Cell Signal Transduction, Second Messengers, and Protein Phosphorylation in Health and Disease

Cell Signal Transduction, Second Messengers, and Protein Phosphorylation in Health and Disease

Edited by

A. Martin Municio
Royal Academy of Sciences of Spain
Madrid, Spain

and

M. Teresa Miras-Portugal
Complutense University of Madrid
Madrid, Spain

Plenum Press • New York and London

Library of Congress Cataloging-in-Publication Data

On file

Proceedings on a symposium on Cell Signal Transduction, Second Messengers, and Protein Phosphorylation in Health and Disease, held July 5–9, 1993, in El Escorial, Spain

ISBN 0-306-44814-9

© 1994 Plenum Press, New York
A Division of Plenum Publishing Corporation
233 Spring Street, New York, N. Y. 10013

Printed in the United States of America

PREFACE

> *It is the great glory as it is also the great threat of science that everything which is in principle possible can be done if the intention to do it is sufficiently resolute.*
>
> Peter Medawar, *"The Threat and the Glory"*

An international symposium on "Cell Signal Transduction, Second Messengers, and Protein Phosphorylation in Health and Disease" was held at El Escorial (Spain) from July 5-9, 1993 as a summer course of the Complutense University in Madrid. The lectures were delivered by renowned scientists from Europe, America, and Asia and attended by a large number of young scientists and graduate students from many countries.

During evolution multicellular organisms have developed the most sophisticated and heterogeneous signals to maintain in harmony their multiple functions. The latest and most controversial aspects and developments in signal transduction were the main focus of this course.

The communication among participants was extremely fluid, alive, and warm. This allowed the understanding of the key steps in cellular communication, from their original and historical sources to the main present hypothesis in the borderline of the latest scientific discoveries in this field. Without any doubt, the special atmosphere of the place, the monuments and the old granite stones, the "patio" with the fountain and the rose garden were responsible for the cordial meeting.

This book comprises the manuscripts of the participants and we hope it will contribute to our knowledge of cellular signal transduction and be of value to a wider scientific community.

Angel Martin Municio
Maria Teresa Miras-Portugal

CONTENTS

CELL SIGNAL TRANSDUCTION, SECOND MESSENGERS AND PROTEIN PHOSPHORYLATION IN HEALTH AND DISEASE

Angel Martín Municio

Royal Academy of Sciences
Valverde 22
28004 Madrid, Spain

GENERAL INTRODUCTION

All cells possess the capacity to process information from their surroundings. Target cells in specialized sensory organs can be stimulated by external signals such as light, odorants and chemical messengers between neighboring or distant cells. The interaction of these factors with specific receptors at the cell surface constitutes the first step in a cascade of biochemical events that underlies transmembrane signaling.

Over the past twenty five years, a series of independent biochemical events prepared the emergency of the remarkable and complex field of signal transduction through intermediates to the cell's interior. It was the time in which the enzyme regulation by covalent modification, conformational changes, molecular interactions, alosteric models, oligomeric proteins, cooperativity, signal amplification, membrane fluidity and bilayer membrane models, and second messengers of the hormonal activity, were introduced in the areas of biochemistry and molecular biology. In that time, a growing list of enzymes in the areas of carbohydrate, lipid, amino acid, and protein metabolism emphasized the importance of regulation by covalent modification. The most frequent modification pathway was already in that time the one involving phosphorylation and dephosphorylation; and the enzymes that catalyzed these transformations were known as protein kinases and protein phosphatases, respectively. Protein phosphorylation and dephosphorylation is one of the fundamental mechanisms of signal integration in eukaryotic cells. It can be accomplished through the reversible phosphorylation of the substrates of protein kinases and phosphatases, through the phosphorylation of these enzymes, or through the phosphorylation of proteins that control the levels of second messengers.

The interconversion of an enzyme from the phosphorylated to the dephosphorylated

Cell Signal Transduction, Second Messengers, and Protein Phosphorylation in Health and Disease
Edited by A.M. Municio and M.T. Miras-Portugal, Plenum Press, New York, 1994

1

forms was usually associated with a marked change in the activity of the substrate enzyme, sometimes in one direction and sometimes in the other. Thus, in glucose metabolism, phosphorylation reciprocally activates glycogen breakdown and inhibits glycogen synthesis, so that the control mechanism of covalent modification acts to regulate energy flux through two otherwise competing systems. There are some unique advantages to enzyme regulation by covalent modification. It provides a means for modulating the extent or direction of energy flux by changing the proportion of an enzyme in the physiologically active form without the waste of removing or replacing the total peptide structure.

During the last decade, this plain protein kinase enzyme activity has evolved up to elaborate networks having multiple potencialities for the amplification of single signals that eventually lead to the regulation of diverse physiological processes such as development, differentiation, morphogenesis, mitogenesis, secretory activity, T-cell activation, motion, gene transcription and cell cycle.

The molecular basis of signal transduction across the cell membrane has long been, and it is nowadays, a subject of the greatest interest, in which are included important biological mechanisms such as cell proliferation and differentiation, and cell-cell communication. Thus, signal transduction is the process by which extracellular molecules, such as cytokines, antigens, polypeptide hormones and neurotransmitters bind to receptors on the cell surface and generate intracellular regulatory signals that are transmitted to different systems of the cell. In other words, the acquisition and subsequent release of signalling information is the process of signal transduction.

Initial emphasis on signal transduction sought to identify the ingredients of the signalling pathways from the molecules interacting outside the cell to the inside functional-protein targets. Emphasis that has moved recently to the participants in pathways that trigger gene transcription.

The architecture of the signalling systems exhibits interactions between the components of a single pathway as well as complex crossing relationships due to interactions between the participants of various pathways and to the presence of common factors involved in different processes. Examples of this complexity are offered by the protein phosphorylation cascades in which many receptors for growth factors are protein tyrosine kinases. Also, in the cases when protein kinase C is activated in the plasma membrane by diacylglycerol generated by agonist-stimulated hydrolysis of polyphosphoinositides. Individual transducing proteins work also in a variety of ways either responding to acquired signals by amplifying them, or damping them down and processing them before they will be transmitted to downstream targets. On the other hand, the heterogeneity and differential tissue expression of the signalling proteins contribute to the complexity of the understanding of a large number of cellular functions.

The signal transduction processes consist of a series of external signals -cytokines, growth factors, antigens, polypeptide hormones, neurotransmitters, nitric oxide-, specific receptor proteins or ion channels, GTP-binding proteins, second messenger-generating enzymes -phospholipases, phosphoinositidase, sphingomyelinase-, soluble intracellular second messengers -Ca^{2+}, cAMP, cGMP, inositol phosphates, diacylglycerols, cyclic ADP-ribose-, protein kinases, regulatory proteins and target functional proteins. Thus, at the molecular level, the situation appears every day more clearly complicated because of the multiple chemical nature of signals and the variety of cells involved; the structural varieties of receptors; the linkage of receptor stimulation to the enzyme hydrolyzing activities acting on either phospholipids or GTP; the diversity of degradation products and their participation in the regulatory activity of the series of the membrane and intracellular molecular species of protein kinases; and the series of target functional proteins. On the other side, many of the signalling cascade components -growth factors, receptors, protein kinases able to phosphorylate specific tyrosine or serine (threonine) amino acid residues of target proteins, DNA-binding proteins, GTP-binding proteins- are expression products of oncogenes. It means that the dissection of this type of intracellular signalling pathways controlling cell growth and differentiation is connected with cell malignancy.

The physiopathological consequences of the crosstalks between intracellular signalling systems are evident at multiple levels within a given signalling cascade, including enzyme activities, ligand-receptor and protein-protein interactions, ligand-gated integral ion channel function and gene expression. Furthermore, the present knowledge of transduction models, and the possibility that the activities of all components of each pathway could be modulated, raises the potential design of new pharmacologically-active drugs to act, as either agonists or antagonists, on the final effects in a predetermined extent. It has opened new therapeutic intervention strategies.

CELL SIGNALS

The long-term effects on cellular regulation are originated by transient signal generated by stimulation of cell-surface receptors. Polypeptide hormones and neurotransmitters are the two classical types of signal factors that being transported in the blood stream exert their function on specific cell-surfaces. Main recent interest is based on the complex cellular networks formed by the cytokine-producing cells and the large family of cytokines as protein mediators influencing proliferation, differentiation and the functions of various lineages of target cells. To this complexity contribute also the multiple types of cells on which each cytokine is able to act, the synergism or interference between cytokines, the pleiotropic functions of the targeted cells, and a kind of functional redundancy among distinct cytokines. At the present time, about fifty different cytokines have been identified as signalling molecules, inducing different biological effects on different target cells.

Cytokines are involved in communication between cells, mainly those of the immune system, determining the quality and the magnitude of the immune response. From an evolutive point of view, some cytokines appeared in vertebrates simultaneously to lymphocytes, although other cytokines have been identified as far down the evolutionary scale as starfish. Frequently, synonymy highly affects cytokine nomenclature, difficulting sometimes the interpretation of research results. To the growing number of cytokines belong two main families: *interleukins* (IL) and a series of *growth factors*.

Interleukins are known for their role as intercellular messengers between white blood cells. Thus, a population of T lymphocytes (TH1) secretes IL-2 and interferon-γ (IFN-γ) and promotes cell-mediated immunity, whilst another subset (TH2) secretes IL-4, IL-5, IL-6 and IL-10 and induces humoral immunity, with production of IgG, IgE and IgA. Simultaneously, IL-4 and IL-10 inhibit the development of TH1 subset, whilst IFN-γ supresses the TH2 response. Development of haemopoietic cells from a multipotential stem cell is regulated by sequential actions of a number of interleukins (IL-3, IL-5) and colony stimulating factors (granulocyte-CSF, granulocyte-macrophage-CSF, monocyte-CSF), according to the lineage-specificity. Also, a number of circulat-ing cytokines, including IL-1, IL-2, IL-6 and tumor necrosis factor (TNF) have been shown to directly affect brain or endocrine activity.

Dysregulation of the haemopoietic system can lead to a plethora of diseases, some of them ultimately fatal. The most intensively developed application of cytokines is the desing of therapeutic regimes to boost a compromised haemopoietic system. The pri-mary clinical applications of these factors has been in either the lack of one or more of the cellular components of the haemopoietic system or ablating chemotherapy-induced cytopenias, particularly in bone-marrow transplants and in solid-tumour cytotoxic the-rapy.

A series of cytokines have been shown to be involved in the molecular mechanisms of many human disorders. Among the **neoplasias**: myeloma (IL-6), Burkitt's lymphoma (IL-10), breast carcinoma (TGF-β), squamous-cell carcinoma (G-CSF), osteosarcoma (GM-CSF), myeloid leukaemia (IL-1, GM-CSF, G-CSF, M-CSF), lymphoid leukaemia (IL-2, IL-7, TNF). In **autoimmune diseases**: diabetes (type II) (IL-1, IL-4, GM-CSF), systemic lupus erythematosus (IL-2, GM-CSF, IFN-γ), multiple sclerosis (IL-1, IL-6, TNF). In **inflammation**: rheumatoid arthritis (IL-1, IL-6, TNF, GM-CSF), psoriasis (IL-6, IL-8), erythroderma (IL-6, IL-8), ankylosing spondylitis (IL-1, IL-6). In **allergic asthma** (IL-1, IL-4, IL-5), **septic shock** (IL-1, IL-6, TNF), **fibrosis** (TGF-β), **aplastic anaemia** (IFN-γ), and **cerebral malaria** (IL-3). Taking into account the participation of these signalling molecules, cytokine and anti-cytokine therapies -anticytokine antibo-dies, soluble cytokine receptors, receptor antagonists, and IL-10- have designed and clinicaly evaluated (1). On the other hand, live vaccine vectors which encode cytokines may increase the safety of viral vectors and permit the manipulation of the immune response so that effective protective immunity can be induced (2).

In this respect, it is important to underline that the tumor necrosis factor (TNF) genes are the only ones located within the major histocompatibility complex (MHC) that are known to code for a cytokine. Despite a striking association between different MHC alleles and various autoimmune diseases, the role of the MHC class I or class II molecules in many such diseases remains largely unknown. It is even questionable whether the MHC genes are the predisposing genes to the disease rather than markers for other closely linked genes. Thus, it would be interesting that the production of TNF-α by mouse peritoneal macrophages, induced by lipopolysaccharide and gamma interferon (IFN-γ), varies from strain to strain and is MHC class II associated (3). Furthermore, the presence of a given cytokine during the disease process does not necessarily imply a pathological role. In this connection, TNF-α has been claimed both to exacerbates autoimmune pathology and exerts a beneficial role (4).

Recently, TNF-α has been associated with the quintessential anabolic state, obesity, and with derangements of glucose homeostasis that accompany this disorder (5). Also, TNF-α has been associated with the most extreme states of both catabolism and anabo-lism (6). Great strides are being made in untangling another signal-transduction path-way which starts with the binding of interferons (IFNs) to their cell-surface receptors. The end results of which is tyrosine phosphorylation and activation of transcription-factor subunits in the cytoplasm, which then translocate into the nucleus and induce transcription (7,8).

RECEPTORS

The biological activity of the growing number of signalling molecules requieres the existence of signal-transducing receptors in cell membranes. From the point of view of their structure, the receptors could be clasified as follows (9):

1. *Single-hydrophobic domain polypeptides.* These polypeptide chains can have either one or no transmembranar amino acid sequence. In the second case, the polypeptide chain is anchored to the membrane trough a glycolipid portion. Concerning the subunit composition, these receptor can be monomers, homodimers, heterodimers, heterotrimers, or heterotetramers.

As transduction system, the binding subunits can exhibit some ligand-stimulated enzyme activity, such as tyrosine kinase (insulin and mitogenic growth factors) or guanylate cyclase (natriuretic peptides). The receptors of many cytokines, growth hormone, prolactin, and neurotrophins, do not show any enzyme activity after ligand binding.

2. *Seven-hydrophobic domain polypepetides* structured as single monomers, homodimers or heterodimers. Each subunit carries a G protein recognition sequence on its intracellular face. The G protein interacts directly on a channel (atrial muscarinic, neuronal α_1-adrenergic), or via a diffusable messenger (PAF, eicosanoids, IL-8, neuropeptides and small transmitters). Some ligands act as proteases to form a self-activating receptor (thrombin).

3. *Channel-enclosing oligomers*, having homomeric or heteromeric subunits. These receptors perform fast signalling since their transduction is independent of any membrane-diffusible or intracellular factor. In addition to the usual way of receptor activation by a presynaptically released transmitter (ACh, GABA$_A$, glycine, glutamate, 5-HT$_3$, ATP), the transmitter molecule may arrive on the intracellular side. The second case includes those receptors where the signalling occurs across an organelle membrane, like the ryanodine receptor and most of the inositol 1,4,5-triphosphate receptors, in which the transference of Ca^{2+} is produced.

The already mentioned functional redundancy showed by different cytokines could be explained by the molecular structure of their receptors. In fact, the high affinity receptors for a group of cytokines with similar function share a common subunit with a critical role in signal transduction. Thus, the gp130 subunit is shared by IL-6, leukaemia inhibitory factor (LIF), oncostatin M and ciliary neurotrophic factor (CNTF) receptors (10).

The clonal activation of antigen-specific T lymphocytes in response to antigen displayed on the surface of an antigen presenting cell (APC) involves the dynamic interaction of multiple receptor-coreceptor pairs necessary for both signal transduction and adhesion. The clonotypic T-cell receptor (TCR) recognizes antigenic peptide in the context of major histocompatibility complex on the APC, and the mechanims by which ligand binding to the T-cell receptor triggers the T-lymphocyte activation has long been one of the most fascinating endeavours in cellular biology. At present, the precise contribution of the diverse surface molecules for the efficient activation is not yet well understood (11). In the TCR signalling mechanism there are several questions involved such as the TCR structure and its modular organization, the antigen receptor signals, the TCR-coupled signalling pathways as well as the TCR-coupled protein phosphorylation, and, finally, the production of a variety of lymphokines. As a consequence of the signal transduction events, T-cells produce different cytokines; among them, interleukins -IL-2, IL-3, IL-4, IL-5, IL-6, IL-10-, granulocyte-macrophage colony-stimulating factor (GM.CSF), tumor necrosis factor α (TNF-α), and interferon-γ (IFN-γ). The coordination of these factor is crucial for the regulation of the immune response.

The complexity of the signal transduction events initiated at the plasma membrane of the T-cell, and having one of their targets in the control of lymphokine gene expresion in the nucleus, offers one of the most interesting models of interactions between signalling systems and spatiotemporal aspects. This complexity results initially from the modular organization of TCR as a eight-subunit transmembranar complex (12): the $\alpha\beta$ heterodimer contains a binding site for antigen complexed to MHC class I or II molecules and dictates ligand-binding specificity, triggering a minimum of two transduction pathways, involving the cytoplasmic tails of CD3-ϵ and -ψ. Effectively it interacts noncovalently with a series of CD3 polypeptides, three of them -γ,δ and ϵ- evolutionarily related showing immunoglobulin-like extracellular domains, and another one, the chain ψ, as a distinct gene product with an external region of nine amino acids. These four different polypeptide chains are grouped as CD3 dimers: $\gamma\epsilon$, $\delta\epsilon$ and $\psi\psi$, are responsible for both efficient receptor expression and signal transduction.

Current data suggest that TCR is coupled to the activation of multiple non-receptor protein kinases; among them, p56 and p59 of the *src* family that mediate many of the phosphorylation events induced by TCR crosslinkage. Tyrosine phosphorylation initiates a signal transduction cascade by modulating the activities of TCR-induced phospholipase C (PLC-γ_1), required for phosphoinositide hydrolysis, and other intracellular signal-generating enzymes. Activation of PLC causes the hydrolysis of phosphatidylinositol 4,5-biphosphate (PIP$_2$) resulting in the formation of second messengers, diacylglycerols and inositol triphosphate. These second messengers are responsible, respectively, for the activation of protein kinase C (PKC) and the increase of intra-cellular Ca^{2+}. It is the reason why tyrosine kinase inhibitors prevent lymphokine gene expression (13). However, the mechanims through which the activation of protein kinases, including PKC, are transduced to the nucleus remain unclear although a number of intermediate targets have been identified. Thus, the accumulation of p21^{ras} in T-cells (14) suggests that this protooncogene product participates in signal relay from the plasma membrane to the cytoplasm.

The regulation of signalling initiated by receptors linked to phosphoinositidase C can be interpreted through mechanisms such as the receptor internalization and eventual downregulation that occur in long-term agonist treatment. This interpretation is similar to that of other receptors linked to adenylate cyclase. However, a rapid desensitization, associated with receptor phosphorylation, regulates receptors linked to phosphoinositidase C (15).

Tyrosine phosphorylation has been shown to be a pivotal event in the activation of latent cytoplasmic transcription factors thar recognize similar DNA elements. Some growth factors, as the epidermal growth factor, and cytokines, as interferon-γ, and IL-6, trough their respective unrelated cell surface receptors, although eliciting distinct biological responses in target cells, activate a common nuclear signal transduction pathway (16).

Besides the clonotypic T-cell receptors already described, the different cell types of the lymphoid-myeloid lineage express constitutively a family of proteins collectively

6

known as Fc receptors (FcR), that play a variety of roles both in the initiation of the immune response and its consequences (17,18). Each immunoglobulin heavy chain class has the corresponding Fc receptor that is expressed on cells of all the major hematopoietic lineages with the exception of the erythroid lineage cells. B-cells, NK cells, macrophages, mast cells, granulocytes and platelets all constitutively express high levels of FcR and the levels of expression are modulated by the state of activation (19). However, FcR are not readily detected on conventional T cells under normal resting conditions. But, recent information (17) identifies a multilevel relationship between Fc receptors and clonotypic TCR, suggesting that this relationship might account for the restricted expression of Fc receptors on T cells. Thus, when occupied by immunoglobulins, FcRs become members of a cognate recognition system which shares with TCRs and BCRs a common usage of associated triggering molecules and signalling pathways.

All findings indicate that the mechanisms of signal transduction from antigen-receptor complexes on several types of lymphoid cells involves enhanced phosphorylation of intracellular proteins on tyrosine residues. The list of known tyrosine phosphorylated polypeptides is grown over the years, now including the substrates CD45, $PLC\gamma1$, $PLC\gamma2$, TCR-ψ, FcϵR, mitogen-activated protein kinase, ras GTPase activating protein and src-related kinases ($p56^{lck}$, $p59^{fyn}$, $p50^{csk}$). Obtained results contribute to unravel the regulatory mechanisms of these proteins involved in lymphocyte activation (20).

The emerging importance of protein phosphatases is illustrated by recent works on T cells. Stimulated T cells lacking the tyrosine phosphatase CD45 are unable to generate the phosphatidyl inositol-derived second messengers inositol triphosphate and Ca^{2+}. In fact, the transmembranar CD45 tyrosin phosphatase is required for normal TCR-mediated signalling as it has been shown by transfecting an active portion of murine CD45 into CD45$^-$ T cells and the recovery of the TCR signalling abnormalities observed in the absence of CD45, including TCR-mediated enhancement of tyrosine kinase activity and Ca^{2+}(21). Thus, tyrosine kinases and phosphatases do not simply oppose each other's action; rather they work in concert to maintain a fine balance of effector activation needed for the regulation of cell growth and differentiation. As another example, phosphotyrosine phosphatase 1D is phosphorylated on tyrosine in cells overexpressing the PDGFβ receptor kinase and this tyrosine phosphorylation correlates with an enhancement of its catalytic activity (22).

One of the most interesting features in the whole pattern of signal transduction is that of receptors exhibiting enzyme activity of tyrosine kinases (RTKs). Many growth factor receptors are tyrosine kinases, containing a cytoplasmic catalytic domain capable of both autophosphorylation and phosphorylation of cellular substrates after ligand activation. Autophosphorylated receptors serve as docking sites for proteins than contain Src homology region 2 (SH2) domains. The signalling molecules bind RTKs through SH2 domains (23) inducing an autophosphorylation of tyrosine residues. Thus, phosphotyrosine residues of receptors serve as highly selective binding sites for some cytoplasmic signalling factors. For instance, when the platelet-derived growth factor β (PDGFβ) activates the corresponding receptor, its intracellular region binds phosphatidylinositol 3-kinase, GTPase-activating factor, phospholipase Cγ and c-src (24). The PDGFβr results suggest that Ras is downstream of phosphatidylinositol 3-kinase and associate with the enzyme (25). The insulin receptor is another example of a RTK in which ligand binding activates a phosphatidylinositol 3-kinase although this enzyme does not physically associate with the receptor (26). Also, signal transduction from interleukin-3, that supports the proliferation and differentiation of early progenitors and cells committed to several myeloid lineages, involves tyrosine phosphorylation, although the cytoplasmic domain lack detectable catalytic domains. This fact has been attributed to a tyrosin kinase that associates with the receptor and is activated by IL-3 binding (27). This tyrosin kinase belongs to the JAK family (*just another kinase*) that shows an extensive similarity in the N-terminal region. Full length cDNA clones for murine JAK1 and JAK2 tyrosine kinase have been obtained (27). Also, JAK1 participates in the signal transduction pathways of interferon-α/β and interferon-γ, and the reciprocal interdependence between JAK1 and JAK2 in the interferon-γ pathway may reflect a requirement for these kinases in the correct assembly of interferon receptor complex (28).

Most growth factors stimulate a group of intracellular protein kinases that includes Raf1 kinase, mitogen-activated protein (MAP) kinases and protein kinase C. Raf1 kinase appears to be essential for both serum-induced proliferation of fibroblasts and activation of specific genes, such as *c-fos* (26).

G PROTEINS AND PHOSPHOLIPASES

G proteins belong to a large superfamily of homologous guanine nucleotide binding proteins involved in a cascade of some signal-transducing mechanisms (29,30). Seven-transmembrane-domain-containing receptors transduce a wide variety of signals across the plasma membrane including those triggered by neurotransmitters, hormones, pheromones, odorants and chemoattractans. Likely, effectors include phospholipases A_2 and C and a plethora of ion channels, transporters and exchangers. These proteins bind to, and hydrolyse, GTP, and participate in the regulation of key cellular events, such as transcription, transmembrane signalling, protein trafficking, proliferation and differentiation (29,31). They become active on binding GTP but their intrinsic ability to hydrolyse GTP eventually converts them to inactive GDP-bound form, and they remain in this state until the GDP es exchanged for GTP.It has been previously mentioned the involvement of *src*-related protein tyrosine kinases (PTKs) in the regulation of an isoform of phospholipase C, PLC-γ1, that generates diacylglycerols and inositol phosphates, via the antigen receptors on T and B cells. Another phospholipase C isoform, PLC-β, is activated through the same antigen receptors on lymphocytes but mediated by G proteins. The participation of G proteins in the regulation of this activity was initially suggested by specific requirements for guanine nucleotides for stimulation by hormones and the more direct effects of non-hydrolisable guanine nucleotides and alu- minium fluoride, universal activator of the heterotrimeric G proteins. An additional reagent that implicated a role for G proteins was the toxin from *Bordatella pertussis*, which ADP-ribosylates the α-subunits of Gi and Go proteins and attenuates their function. The action of several hormones was abolished by this toxin, although in many cases, the toxin had only a partial effect.

The relative roles of PTKs and G proteins in the regulation of antigen receptor-mediated signalling in T and B cells is controversial, although they need not be mutually exclusive (32). The controversy has focussed on the fact that the antigen receptors on T and B cells contain only two single transmembrane-spanning proteins with short cytoplasmic domains; the possibility exists, however, that receptors with single transmembrane-spanning regions as the receptors for insulin, insulin-like growth factor, and tumor necrosis factor-α, may be G protein regulated. Nevertheless, the emergence of the G protein-coupled receptor family, with seven transmembrane-spanning domains, sparked off the controversy concerning the role of G proteins in the coupling of the lymphoid antigen receptors to PLC (33). In this connection, the hypothesis that G proteins and PTKs may interact to regulate antigen receptor-PLC coupling has been supported by experimental evidence (34). The G proteins that regulate phosphatidylinositol 4,5-biphosphate hydrolysis can be separated into two distinct classes on the basis of their inhibition by pertussis toxin: the pertussis-sensitive G proteins thought to involve members of the Gi/Go family, and the toxin-insensitive Gq class, implicated in regulation of PLC-β (29-31). Another members of the GTPase superfamily are the elongation factor, EF-Tu, that participates in the biosynthesis mechanism of polypeptide chains, and a series of monomeric proteins, of 20-25 kDa, as the p21^{ras}. In contrast to the large heterotrimeric G proteins, ras-related G proteins are monomeric and of small molecular mass (22-30 kDa). They share a number of structural and sequence homologies which account for their primary biochemical function, the binding and hydrolysis of GTP; similarly to the trimeric G proteins, these proteins exist in the cell as one of two interconvertible conformational states: an active form, when bound to GTP, and an inactive form, when bound to GDP.

The mammalian ras family of G proteins is composed of about fifty genes which, based on sequence similarities, can be subdivided into three major subgroups: Ras, Rho and Rab. Besides the p21^{ras}, proteins of the Rho and Rab subgroups are likely to play a pivotal role in the final steps of the signal transduction cascade that elicits the various effector functions of lymphocytes. The Rho proteins play a central role in the modulation of cellular functions involving the actin cytoskeleton such as in the establishment of cell polarity and morphology (35). Thus, Rho A, Rho C and Rac1 controls actin cytoskeleton; Rac 2 activates NADPH oxidase; Rho B controls the early induction by growth factors and Rho G the late-response gene (36).

The G proteins that mediate regulation of several effector molecules by hormones constitute a large family of highly homologous proteins. G proteins are heterotrimers, composed of three distinct subunits: α (39-46 kDa), β (37 kDa) and γ (8 kDa). The α subunits have a single, high-affinity binding site for GDP or GTP. The GDP-bound form of α binds tightly to $\beta\gamma$ ($\alpha_{GDP}\beta\gamma$) and is inactive and stimulated by a ligand-activated receptor to exchange GDP for GTP. In the active form α_{GTP} dissociates from the $\beta\gamma$ and both -α_{GTP} and $\beta\gamma$- are then able to interact specifically with cellular effector molecules to evoke the cellular response. The β and γ subunits exist as a tightly associated complex that functions as a unit. Until now about twenty different α subunits have been described and used traditionally to define the purified heterotrimeric proteins: G$_s$ (ubiquitous, olfactory neuroepithelium), G$_i$ (brain, retinal rods and cones, taste buds, platelets), G$_q$ (T and B cells, myeloid cells, lung, kidney, liver), and G$_{12}$ (ubiquitous). Each one of these families interacts with particular receptors and effectors; among them, adenylate cyclase, phospholipases C (β_1, β_2, β_3), phospholipase A$_2$, and Na$^+$-, K$^+$-, Ca^{2+}-channels (37). Frequently, the α-subunits can account for the primary activity of the G proteins. Thus, G$_s\alpha$ stimulates adenylate cyclase, and transducin α (G$_t\alpha$) activates a cyclic GMP-dependent phosphodiesterase in retina (38). The complex $\beta\gamma$, tightly associated under physiological conditions, may directly interact with effector molecules. Four different β polypeptide sequences are known and all have similar sequences, and five sequences of γ cDNA have been established. Because of the apparent sequence heterogeneity in the γ subunits, functional differences of the $\beta\gamma$ complexes have been attributed to the γ subunits (39,40). Thus, among γ subunits the γ_3 subtype was found to be required for coupling of the somatostatin receptor to voltage-sensitive calcium channels, whereas the γ_4 subtype was found to be required for coupling of the muscarinic receptor to those channels (41). Furthermore, two of the four subtypes of β subunits tested, β_1 and β_2, are selectively involved in the signal transduction cascades from muscarinic M_4 and somatostatin receptors, respectively, to voltage-dependent Ca^{2+} channels (42).

The new roles for G protein $\beta\gamma$-dimers have been recently reviewed (43). The summarized effectors to be regulated by the complex $\alpha\beta$ are: K^+ channel ($I_{K,ACh}$), phospholipase A_2, pheromone response, adenylate cyclase (types I, II, IV), phospholipases β1-3 (44), muscarinic receptor kinase and β-adrenergic receptor kinase (45). Also, $\beta\gamma$ subunit can regulate receptor function by enhancing the receptor's ability to interact with the α subunit or by bringing a receptor-specific kinase to the receptor to desensitize it (43).

The modulation of the G protein function has been studied in the two monomeric interconvertible conformational states of the ras-related family: GTP-bound, active, and GDP-bound, inactive. The GTP/GDP cycle of ras-like proteins is regulated by two classes of proteins: (1) GTPase activating proteins that stimulate the low intrinsic GTPase activity of ras-G proteins, inactivating them; (2) exchange factors such as the GDP dissociation stimulators that promote the exchange of GDP for GTP, thus returning the G protein to an activated state, or the GDP dissociation which prevent this exchange (36). Also, the ability of the $\beta\gamma$ subunit to interact with the α subunit in the retinal G protein has been found to be modulated by the phosphoprotein phosducin. Phosducin inhibits *in vitro* the retinal cyclic GMP phosphodiesterase, probably by binding to $\beta\gamma$ and preventing the heterotrimer formation (46). Arrestins have been implicated in the regulation of many G protein-coupled receptor signalling cascades. Mutations in two *Drosophila* photoreceptor-specific arrestin genes, *arrestin 1* and *arrestin 2*, were studied and results demonstrate the fundamental requirement for members of the arrestin protein family in the regulation of G protein-coupled receptors (47).

Given these numerous physiological roles of G protein-coupled receptors in neurotransmission, endocrine signalling, olfaction, vision and chemotaxis, it might expected that abnormalities of such receptors underlie a variety of human patologies. Several mutations that disturb the structure and functions of hormonal receptors have been shown to be associated with states of hypo and hyper function of the affected receptors. Also, mutations in the intracellular loops of the prototypic G protein-coupled receptors involved in coupling the receptors to G proteins lead to constitutive activation of G protein-mediated phospholipase C activation (48). The molecular defects responsible for an uncommon form of hyperthyroidism and for familial precocius puberty, have been identified with mutations in G protein-coupled receptors (49,50). All these facts have been recently described and related with the possible therapies for such illnesses (51).

Interleukin 8, one of the most potent chemoattractants for neutrophils, mediates cytokine-induced transendothelial neutrophil migration, induces angiogenesis, and triggers a variety of other effects associated with the inflammatory response. The receptors for IL-8, designated as α and β, appear to couple to G proteins and activate phospholipase C in neutrophils (52). These findings suggest that IL-8 acts through signal-transducing pathways that are limited to specific heterotrimeric G proteins and effectors and provide suitable targets for the development of new anti-inflammatory agents.

Because the trafficking of some intracellular organells involves G proteins and they are associated with synaptic cells, the GTP-binding proteins may regulate synaptic vesicle traffic (53). Squid giant synapse has been used to examine how the activation of G proteins affects the release of neurotransmitters and the distribution of synaptic vesicles within the terminal. A model has been proposed (54) that combines previous schemes for synaptic vesicle trafficking and small molecular weight protein cycling. According to this model, GTPγS substitutes for GTP in binding to one of these proteins, such as Rab3A, that is responsible for docking synaptic vesicles at the active zone. Also, G proteins -either heterotrimeric G proteins or small molecular weight proteins- regulate every step in the trafficking of other intracellular organelles and it seems possible that GTP-binding proteins regulate every step in the transit of a synaptic vesicle through its cycle.

The neuropeptide galanin of 30 amino acids in human, is synthesized as a prepro-hormone of unknown function. Galanin regulates K^+ channels, adenylate cyclase and phospholipase C by acting at Gi/Go protein-coupled receptors. The N-terminal fragment 1-16 acts synergistically with morphine in the somatosensory system and have potential analgetic applications; also, antagonists may be useful therapeutic agents in neurology and endocrinology (55).

SECOND MESSENGERS

It has been previously mentioned that the interaction of external messengers with specific receptors at the cell surface represent the first step in a cascade of molecular events that underlies transmembrane signalling. In many cases, stimulation of these receptors result in activation of effector proteins either ion channels or enzymes, which mobilize chemical second messengers that initiate characteristic actions within the cell. Thus, phospholipase C-β activation, through G protein-coupled receptor, and phospholipase C-γ activation, tyrosine kinase mediated, produce intracellular second messengers, namely inositol 1,4,5-triphosphate (IP_3) and diacylglycerols (DAGs). IP_3 mobilizes intracellular stores of calcium whereas DAGs are physiological activators of protein kinase C. In many cells, IP_3 mediates the effects of receptors linked to phosphoinositide hydrolysis on intracellular Ca^{2+} mobilization. The sequence homology and tertiary structural features of IP_3 receptors are shared with those of the ryanodine receptors (the channels responsible for Ca^{2+} mobilization from the SR); both receptors are very large homotetrameric Ca^{2+}-channels and shared also many allosteric regulators and means of communication with plasma membrane Ca^{2+}-channels (56). The relationship between IP_3 receptors and Ca^{2+} has been recently reviewed (57).

The expanding number of inositol phosphate metabolites has prompted consideration that inositol phosphates other than IP_3 have important cellular functions. Thus, function of IP_4 has been related to the cellular regulation of Ca^{2+} fluxes (58); IP_5 regulates the affinity of avian hemoglobin for oxygen (59), whereas both, IP_5 and IP_6, have been described as neurotransmitters (60). Other peculiar inositol phosphates have recently emerged: the pyrophosphates IP_5P and IP_6P found in the slime-mould *Dictyostelium discoideum* and cell lineages. They behave in a complex and dynamic way, showing a particularly interesting rapid turnover (61).

On the other hand, it has been mentioned previously that adenylate cyclase, as well as guanylate cyclase, belong to the effector enzymes activated by extracellular agonist-coupled receptors; this activity causes the formation of an intracellular messenger molecule, cyclic AMP and cyclic GMP, that regulates some characteristic biochemical functions of the target tissue. It is one of the basic principles of hormone action emerged from research three decades ago, under the guidance of C.F.Cori and G.T.Cori, and of E.G. Krebs and E.H.Fischer. It is, surely, the right moment to honor these hystorical names of Science and, particularly, of the topic of this symposium. The presence in it of professor Fischer has permanently given us a double lesson: those of his scientific teaching and his amicable "togetherness".

Thus, the activity of phospholipase C on phospholipids, phosphatidylinositol and phosphatidylcholine, contributes to the generation of DAGs, whereas hydrolysis by phospholipase A_2 produces unsaturated free fatty acids and lysophospholipids. Phospholipase D generates phosphatidic acid that yields further DAGs by the action of phosphatidic acid phosphohydrolase. All these lipid metabolites are produced in response of particular signal-induced degradation of membrane phospholipids, and play a role in the activation of protein kinase C (62-65) that have multiple functions within cells. The interaction among phospholipases and Ca^{2+} mobilization has been recently described (66), showing the distinct responses of the protein kinase C family to Ca^{2+} and the phospholipid degradation products. The concurrence of phospholipases D and A_2 on phospholipids yields monoacylglycerol 3-phosphate (lysophosphatidic acid) which could be an intracellular messenger, possibly acting through a G protein-coupled receptor, and with a role in cell growth and motility (67).

On the other side, the second messenger cyclic AMP exerts nearly all its effects by activating the ubiquitous cAMP-dependent protein kinase (PKA), whereas cyclic GMP activates a specific cGMP-dependent protein kinase (PKG), located predominantly in smooth muscle and the cerebellar region of the brain (68).

Besides the already described interactions of calcium ions with other messengers, Ca^{2+} binds to calmodulin, triggering conformational changes in the polypeptide chain structure that allow it to activate many enzymes, including a number of protein kinases. The calmodulin-dependent protein kinases can have either a broad (69) or a restricted specificity, such as those acting on myosin light chain (70), phosphorylase kinase (71) and elongation factor 2 (72), that regulate muscle contraction, glycogenolysis and protein synthesis, respectively. The interactions between both second messengers, cAMP and Ca^{2+}, concerning glycogenolysis in mammalian skeletal muscle and the regulation of cardiac muscle contractility, have been also reviewed (73).

PROTEIN KINASES AND PROTEIN PHOSPHATASES

It has been said (74) that 2-3% of all eukaryotic genes may code for protein kinases. In fact, several hundreds of protein kinases and protein phosphatases, regulated by second messengers, have already been identified enabling a diversity of responses to many physiological stimuli. Phosphorylation, or dephosphorylation, of either serine or threonine residues, and occasionally tyrosine, trigger conformational changes in the protein substrates leading to the physiological responses that are evoked by particular agonist-coupled receptors. Thus, protein phosphorylation and dephosphorylation is one of the major mechanisms of signal integration in eukaryotic cells, and the diversity of effects produced arises from the variety of agonists, the pleiotropic actions of protein kinases and phosphatases, and the cellular presence of the particular protein substrates (75).

Protein kinase C activity is sustained by DAGs, which is essential for maintaining cellular responses. A series of species and subspecies of PKC have been defined: cPKC (α, βI, βII, γ), nPKC (δ, ϵ, η, θ) and aPKC (ζ, λ) (66), showing distinct physicochemical and regulatory properties, as well as differential tissue expression with specific intracellular localization (76). All members of the PKC family so far described are dependent on phosphatidylserine, but exhibit different requirements of phospholipid metabolites and Ca^{2+}. All subspecies of cPKC have similar molecular mass and are activated by the same series of effectors: Ca^{2+}, DAGs, FFA and LysoPC. However they respond differently to the various combinations of Ca^{2+} and the phospholipid degradation products with respect to the extent and duration of the response, suggesting their unique functions in cell signalling (66). Also, they show a different tissue expression; whereas the distribution of PKCα is universal, that of PKCγ is restricted only to brain.

Protein kinase C participates in the regulation of important mechanims of molecular biology. Thus, long-term potentiation (LTP) is a use-dependent increase in synaptic strength considered as cellular mechanism contributing to memory formation in mammals. Recent results suggest that inhibitors of a number of protein kinases block both the induction and the maintenance of LTP. Trying to identify the particular protein kinase involved in this mechanism a peptide substrate selective for PKC has been found. It corresponds to the phosphorylation site of a neural protein, neurogranin-(28-43), that is an endogenous selective substrate for PKC. Direct biochemical evidence for activation of PKC in both the induction and maintenance phases of LTP has been associated to the phosphorylation of the kinase (77). These molecular mechanisms of induction and maintenance of LTP are accompanied by the translocation of PKC (78). The involvement of protein kinases in cognitive processes has been recently reviewed (79). Also, activation of protein kinase C reduced β/A4 peptide, which is derived from the Alzheimer amyloid protein precursor, in a proportion of 50-80% (80). It offers the possibility to influence the production of β/A4 peptide, through the activation of PKC.

Many other physiopathological processes are participated by protein kinase C. Thus, PKC regulates the GTP-dependent binding of the two coat proteins ADP-ribosylation factor and β-COP to Golgi membranes in rat basophilic leukaemia cells, suggesting that the secretory traffic would be modulated by membrane receptors and messengers (81). An eye-specific PKC has been identified in *Drosophila*; its activation by cytosolic calcium and DAGs is required for adaptation (82). Among the various molecular interpretations of the mechanisms of anaesthesia, the inhibition of the regulatory subunit of PKC has been reported (83).

It is also interesting to underline that some functions, such as that of voltage-gated sodium channels, responsible for action potential generation in mammalian brain neurons, is regulated by phosphorylation of both protein kinase C and cAMP-dependent protein kinase. This convergent modulation of sodium channels reequired phosphorylation of serine 1056 by PKC accompanied by phosphorylation of additional sites by PKA (84).

Despite the identification of hundreds of distinct protein kinases, few interconnections between these kinases have been firmly established. Some of these connections are those exhibited in the mitogenic signalling routes. Many extracellular mitogens, such as platelet-derived growth factor, epidermal growth factor and nerve growth factor, induce autophosphorylation of their respective receptors on tyrosine residues by activation of intrinsic catalytic kinase domains. This is finally translated into phosphorylation of serine and threonine residues in proteins throughout the cell, such as the ribosomal protein S6. In the protein kinase cascades involved in growth factor and tumor promoter induced signalling, the mitogen-activated protein (MAP) kinases, serine/threonine kinases, are highly versatil as transducer for multiple intracellular signalling pathways (growth factors, differentiating agents, neurotransmitters, heat shock) in virtually all cell types (85). MAPK exists in a dephosphorylated form in quiescent cells or in unstimulated cells and becomes activated by combined tyrosine and threonine phosphorylation catalysed by a MAP kinase kinase (86). The MAP kinase cascade (signalling factor → ras p21 → MAPKKK → MAPKK → MAPK) appears to be conserved in various signal transduction pathways from yeasts to vertebrates. Forms of Ras activated either by mutation (Gly12Val) or by binding of guanylyl-imidodiphosphate (GMP-PNP) interact with active MAP kinase kinase. Raf 1 protein and its complexes with MAPKK are dependent upon the activity of Ras (87). The normal cellular homologue of the acutely transforming oncogene v-raf is c-raf-1, which encodes a serine/threonine protein kinase that is activated by many extracellular stimuli and use MAPK as the immediate physiological substrate (88). Thus, activated Raf-1 triggers a protein kinase cascade by direct phosphorylation of MAPKK; however, the kinase Raf-1 can be activated by treatment of cells with mitogens and by the protein kinase C activator 12-O-tetradecanoyl-phorbol-13-acetate. PKCα induces Raf-1 phosphorylation at several sites, including a serine residue at position 499 (89). Insulin-induced activation of MAPKs was enhanced by overexpression of growth factor receptor-bound protein 2 (GRB2). This overexpression led to increased formation of a complex between the guanine nucleotide-releasing factor Sos and GRB2. In response to insulin stimulation, this complex bound to tyrosine-phosphorylated IRS-1 (insulin receptor substrate-1) and Shc (90).

Two highly related mammalian MAPKs, $p44^{mapk}$ and $p42^{mapk}$, have been cloned and found to be ubiquitously expressed in vertebrates (91-93). A unique feature of this family of protein kinases is that they require dual phosphorylation on both tyrosine and threonine residues to become fully aactive. Activation of of these MAPKs play a pivotal role in integrating and transmitting signals required for growth and differentiation; particularly, it is essential for G_o-arrested fibroblasts to enter the cell cycle (94).

The cdc2 gene was initially identified in the fission yeast Schizosaccharomyces pombe because its mutation has the property of causing the cell cycle to stop at one of two discrete points. This gene encodes a 34 kDa protein serine/threonine kinase, that has been highly conserved through evolution. In animal cells there is a family of cdc2-related proteins that each regulate different parts of the cell cycle (95). These proteins all need to bind a cyclin to become active protein kinases, and thus they are called cyclin-dependent kinases. The cyclin-dependent protein kinases (CDKs) are crucial regulators of the timing and coordination of eukaryotic cell cycle events. Cell-cycle dependent oscillations in CDK activity are induced by complex mechanisms that include binding to positive regulatory subunits and phosphorylation at positive and negative regulatory sites. Human CDK2 is associated with at least two different cyclins: cyclin A, implicated in the progress of DNA replication, and cyclin E, implicated in the initiation of DNA replication. Both, cyclins A and E, have been found complexed with the retinoblastoma-related protein p107, and with the transcription factor E2F. The D-type cyclins link the cell cycle, signal transduction and oncogenesis. Their properties and kinase partners have been recently reviewed (96). The effects of transforming growth factor-β on the cell cycle did not support activation of endogenous cyclin-dependent protein kinases by exogenous cyclins. These features correlate with the inhibition of retinoblastoma protein phosphorylation and suggest that mammalian G1 cyclin-dependent protein kinases are targets for negative regulators of the cell cycle (97). Studies with Xenopus oocytes indicate that metaphase arrest is a result of cooperation between a proto-oncogen kinase and a cyclin-dependent kinase and illustrate the interaction of a cell growth regulator with a cell cycle control factor (98). The crystal structures of the human CDK2 and its Mg^{2+}ATP complex have been determined to 2.4 A resolution (99). The structure is bilobate, like that of the cAMP-dependent protein kinase, but contains a unique helix-loop segment that interferes with ATP and protein substrate binding and probably plays a key part in the regulation of all cyclin-dependent kinases. Also, the crystal structure of the catalytic subunit of cAMP-dependent protein kinase, complexed with ATP and a 20-residue inhibitor peptide has been correlated with chemical and genetic data. The striking convergence of the structure with the biochemical properties and genetics provides the molecular basis for understanding the function of this enzyme, as well as an explanation for the highly conserved residues that are scattered throughout the molecule (100). A plethora of studies deal at present time with protein kinase regulation of mitosis in eukaryotes. Thus, the differential phosphorylation of c-Abl in the cell cycle is determined by an equilibrium between *cdc2* kinase and protein phosphatase activities; two interphase and all mitotic c-Abl sites are phosphorylated by *cdc2* kinase (101). The p34^{CDC28} is a universal regulator of mitosis and its activity is required for entry into the DNA replicative phase (S phase) and for the initiation of mitosis (M phase) (102). In mammals, the macrolide rapamycin inhibits interleukin-2 receptor induced S phase entry and subsequent T-cell proliferation, resulting in immunosupression; rapamycin completely and rapidly inhibits interleukin-2-induced phosphorylation and activation of p70 S6 kinase (103).

Many other important substrates for kinases have been described; among them myotonin-protein kinase and β-adrenergic receptor kinase. The genetic basis of myotonic muscular dystrophy include mutational expansion of a repetitive trinucleotide sequence $(CTG)_n$ located in the 3' untranslated portion of mRNA from a gene designated as myotonin-protein kinase. The gen product exhibits an extensive homology to protein kinase catalytic domains (104,105). Two other diseases caused by triplet repeat expansion are Kennedy syndrome and Fragile X syndrome. Decreased levels of the messenger RNA and protein expression are associated with the adult form of myotonic dystrophy (106). The signal transduction in cilia of olfactory receptor neurons involves the generation of cAMP and IP_3; odorants induce a rapid and transient elevation of cAMP, which activates a nonspecific cation channel and produces membrane depolarization. In the agonist-dependent desensitization in olfaction participates β-adrenergic receptor kinase (107).

Protein kinase A (cAMP-dependent protein kinase), as well as calcium-calmodulin-dependent protein kinase (CaM kinase) participate in numerous physiopathological processes. The glutamate-gated ion channels mediate most excitatory synaptic transmission in the central nervous system and play crucial roles in synaptic plasticity, neuronal development and some neuropathological conditions. This glutamate receptor function can be directly modulated by protein phosphorylation, suggesting that a dynamic regulation of excitatory receptors could be associated with some forms of learning and memory in the mammalian brain (108). Protein phosphorylation of glutamate receptors, transiently expressed in mammalian cells, by cAMP-dependent protein kinase has been suggested to regulate their function, possibly playing a prominent role in certain forms of synaptic plasticity such as long-term potentiation and long-term depression (109). The dependence of glutaminergic synapses from calmodulin-dependent protein kinase II has been also reported (110). Also, a large, voltage- and frequency-dependent potentiation of skeletal muscle L-type Ca^{2+} currents by trains of high-frequency depolarizing prepulses, which is caused by a shift in the voltage-dependence of channel activation to more negative membrane potentials and requires phosphorylation by cAMP-dependent protein kinase in a voltage-dependent manner (111).

In the yeast *Saccharomyces cerevisiae*, addition of glucose to starved cells triggers a transient rise in the intracellular level of cyclic AMP that induces a protein phosphorylation cascade in which the signal is processed by the Cdc25/Ras/adenylate cyclase pathway. It has been shown that, in response to glucose, the phosphoprotein Cdc25 is hyperphosphorylated rapidly by a cAMP-dependent protein kinase. These results have been considered of general significance because of the highly conserved sequence of Ras-guanyl nucleotide exchange factors from yeasts to mammals (112).

It has been also reported that the virulence plasmid of *Yersinia pseudotubereculosis* encodes a secreted protein kinase (YpkA) with high homology to eukaryotic serine/threonine protein kinases. Pathogenicity results presumably by interfering with the signal transduction pathways of the target cell (113). During neurogenesis in *Drosophila*, groups of equipotential, neurally competent cells has to choose between epidermal and neural fates. *Shaggy* is required for the lateral signal and encodes serine/threonine protein kinases with homology to the glycogen synthase kinase-3 (GSK-3) that act in signal transduction in vertebrates (114). Recent results indicate that *shaggy*/GSK-3 is part of a signalling pathway downstream of *Notch* (115).

An interesting method of labelling both catalytic and regulatory subunits has been described (116) for monitoring the compartmentation of the dissociated subunits. Cyclic AMP-dependent protein kinase was labeled with fluorescein and rhodamine on catalytic and regulatory subunits, respectively, and injected into *Aplysia* sensory neurons; labelling was followed by confocal fluorescence microscopy. Also, fluorescence emission anisotropy has been used for monitoring dansylated calmodulin and its affinity for Ca^{2+}-calmodulin-dependent protein kinase. Autophosphorylation markedly slowed the release of bound calcium-calmodulin; in other words, calmodulin is trapped by autophosphorylation (117).

A new type of protein kinases uses DNA as a signal to act enzymatically. DNA-activated phosphorylation was first observed in lysates of rabbit reticulocytes, and, subsequently, in ex tracts from cultured human cells and in oocytes, eggs, or embryos of several marine invertebrates (118). DNA-activated protein kinase is a nuclear serine/threonine protein kinase that is activated *in vitro* by DNA fragments. Its cellular targets are nuclear, DNA-binding, regulatory proteins, such as transcription factors cJun, cFos, cMyc, and Sp1, SV40 large tumor antigen, replication factor A, tumor supressor protein p53, topoisomereases I and II, polyomavirus VP1; and also non-DNA-binding proteins as heat shock protein 90 and microtubule-associated protein tau (119). The coordination of nuclear processes and the modulation of checkpoint mechanisms activated by DNA damage have been suggested as the roles to be accomplished for these kinases.

The broad specificities displayed by many protein kinases and phosphatases *in vitro*, including towards non-physiological protein substrates, allow to suggest that their activities be strictly regulated *in vivo*. The targetting subunit hypothesis establishes the existence of targetting subunits as parts of protein kinases or protein phosphatases which directs the catalytic subunit to the target locus; these targetting subunits specify the location, catalytic and regulatory properties of kinases and phosphatases, ensuring the fidelity of protein phosphorylation (74). Thus, targetting subunits provide a means of introducing specificity to pleiotropic protein kinases and phosphatases, tailoring their properties to the needs of a particular subcellular compartment, locus or cellular function. There is evidence that at least two classes of protein kinases (PKA and cyclin-dependent protein kinases) are regulated by types of targetting subunits. These two protein kinases are able to phosphorylate many proteins *in vitro* and their localization to particular sites *in vivo* by targetting subunit types may similarly serve to restrict their actions to physiologically relevant substrates.

Another family of proteins, known as the 14-3-3 family, identified as a series of acidic brain proteins, widely distributed in most mammalian tissues and found also in a wide range of other eukaryotic organisms including plants, insects, amphibians and yeasts. Initially, the first function attributed to the 14-3-3 proteins was the activation of tyrosine and tryptophan hydroxylases, the rate-limiting enzymes involved in catecholamine and serotonin neurotransmitter biosynthesis, respectively (120). PKC, cAMP-dependent protein kinase and 14-3-3 protein phosphorylate an identical site on tyrosine hydroxylase; the Ca^{2+}-calmodulin-dependent protein kinase II phosphorylates an additional unique site. The outstanding feature of this family is the extraordinarily high sequence conservation observed. The acidic proteins of 29-33 kDa from sheep brain are potent inhibitors of PKC (KCIP-1). Members of this protein family may regulate the subclass of PKC isoforms (α, β and γ) that are Ca^{2+} dependent and translocate to the plasma membrane as part of their activation mechanisms (121).

Kinases are deactivated and dephosphorylated by a single type-2A protein phosphatase. Also, the catalytic subunit of type-2A protein phosphatase was phosphorylated by tyrosine-specific protein kinases and enhanced in the presence of the phosphatase inhibitor okadaic acid, consistently with an autodephosphorylation reaction. Phosphorylation was catalyzed by epidermal growth factor receptors, insulin receptors, $p60^{v\text{-}src}$, and $p56^{lck}$ (122). A mouse phosphoprotein phosphatase, Syp, containing two Src homology 2 domains (SH2) has been identified (123); it bound to autophosphorylated epidermal growth factor and platelet-derived growth factor receptors through its SH2 domains and was rapidly phosphorylated on tyrosine in both ligand-stimulated cells. Syp may function in mammalian embryonic development and as a common target of both receptor and non-receptor tyrosine kinases.

It has been previously mentioned that many extracellular ligand receptors are tyrosine kinases that catalyzes autophosphorylation and phosphorylation of a number of cytoplasmic proteins. Although the first tyrosine kinase was discovered as the protein product of the Rous sarcoma virus oncogene, $p60^{v\text{-}src}$, the normal cellular forms of these enzymes occur either as transmembrane receptors or as cytoplasmic non-receptor proteins associated with the inner surface of the plasma membrane. Subsequent transmission of the ligand-induced signals depends on the recognition of the phosphorylated tyrosines by distinctive domains containing about 100 amino acid residues (*src* homology-2, SH2). These domains complement the action of the catalytic kinase activity by communicating the phosphorylation states of signal transduction proteins to elements of the signalling pathway. Three-dimensional structures of complexes of the SH2 domain of the *v-src* oncogene product with two phosphotyrosyl peptides have been determined by X-ray crystallography. A central antiparallel β-sheet is flanked by two α-helices, with peptide binding mediated by the sheet, intervening loops and one of the helices. The structure also defines those elements of the sequence that are required for maintaining the integrity of the three-dimensional fold, and hence provides a basis for the design of proteins (124).

Among the roles of tyrosine phosphorylation by the concerted actions of protein tyrosine kinases and protein tyrosine phosphatases, it is interesting to mention the recent studies on the pathogenic mechanisms of several bacterial genera, including the *Yersinia*, *Salmonella* and *Escherichia*, revealing novel strategies of infection that involve the signal transduction processes of the mammalian host. Tyrosine phosphorylation is implicated in the bacterial-host cell interactions (125).

Entry into mitosis in *Schizosaccharomyces pombe* is negatively regulated by the *wee1+* gene, which encodes a protein kinase with serine-, threonine-, and tyrosine phosphorylating activities (126). The *wee1+* kinase negatively regulates mitosis by phosphorylating p34^{cdc2} on tyrosine 15, thereby inactivating the p34^{cdc2}- cyclin B complex. Biochemical evidence on the stimulation of the Ras guanine nucleotide exchange factor activity by epidermal and platelet-derived growth factors in quiescent NIH 3T3 cells has been described (127). The results of these studies suggest that the guanine nucleotide exchange factor plays a major role in the Ras activation in cell proliferation initiated by growth factor receptors (TPKs) and malignant transformation by oncogenic TPKs and that tyrosine phosphorylation of either the exchange factor or a tightly bound protein may mediate the activation of the exchange factor by these TPKs. The mitogen-induced gene *PAC-1* encodes the sequence of the enzymatic site of known protein phosphotyrosine phosphatases; PAC-1 is similar to a phosphatase induced by mitogens or heat shock in fibroblasts, and a vaccinia virus-encoded serine-tyrosine phosphatase (128).

Interferon-α-stimulated gene factor 3(ISGF3), a transcriptional activator, contains three proteins that reside in the cell cytoplasm until they are activated in response to IFN-α. Treatment of cells with IFN-α caused these three proteins to be phosphorylated on tyrosine and to translocate to the cell nucleus where they stimulate transcription through binding to IFN-α-stimulated response elements in DNA (129). Interferon-γ regulates also gene expression by tyrosine phosphorylation of several transcription factors that have the protein of interferon-stimulated gene factor 3 as a common component (130). Intraperitoneal injection of epidermal growth factor into mice resulted in the appearance in liver nuclei of three tyrosine phosphorylated proteins (84, 91 and 92 kDa) within minutes. Administration of interferon-γ resulted in the appearance in liver nuclei of two tyrosine phosphorylated proteins (84 and 91 kDa). In both cases, the 84 and 91 kDa proteins were identified as the IFN-γ activation factors (131). Several cytokines other than interferons can activate putative transcription factors by tyrosine phosphorylation (130).

Transient tyrosine phosphorylation of an overlapping but distinct sets of proteins as growth signals has been detected by interaction of interleukin-3 and interleukin-2 with their membrane receptors (132).

Tyrosine kinases are important also in the signalling pathways regulating the orderly expression of cell surface markers and responses to specific activation signals of B cells. The cytoplasmic tyrosine kinase, the Bruton's tyrosine kinase, is crucial for B cell development and loss of kinase activity results in a human immunodeficiency, X-linked agammaglobulinemia, characterized by a failure to produce B cells (133). In the murine K-linked immunodeficiency (XID), B cells are present but respond abnormally. In murine XID, a mutation in the amino terminal region of these tyrosine kinase interferes with normal B cell signalling.

As a final, I should like to emphasize the wonderful variety of signals, the complexity of cascades, the crosstalks between cascades, the diversity of steps and branches, the enormous heterogeneity of responses, the great evolutionary flexibility, and the important physiopathological implications. And, we are here for thinking on the answers to many crucial questions on molecular and cell biology of signal transduction, as what causes cells to begin and to stop dividing and the many disturbances that the regulatory mechanisms can rise. Nevertheless, twenty five centuries ago, the Greek philosopher Socrates advised: "Only the good questions will allow us to keep the intelectual honesty". Thus, I am sure of the level of the questions thaat will be raised during these days. Let me add that I am also sure of the answers.

* * * *

As the Director of this Symposium organized as a part of the Summer Courses of the UNIVERSIDAD COMPLUTENSE in this place of EL ESCORIAL, my first duty and great pleasure is to welcome our foreign lecturers to our country and to this old University. On behalf of the Spanish scientific community, I thank you all for coming to participate in this course on one of the most exciting topics of the present molecular and cellular biology. We do hope that, at the same time, you will enjoy your stay with us and have the opportunity to know our historical places and scientific environment.

In addition, in the last decades, the scientific organization of this country, and, mainly, the efforts of the Spanish scientists have achieved a remarkable international position. The participation of a short and selected number of our scientists as lecturers in this Symposium is also a good signal of that. They represent to many others doing research at the present time on the topics covered by the wide umbrella of "signal transduction". They are contributing for filling pieces in the international and beautiful puzzle of how the cells respond to the many signals they receive from the environment.

Thank you, and welcome!

REFERENCES

1.   Mire-Sluis, A. (1993) *Tibtech*. **11**, 74-77.
2.   Ramshaw, I.A., Ruby, J. and Ramsay, A. (1992) *Tibtech*. **10**, 424-426.
3.   Jacob, C.O. and McDevitt, H.O. (1991) In *Molecular Autoimmunity*, pp.7-29. Academic Press.
4.   Jacob, C.O. (1992) *Immunol. Today* **13**, 122-125.
5.   Hotamisligil, G.S., Shargill, N.S. and Spiegelman, B.M. (1993) *Sience* **259**, 87-91.
6.   Spiegelman, B.M. and Hotamisligil, G.S. (1993) *Cell* **73**, 625-627.
7.   Müller, M. *et al.* (1993) *Nature* **366**, 129-135.
8.   Watling, D. *et al.* (1933) *Nature* **366**, 166-170.
9.   Barnard, E.A. (1992) *TIBS* **17**, 368-382.
10. Miyajima, A., Hara, T. and Kitamura, T. (1992) *TIBS* **17**, 378-382.
11. Fraser, J.D., Straus, D. and Weiss, A. (1993) *Immunol. Today* **14**, 357-362.
12. Abraham, R.T., Karnitz, L.M., Sacrist, J.P. and Leibson, P.J. (1992) *TIBS* **17**, 434-438.
13. June, C.H., Fletcher, M.C., Ledbetter, J.A. *et al.* (1990) *Proc. Natl. Acad. Sci.* **87**, 7722-7726.
14. Downward, J. *et al.* (1990) *TIBS* **346**, 719-723.
15. Wojcikiewicz, R.J.H., Tobin, A.B. and Nahorski, S.R. (1993) *TIPS* **14**, 279-285.
16. Sadowski, H.B., Shuai, K., Darnell, J.E. and Gilman, M.Z. (1993)*Science* **262**, 1739-1744.
17. Sandor, M. and Lynch, R.G. (1993) *Immunol. Today* **14**, 227-231.
18. Beaven, M.A. and Metzger, H. (1993) *Immunol. Today* **14**, 222-226.
19. Lynch, R.G. and Sandor, M. (1990) In *Fc Receptors and the Action of Antibodies* (Metzger, H., ed.), pp. 305-334.
20. Mustelin, T. and Burn, P. (1993) *TIBS* **18**, 215-22.

21. Volaarevic, S., Niklinska, B.B., Burns, C.M., June, C.H. *et al.* (1993) *Science* **260**, 541-544.
22. Vogel, W., Lammers, R., Huang, J. and Ullrich, A. (1993) *Science* **259**, 1611-1614.
23. Matsuda, M., Mayer, B.J. and Hanafusa, H. (1991) *Mol.Cell.Biol.* **11**, 1607-1613.
24. Kazlauskas, A., Kashishian, A., Cooper, J.A., and Valius, M. (1992) *Mol. Cell. Biol.* **12**, 2534-2544.
25. Sjolander, A., Yamamoto, K., Huber, B.E. and Lapetina, E.G. (1991) *Proc. Natl. Acad. Sci.* **88**, 7908-7912.
26. Pazin, M.J. and Williams, L.T. (1992) *TIBS* **17**, 374-378.
27. Silvennoinen, O., Witthuhn, B.A., Quelle, F.W. *et al.* (1993) *Proc. Natl. Acad. Sci.* **90**, 8429-8433.
28. Müller, M., Briscoe, Laxton, C., Guschin, D. *et al.* (1993) *Nature* **366**, 129-135.
29. Harnett, M.M. and Klaus, G.G.B. (1988) *Immunol. Today* **9**, 315-320.
30. Cook, S.J. and Wakelam, M.J.O. (1992) *Rev. Physiol. Biochem. Pharmacol.* **119**, 13-45.
31. Simon, M.I., Strathman, M.P. and Gautam, N. (1991) *Science* **252**, 802-808.
32. Harnett, M. and Rigley, K. (1992) *Immunol. Today* **13**, 482-486.
33. Dohlmann, H.G., Caron, M.G. and Lefkowitz, R.J. (1987) *Biochem. J.* **26**, 2657-2664.
34. Melamed, I., Wang, G. and Roifman, C.M. (1992) *J. Immunol.* **149**, 169-174.
35. Yang, Z. and Watson, J.C. (1993) *Proc. Natl. Acad. Sci.* **90**, 8732-8736.
36. Chavrier, P., Gorvel, J.P. and Bertoglio, J. (1993) *Immunol. Today* **14**, 440-444.
37. Hepler, J.R. and Gilman, A.G. (1992) *TIBS* **17**, 383-387.
38. Sternweis, P.C. and Smrcka, A.V. (1992) *TIBS* **17**, 502-506.
39. Tamir, H., Fawzi, A.B., Tamir, A., Evans, T. and Northup, J.K. (1991) *Biochemistry*, **30**, 3929-3934.
40. Hildebrandt, J.D. *et al.* (1985) *J. Biol. Chem.* **260**, 14867-14871.
41. Kleuss, C., Scherübl, H., Hescheler, J., Schultz, G. and Wittig, B. (1993) *Science* **259**, 832-834.
42. Kleuss, C., Scherübl, H., Hescheler, J., Schultz, G. and Wittig, B. (1992) *Nature*, **358**, 424-426.
43. Clapham, D.E. and Neer, E.J. (1993) *Nature* **365**, 403-406.
44. Katz, A., Wu, D. and Simon, M.I. (1992) *Nature* **360**, 686-689.
45. Pitcher, J.A., Inglese, J., Higgins, J.B., Arriza, J.L. *et al.* (1992) *Science* **257**, 1264-1267.
46. Bauer, P. *et al.* (1992) *Nature* **358**, 73-75.
47. Dolph, P.J., Ranganathan, R., Colley, N.J. *et al.* (1993) *Science* **260**, 1920-1916.
48. Cotecchia, S., Exum, S., Caron, M.G. and Lefkowitz, R.J. (1990) *Proc. Natl. Acad. Sci.* **87**, 2896-2900.
49. Parma, J. *et al.* (1993) *Nature* **365**, 649-651.
50. Shenker, A. *et al.* (1993) *Nature* **365**, 652-654.
51. Lefkowitz, R.J. (1993) *Nature* **365**, 603-604.
52. Wu, D., LaRosa, G.J. and Simon, M.I. (1993) *Science* **261**, 101-103.
53. Sudhof, T.C. and Jahn, R. (1991) *Neuron* **6**, 665-668.
54. Hess, S.D., Doroshenko, P.A. and Augustine, G.J. (1993) *Science* **259**, 1169-1172.
55. Bartfai, T., Fisone, G. and Langel, U. (1992) *TIPS* **13**, 312-317.
56. Taylor, C.W. and Richardson, A. (1991) *Pharmacol. Ther.* **51**, 97-137.
57. Taylor, C.W. and Marshall, C.B. (1992) *TIBS* **17**, 403-407.
58. Berridge, M.J. and Irvine, R.F. (1989) *Nature* **341**, 197-205.
59. Isaacks, R.E. and Harkness, D.R. (1980) *Am. Zool.* **20**, 115-129.
60. Vallejo, M., Jackson, T. and Hanley, M.R. (1988) *Nature* **330**, 656-658.
61. Menniti, F.S., Oliver, K.G., Putney, J.W. and Shears, S.B. (1993) *TIBS* **18**, 53-56.
62. Exton, J.H. (1990) *J. Biol. Chem.* **265**, 1-4.
63. Billah, M.M. and Anthes, J.C. (1990) *Biochem. J.* **269**, 281-291.
64. Axelrod, J., Burch, R.M. and Jelsema, C.L. (1988) *Trends Neurosci.* **11**, 117-123.
65. Yoshida, K., Asaoka, Y. and Nishizuka, Y. (1992) *Proc. Natl. Acad. Sci.* **89**, 6443-6446.
66. Asaoka, Y., Nakamura, S., Yoshida, K. and Nishizuka, Y. (1992) *TIBS* **17**, 414-417.
67. Durieux, M.E. and Lynch, K.R. (1993) *TIPS* **14**, 249-254.
68. Naim, A.C., Hemmings, H.C. and Greengard, P. (1985) *Annu. Rev. Biochem.* **54**, 931-976.

69. Ohmstede, C.A., Jensen, K.F. and Sahyoun, N. (1989) *J. Biol. Chem.* **264**, 5886-5875.
70. Stull, J.T. (1988) *Mol. Asp. Cell. Reg. (Vol.5)* (Cohen, P. and Klee, C.B., eds), pp. 91-122. Elsevier.
71. Cohen, P. (1988) *Mol. Asp. Cell. Reg. (Vol.5)* (Cohen, P. and Klee, C.B. eds), pp. 123-144. Elsevier.
72. Ryazanov, A.G., Shestakova, E.A. and Watapor, P.G. (1988) *Nature* **334**, 170-173.
73. Cohen, P. (1992) *TIBS* **17**, 408-413.
74. Hubbard, M.J. and Cohen, P. (1993) *TIBS* **18**, 172-177.
75. Cohen, P. (1988) *Proc. R. Soc. London Ser. B* **234**, 115-144.
76. Nishizuka, Y. (1988) *Nature* **334**, 661-665.
77. Klann, E., Chen, S. and Sweatt, D. (1993) *Proc. Natl. Acad. Sci.* **90**, 8337-8341.
78. Sacktor, T.C., Osten, P., Valsamis, H. *et al.* (1993) *Proc. Natl. Acad. Sci.* **90**, 8342-8346.
79. Schwartz, J.H. (1993) *Proc. Natl. Acad. Sci.* **90**, 8310-8313.
80. Duxbaum, J.D., Koo, E.H. and Greengard, P. (1993) *Proc. Natl. Acad. Sci.* **90**, 9195-9198.
81. De Matteis, M.A., Santini, G., Kahn, R.A., Di Tullio, G. and Luini, A. (1993) *Nature* **364**, 818-821.
82. Hardie, R.C., Peretz, A., Suss-Toby, E., Rom-Glas, A. *et al.* (1993) *Nature* **363**, 634-637.
83. Slater, S.J., Cox, K.J.A., Lombardi, J.V. *et al.* (1993) *Nature* **364**, 82-84.
84. Li, M., West, J.W., Numann, R., Murphy, B.J. *et al.* (1993) *Science* **261**, 1439-1441.
85. Hoshi, M., Nishida, E. and Sakai, H. (1988) *J. Biol. Chem.* **263**, 5396-5401.
86. Nishida, E. and Gotoh, Y. (1993) *TIBS* **18**, 128-131.
87. Moodie, S.A., Willumsen, B.M., Weber, M.J. and Wolfman, A. (1933) *Science* **260**, 1658-1661.
88. Myriakis, J.M., App, H., Zhang, X., Banerjee, P. *et al.* (1992) *Nature* **358**, 417-421.
89. Kolch, W., Heidecker, G., Kochs, G. *et al.* (1993) *Nature* **364**, 249-252.
90. Skolnik, E.Y., Batzer, A., Li, N., Lee, C.H. *et al.* (1993) *Science* **260**, 1953-1955.
91. Boulton, T.G., Yancopoulos, G.D., Gregory, J.S., Slaughter, C., *et al.* (1990) *Science* **249**, 64-67.
92. Boulton, T.G., and Cobb, M.H. (1991) *Cell Regul.* **2**, 357-371.
93. L'Allemain, G., Her, J.H., Wu, J., Sturgill, T. and Weber, M. (1992) *Mol. Cell. Biol.* **12**, 2222-2229.
94. Pagès, G., Lenormand, P., L'Allemain, G., Chambard, J.C. *et al.* (1993) *Proc. Natl. Acad. Sci.* **90**, 8319-8323.
95. Osada, S. *et al.* (1990) *J. Biol. Chem.* **265**, 22434-22440.
96. Pines, J. (1933) *TIBS* **18**, 195-197.
97. Koff, A., Ohtsuki, M., Polyak, K. *et al.* (1993) *Science* **260**, 536-539.
98. Gabrielli, B.G., Roy, L.M. and Maller, J.L. (1993) *Science* **259**, 1766-1769.
99. De Bondt, H.L., Rosenblatt, J., Jancarik, J. *et al.* (1993) *Nature* **363**, 595-602.
100. Taylor, S.S., Knighton, D.R., Zheng, J. *et al.* (1993) *TIBS* **18**, 84-89.
101. Kipreos, E.T. and Wang, J.Y.J. (1990) *Science* **248**, 217-220.
102. Mendenhall, M.D. (1993) *Science* **259**, 216-219.
103. Kuo, C.J., Chung, J., Florentino, D.F. *et al.* (1992) *Nature* **358**, 70-73.
104. Fu, Y.H. (1992) *Science* **255**, 1256-1258.
105. Brook, J.D. *et al.* (1992) *Cell* **68**, 799-808.
106. Fu, Y.H. *et al.* (1993) *Science* **260**, 235-238.
107. Dawson, T.M., Arriza, J.L., Jaworsky, D.E. *et al.* (1993) *Science* **259**, 825-829.
108. Wang, L.Y., Taverna, F.A., Huang, X.P. *et al.* (1993) *Science* **259**, 1173-1175.
109. Raymond, L.A., Blackstone, C.D. and Huganir, R.L. (1993) *Nature* **361**, 637-641.
110. McGlade-McCulloh, E., *et al.* (1993) *Nature* **362**, 640-642.
111. Sculptoreanu, A., Scheuer, T. and Catterall W.A. (1993) *Nature* **364**, 240-243.
112. Gross, E., Goldberg, D. and Lavitzki, A. (1992) *Nature* **360**, 762-765.
113. Galyov, E.E., Hakansson, S. *et al.* (1993) *Nature* **361**, 730-732.
114. Woodgett, J.R. (1991) *TIBS* **16**, 177-181.
115. Ruel, L., Bourouis, M. *et al.* (1993) *Nature* **362**, 557-559.
116. Bacskai, B.J., Hochner, B., Mahaut-Smith, M. *et al.* (1993) *Science* **260**, 222-226.
117. Meyer, T., Hanson, P.I., Stryer, L. and Schulman, H. (1992) *Science* **256**, 1199-1202.
118. Walker, A.I., Hunt, T., Jackson, R.J. and Anderson, C.W. (1985) *EMBO J.* **4**, 139-145.

119. Anderson, C.W. (1933) *TIBS* **18**, 433-437.
120. Ichimura, T. *et al.* (1987) *FEBS Lett.* **219**, 79-82.
121. Aitken, A., Collinge, D.B., van Heusden, B.P.H. *et al.* (1992) *TIBS* **17**, 498-501.
122. Chen, J., Martin, B.L. and Brautigan, D.L. (1992) *Science* **257**, 1261-1264.
123. Feng, G.S., Hui, C.C., and Pawson, T. (1993) *Science* **259**, 1607-1610.
124. Waksman, G., Kominos, D., Robertson, S.C., Pant, N. *et al.* (1992) *Nature* **358**, 646-653.
125. Bliska, J.B. and Falkow, S. (1993) *TIG* **9**, 85-89.
126. Parker, L.L. and Piwnica-Worms, H. (1992) *Science* **257**, 1955-1957.
127. Li, B.Q., Subleski, M., *et al.* (1993) *Proc. Natl. Acad. Sci.* **90**, 8504-8508.
128. Rohan, P.J., Davis, P., Moskaluk, C.A. *et al.* (1993) *Science* **259**, 1763-1767.
129. Schindler, C., Shuai, K., Prezioso, V.R. and Darnell, J.E. (1992) *Science* **257**, 809-812.
130. Larner, A.C., David, M., Feldman, G.M. *et al.* (1993) *Science* **261**, 1730-1733.
131. Ruff-Jamison, S., Chen, K. and Cohen, S. (1933) *Science* **261**, 1733-1736.
132. Chiba, T., Nagata, Y., Machide, M. *et al.* (1993) *Nature* **362**, 646-649.
133. Rawlings, D.J., Saffran, D.C., Tsukada, S. *et al.* (1993) *Science* **261**, 358-361.

TYROSINE PHOSPHATASES IN CELL CYCLE AND TRANSFORMATION

Edmond H. Fischer

Department of Biochemistry
University of Washington,
Seattle, Washington, 98195

SUMMARY

The first part of this article is devoted to a historical overview of cellular regulation by protein phosphorylation. The field originated with a study of glycogen phosphorylase and the cascade system by which this enzyme can be activated. Reversible protein phosphorylation can be considered today as one of the most prevalent mechanism of control of biological systems. The second part will be devoted to protein tyrosine phosphatases (PTPs), an expanding family of intracellular and receptor-linked enzymes, and their involvement in cell cycle and transformation. Most transmembrane forms contain two catalytic domains but highly variable external segments. Likewise, the low Mr PTPs display a great variety of regulatory segments either preceding or following highly conserved catalytic domains. Deletion of the regulatory segment from a T-cell PTP and overexpression of the truncated enzyme in BHK cells results in multinucleation and asynchronous nuclear division. Furthermore, growth on soft agar and tumor formation in nude mice is inhibited. Similar results are obtained with rat-2 cells transformed with v-fms. The data indicate that, depending on the type of phosphatases involved and their localization within the cells, PTPs can act either synergistically or antagonistically with the tyrosine kinases to elicit a particular physiological response.

HISTORICAL OVERVIEW

Fifty years ago, very little was known about cellular regulation and nothing, of course, about a possible involvement of protein phosphorylation/dephosphorylation in these processes. In fact, this is how they were described around 1950 in the Hawk, Oser and Summerson, a classical source of biochemical information:

> "These proteins contain phosphate bound in ester linkage to the hydroxy amino acids serine and threonine. They include casein from milk, ovovitellin from egg yolk, and other proteins associated with the feeding of the young; also the proteolytic enzyme pepsin".

But in the mid '50's, a first indication was obtained showing that protein phosphorylation could serve as a means to regulate the activity of glycogen phosphorylase, the enzyme that catalyzes the first step in the degradation of glycogen. The muscle enzyme was discovered by Parnas in Poland and Carl and Gerti Cori in the U.S. In the 30's, this enzyme was thought to have an absolute requirement for AMP for activity (Fig. 1). But in the early 40's, Arda Green in Cori's lab crystallized it in a form that was active

Cell Signal Transduction, Second Messengers, and Protein Phosphorylation in Health and Disease
Edited by A.M. Municio and M.T. Miras-Portugal, Plenum Press, New York, 1994

23

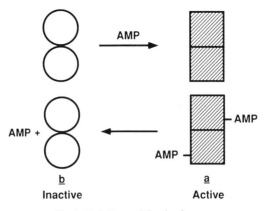

Fig. 1. Early history of phosphorylase.

without added AMP (1,2). They called this form phosphorylase a, and logically assumed that it contained covalently-bound AMP. They further thought that it had to be the native form of the enzyme because, when left standing in crude extracts, it was rapidly converted to the old species which they called phosphorylase b. However, they could never find any AMP released in the reaction, and no AMP or adenine or ribose was detected in the "native" enzyme, using the most sensitive microbiological assays available at that time. They knew that the enzyme existed in two forms but did not know how these two forms differed. And they actually dropped the problem.

It is in the mid-50's that, with Ed Krebs, we decided to take a crack at this problem. Within a few months, we found that muscle phosphorylase could be activated by a totally different mechanism, namely, by phosphorylation of the protein (3,4). The reaction had to be enzymatic, requiring a kinase which we called phosphorylase kinase; the reverse reaction had to be catalyzed by a phosphorylase phosphatase. It did not occur spontaneously in crude muscle extracts because there is an absolute requirement for Ca2+ or Mn2+. The only possibility then, was that phosphorylase kinase also existed in inactive and active forms, and that Ca2+ was required for the activation reaction (Fig. 2).

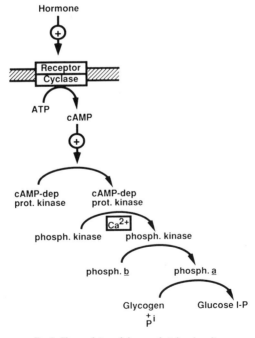

Fig. 2. The regulation of glycogenolysis by adrenaline.

It should be noted that during the same period of time, Earl Sutherland's laboratory arrived at the same conclusion, working on liver phosphorylase. He and Ted Rall and Wosilait (5-7) showed that inorganic phosphate was released when active liver phosphorylase was incubated with a separate liver fraction, also suggesting that interconversion of the two forms might involve a phosphorylation/dephosphorylation reaction. An epochal finding that grew out of these studies was the discovery by Earl of cAMP which, by an unknown mechanism was found to shift the balance between non-activated and activated liver phosphorylase toward the latter form (8). Sutherland generously supplied us with samples of cAMP which allowed us to determine that its action was directed towards the activation of phosphorylase kinase, not directly, but by the activation of a second kinase (a kinase-kinase as we first called it) which turned out to be the cAMP-dependent protein kinase which Ed Krebs and Don Walsh isolated several years later. Therefore, the activation of phosphorylase occurs at the last step of a complex cascade system of signaling events originating at the membrane level as depicted in Fig. 2.

At the time of its discovery, it was not known whether the reversible phosphorylation of glycogen phosphorylase represented a unique occurrence, restricted to the control of that enzyme only or, at best, of carbohydrate metabolism. As it turned out, it is one of the most widespread mechanisms by which cellular processes can be regulated (Fig. 3) (9).

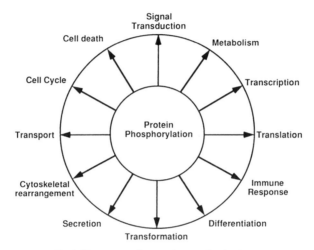

Fig. 3. Ubiquitous involvement of protein phosphorylation.

ALLOSTERIC vs. COVALENT REGULATION

Here, a question should be raised: why two different mechanisms to control the activity of a single enzyme, namely, allosteric regulation by AMP and covalent regulation by phosphorylation of the protein. According to the model proposed by Jacques Monod in the early '60's, under allosteric regulation, the enzyme responds to effectors that are generated during the normal housekeeping operations of the cell and reflect its overall internal condition: whether it is active or quiescent, its metabolic state, its energetic balance, that is, the ratio of AMP/ATP, etc. According to the rule that enzymes are subjected to end-product or feed-back inhibition, phosphorylase would be expected to be inhibited by G6P (an early metabolite of the phosphorylase reaction) and by ATP, the ultimate end product of carbohydrate metabolism. By the same token, it would be activated by AMP as indeed it is. But then, many enzymes of the glycolytic pathway would be subjected to this same type of regulation and all these "doors" would become open at the same time. By contrast, the phosphorylase kinase reaction is highly specific; it allows the activation of that single enzyme without affecting the activity of any other. Furthermore, and this is perhaps one of the major lessons we have learned over the last 30 years, covalent regulation responds mainly to external signals.

A classical example is that of the liver; its prime purpose in regulating glycogenolysis is not to satisfy its own metabolic needs, but to maintain blood glucose levels for the benefit of other organs, particularly the brain and the erythrocyte during fasting. Therefore, the sensory mechanism will be external, and the signals that will be sent to the liver to maintain glucose homeostasis will come from the outside, principally in the form of hormones released from the adrenals and the pancreas.

25

External signals will come in the form of hormones, neurotransmitters, growth factors and other stimuli such as drugs, light, odorants, probably touch in plants. They will act on their specific membrane receptors and, in reactions regulated by G-proteins, cause the release of second messengers or induce the intrinsic protein tyrosine kinase (PTK) activity of the receptors themselves (Fig 4). These second

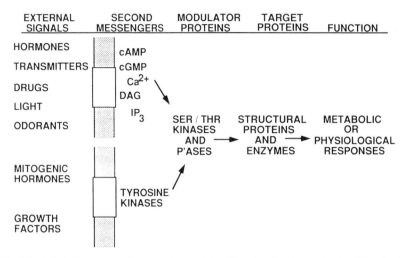

Fig. 4. Control of cellular processes by protein phosphorylation. Cascade system triggered by extracellular stimuli.

messengers or internal signals will act on other kinases, then on target enzymes to finally elicit a metabolic or physiological response. As can be seen, one has here all the elements of a cascade system: enzymes acting on enzymes and providing not only the enormous amplification of an external signal, but also the possibility of bringing about the coordinate or synchronous regulation of different physiological events through the pleiotropic action of some of the enzymes involved.

The upper pathways, such as the one described earlier, are, for the most part well understood. As to the lower pathways, we are just beginning to understand the steps that link a signaling system that originates at the membrane level and relies on tyrosine phosphorylation, and the more widespread Ser/Thr phosphorylation reactions that occur downstream. Some of the enzymes have dual specificity, or are regulated by both tyrosine and Ser/Thr phosphorylation. This dual control would ensure that no accidental initiation of crucial cellular events would occur at inappropriate times, just as one would need two keys to gain access to a safety deposit box.

TYROSINE KINASES

Regarding protein tyrosine phosphorylation, this has represented one of the most exciting developments that has occurred in the last 15 years with the realization that this reaction was intimately implicated in cell growth, proliferation, differentiation and transformation. This brings into play tyrosine kinases of cellular or viral origin (shown by the work of Erickson and his coworkers (10,11), Levinson et al (12), and Tony Hunter and Bart Sefton on Rous sarcoma virus), or associated with hormone and growth factor receptors (as shown by Stanley Cohen for the EGF receptor) (13-15) such as depicted in Fig. 5.

Just as overexpression or mutation of the intracellular tyrosine kinases can lead to cell transformation, mutation of the growth factor receptors can generate oncogenic products. The first to be identified was the retroviral oncogene v-erb B generated by a truncation of the external domain of the EGF receptor, its cellular progenitor (16,17). Many others have since been cloned and characterized.

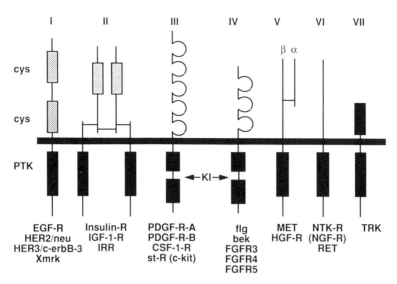

Fig. 5. Growth factor tyrosine kinase receptors.

In most instances, they involve truncation at the amino and/or carboxyl-terminus point mutations here and there or fusion to residues borrowed from other viral or animal genes. With accumulating evidence implicating tyrosine phosphorylation in cell transformation (18,19), it is hardly surprising that several research groups would go after the protein tyrosine phosphatases that catalyze the reverse reaction.

TYROSINE PHOSPHATASES

First evidence for phosphotyrosine dephosphorylation in proteins was obtained by Graham Carpenter and Stanley Cohen using A431 cell membranes overexpressing the EGF receptor - allowing these to undergo autophosphorylation in the presence of EGF and then following their dephosphorylation (13,20). Similar data were obtained by Bart Sefton and Tony Hunter using cells transformed with a temperature-sensitive mutant of Rous sarcoma virus (21). Then followed a flurry of investigations by a number of groups, including our own six years ago carried out in collaboration with Nick Tonks (22-24). When we started, we too assumed somewhat naively that if transformation could be brought about by overexpression of the tyrosine kinases, or mutations that would render them constitutively active, then overexpression of the phosphatases would inevitably block or reverse these reactions. This assumption was incorrect. Within two years, Nick Tonks succeeded in isolating a tyrosine phosphatase in homogeneous form from human placenta (25,26). The enzyme was totally specific for phosphotyrosyl residues and with a specific activity up to 3 orders of magnitude higher than certain tyrosine kinases, suggesting that it had to be tightly regulated to allow for those mitogenic signals that are necessary for normal cell development.

But the surprise came when the amino acid sequence of the enzyme was determined because it showed no homology with any of the other serine/threonine phosphatases (27). However, a search of the data base indicated that it was structurally related to an abundant and already well-known surface antigen, the leukocyte common antigen also designated as CD45 (Fig. 6) (28).

The leukocyte common antigen represents a broad family of membrane-spanning molecules found in all hematopoietic cells except mature erythrocytes (29). Their intracellular moiety is highly conserved and contains two internally homologous domains of approximately 30 kDa each. It is those two domains that are structurally related to the placenta phosphatase. The same basic structure has been cloned from humans, rats, mice, chickens, etc. libraries. CD45 has been implicated in the regulation of lymphocyte function, including cytotoxicity, proliferation and differentiation and in modulating IL2 receptor expression.

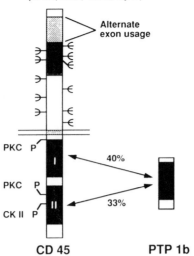

Fig. 6. The leukocyte common antigen CD45.

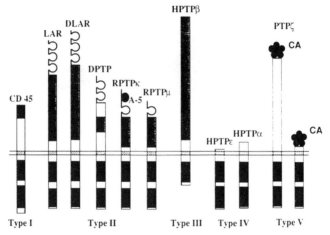

Fig. 7. Tyrosine phosphatase receptors.

Since then, a great variety of receptor forms have been identified by low stringency hybridization or PCR (fig. 7) (22-24). All but a couple display the same double catalytic domains in their cytoplasmic portion but a considerable diversity in their external segments. Some have structural characteristics of cell adhesion molecules such as the LARs first cloned by Haruo Saito and Stu Schlossman that appear to be related to the N-CAMs (Neuro Cell Adhesion Molecules) or fasciclin II, suggesting that they might be involved in homophilic cell-cell interactions and perhaps modulate morphogenesis and tissue development. Others contain many fibronectin type III repeats and might be involved in cell-cell or cell-matrix signaling. Some have very short external domains and their function has yet to be elucidated. Perhaps the most intriguing receptors are the ones that contain, at the end of a long or a short external arm, a globular molecule homologous to carbonic anhydrase except that it contains only one of the three histidines involved in the binding of Zn2+ (30,31). The function of this receptor is unknown. Except for CD45, no ligand has been found for any of these structures.

TYROSINE PHOSPHATASES IN CELL CYCLE AND TRANSFORMATION

Likewise, the low molecular weight, intracellular protein tyrosine phosphatases (PTPs) display a great diversity of structures, either preceding or following a highly conserved catalytic core. These are undoubtedly involved in the regulation and localization of the enzymes (Fig. 8). Some PTPs have segments homologous to cytoskeletal proteins such as band 4.1, ezrin, and talin; others contain two src-homology 2 (SH2) domains which might allow them to interact with phosphotyrosyl residues at sites of autophosphorylation of growth factor receptors. Tyrosine phosphatases are also found as the gene products (YOPs) of virulence plasmids from bacteria of the genus Yersinia (such as Y.pestis responsible for the bubonic plague) (24,32).

Now, I would like to discuss the role the regulatory domain might play in enzyme localization and function, looking at a human T-cell enzyme cloned by Debbie Cool (33). The regulatory domain is entirely hydrophilic until one reaches the last 19 residues that are very hydrophobic, in fact reminiscent of a transmembrane domain. There is also a short stretch of five basic residues that could serve as a nuclear localization signal. So this section was mutated, as well as other segments of putative physiological importance. As a first step, a premature stop codon was introduced after the catalytic domain to delete the entire C-terminal tail. When this is done, the truncated enzyme becomes soluble whereas the wild-type protein is particulate (34,35). Furthermore, large differences in cell cycle progression and transformation are observed.

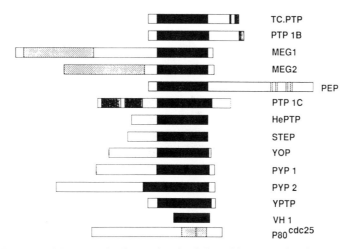

Fig. 8. Intracellular tyrosine phosphatases, aligned on the basis of their conserved, catalytic domains.

In BHK cells in which the wild-type enzyme is overexpressed, one sees no obvious change in cell morphology. By contrast, 60-70% of the cells overexpressing the truncated form are multinucleate (36). Blockage of cell division is not unusual and can be observed under a number of conditions. It can occur by cell fusion particularly under the influence of viruses; with drugs such as cytochalasin or nocodazol or with antibodies to myosin ATPase, since cell division is an actomyosin-dependent process (37,38). But in all these instances, when nuclear division goes on, it goes on synchronously: nuclei divide simultaneously. What is unusual here is that nuclear division that persists is more often than not asynchronous; that is, one nucleus will divide while the other will not, as shown in figure 9 (36). Therefore, one will see cells with nuclei at all phases of mitosis. At this time, we don't know which internuclei signals have been disrupted.

Differences in cell behavior brought about by expression of the wild-type vs. truncated T-cell enzyme are also seen in cell transformation, using the same highly tumorigenic BHK cell line. As expected, the transformed BHK cell line grows readily on soft agar but it grows just as well, if not better, when transfected with the wild-type full-length enzyme. By contrast, overexpression of the truncated form almost abolishes growth under these conditions, as if transformation had been suppressed.

A similar enhancement in tumorigenesis by overexpression of the full-length phosphatase is observed when these cells are injected into nude mice. Tumors produced are more fully developed and highly vascularized. By contrast, tumor formation is greatly reduced (if not suppressed) with BHK cells containing the truncated form. In several animals, no tumor was detected (D. Cool, unpublished data).

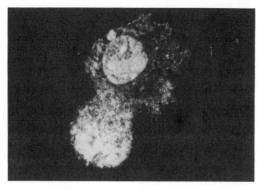

Fig. 9. BHK cells overexpressing the truncated 37 kDa form of the T-cell tyrosine phosphatase showing asynchronous nuclear division. The cell has rounded up around the lower nucleus ready to undergo mitosis. The nucleus on top is in an interphase configuration.

Since the nature of the transforming agent in these BHK cells was not defined, these studies were repeated in embryonic Rat-2 cells transformed with the well-defined viral oncogene product v-fms (39). V-fms was first isolated from a feline sarcoma virus and belongs to the platelet-derived growth factor (PDGF) family of tyrosine kinase receptors (40). Its non-transforming progenitor, the cellular protooncogene c-fms, encodes the receptor for CSF-1, the macrophage colony-stimulating factor. Binding of CSF-1 triggers signaling events that lead to the transcription of CSF-1 genes necessary for mononuclear phagocyte growth, differentiation and survival (41).

Control rat-2 cells display a typical, non-transformed, cobblestone morphology - as opposed to the spindle-shape, stringy v-fms transformed cells (Fig. 10). Cells overexpressing the full-length enzyme

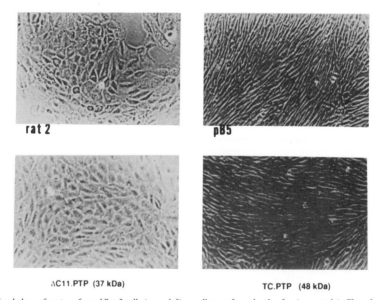

ΔC11.PTP (37 kDa)   TC.PTP (48 kDa)

Fig. 10. Morphology of nontransformed Rat-2 cells (upper left) or cells transformed with v-fms (upper right). The v-fms-transformed cell lines were cotransfected with full-length T-cell tyrosine phosphatase (lower right) or its 37 kDa truncated form (lower left). The data indicate that while the full-length enzyme has not altered the transformed (spindle-shape) morphology, cells containing the truncated enzymes display a nontransformed phenotype.

have the same transformed, stringy appearance while clones containing the truncated enzyme present a morphology similar to the non-transformed phenotype (39). The transformed cells will grow on soft agar, the others will not. To achieve a high level of expression of the tyrosine phosphatase, the enzyme was packaged in retroviruses with which the cells were infected. That the transformed cells had reverted to a non-transformed state was further demonstrated by examining their cytoskeleton. In transformed cells, the array of actin microfilaments are disrupted and they show a loss of focal adhesion, the plaques with which normal cells adhere to their substratum. Actin fibers and focal adhesions are restored in cells expressing the truncated enzyme.

Finally, the same differences were observed in the ability of these two types of cells to form tumors when injected into nude mice. All cells containing the v-fms oncogenes produced large tumors, and co-transfection with the full-length, wild-type enzyme didn't change anything. By contrast, tumor formation was totally abolished in most cells transfected with the truncated/deregulated enzyme. After many passages, though, some clones produced tumors once again but when these were examined, it was apparent that the cells no longer contained the phosphatase. In other words, they had found ways to reject the enzyme or block its transcription or expression, thereby allowing tumors to reappear in the long run. Once again, we don't know the mechanism of this rejection process.

CONCLUSIONS

The above data, plus others I did not have the time to develop, indicate: first, that phosphatases cannot be viewed simply as providing an "off" switch in an "on/off" kinase/phosphatase system. In certain cases and depending on the form of the enzyme involved, tyrosine phosphatases can clearly act synergistically with the kinases to bring about a particular physiological response. An obvious way by which they could do that would be by activating the src family kinases whose activities are repressed by phosphorylation at the C-terminus (Fig. 11). Removal of this phosphate group activates the enzyme many

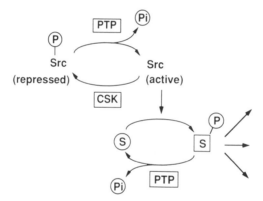

Fig. 11. Possible dual role of protein tyrosine phosphatases (PTP) in signal transduction. Upper reaction, activation of src family tyrosine kinases by dephosphorylation of the C-terminal phosphotyrosyl residue. Lower reaction, dephosphorylation of other tyrosine phosphorylated substrates to return the system to its original state. CSK is the cellular src kinase that represses enzymatic activity by tyrosine pohsphorylation at the C-terminus.

fold. In fact, if that tyrosyl residue were mutated in such a way that it could no longer be phosphorylated, the enzyme would become constitutively active and oncogenic. Second, the factors that would determine whether a phosphatase would enhance or oppose a kinase reaction would seem to depend less on its state of activity than on its subcellular localization. This would suggest that if one wanted to call upon these enzymes to control transformation, one should try to tamper with their localization segments - or whatever binding proteins they might be attached to - rather than with their catalytic domains. Displacement of these enzymes from where they are meant to bind would seem a more promising approach than trying to modulate their catalytic activity.

There are, of course, many questions that need to be answered. To mention a few:
a) What controls the activity of phosphatases under normal conditions? We know that it is easier to achieve high level expression of these enzymes in transformed lines in which the signal pathways are turned on, rather than in normal cells; as if a balance existed between kinases and phosphatases.
b) Do the differences just described between wild-type and truncated enzymes depend only on matters of localization or is enzyme specificity also implicated?

c) Would the same effects be observed with any oncogene, including those that are not tyrosine kinases (such as v-ras, v-raf, v-mos, etc.)?

d) Which are the steps in the signaling pathway that are specifically affected by the enzymes? And finally,

e) If oncogenicity can result from an overexpression of tyrosine kinases, and if the function of certain phosphatases is to oppose the activity of these kinases, could oncogenicity also result from an underexpression or suppression of some of these enzymes? We are trying to address this question using antisense RNA.

These are some of the questions we, and many others, are trying to answer.

ACKNOWLEDGEMENTS

These studies were supported by National Institutes of Health grants #DK0709 and GM42508 and from a grant from the Muscular Dystrophy Association of America.

REFERENCES

1. Cori, G.T. and Green, A.A. (1943) J. Biol. Chem. 31-38
2. Cori, G.T. and Cori, C.F. (1945) J. Biol. Chem. 321-332
3. Fischer, E.H. and Krebs, E.G. (1955) J. Biol. Chem. 216, 121-132
4. Krebs, E.G. and Fischer, E.H. (1956) Biochim. Biophys. Acta 20, 150-157
5. Sutherland, E.W. and Wosilait, W.D. (1955) Nature 175, 169
6. Rall, T.W., Sutherland, E.W., and Wosilait, W.D. (1956) J. Biol. Chem. 218, 483
7. Wosilait, W.D. (1958) J. Biol. Chem. 233, 597
8. Sutherland, E.W. and Rall, T.W. (1958) J. Biol. Chem. 233, 1077-1091
9. Edelman, A.M., Blumenthal, D.K., and Krebs, E.G. (1987) Annu. Rev.   Biochem. 56, 567-613
10. Brugge, J.S. and Erikson, R.L. (1977) Nature 269, 346-347
11. Collett, M.S. and Erikson, R.L. (1978) Proc. Natl. Acad. Sci. USA 75, 2021-2024
12. Levinson, A.D., Oppermann, H., Levintow, L., Varmus, H.E., and Bishop, J.M. (1978) Cell 15, 561-572
13. Ushiro, H. and Cohen, S. (1980) J. Biol. Chem. 255, 8363-8365
14. Carpenter, G. (1987) Annu. Rev. Biochem. 56, 881-914
15. Yarden, Y. and Ullrich, A. (1988) Annu. Rev. Biochem. 57, 443-478
16. Downward, J., Yarden, Y., Mayes, E., Scrace, G., Totty, N., Stockwell, P.,   Ullrich, A., Schlessinger, J., and Waterfield, M.D. (1984) Nature 307, 521-527
17. Ullrich, L., Coussens, L., Hayflick, J.S., Dull, T.J., Gray, A., Tam, A.W.,   Lee, J., Yarden, Y., Libermann, T.A., Schlessinger, J., Downward, J., Mayes, E.L.V. , Whittle, N., Waterfield, M.D., and Seeburg, P.H. (1984)   Nature 309, 418-425
18. Hunter, T. (1991) Methods. Enzymol. 200, 3-37
19. (1989) Oncogenes and the Molecular Origins of Cancer, Cold Spring Harbor Laboratory Press, Cold Spring Harbor, NY
20. Carpenter, G., King, L., and Cohen, S. (1979) J. Biol. Chem. 254, 4884-4891
21. Sefton, B.M., Hunter, T., Beemon, K., and Eckhart, W. (1980) Cell 20,   807-816
22. Fischer, E.H., Charbonneau, H., and Tonks, N.K. (1991) Science 253,   401-406
23. Saito, H. and Streuli, M. (1991) Cell Growth and Differ. 2, 59-65
24. Charbonneau, H. and Tonks, N.K. (1992) Annu. Rev. Cell Biol. 8, 463-493
25. Tonks, N.K., Diltz, C.D., and Fischer, E.H. (1988) J. Biol. Chem. 263,   6731-6737
26. Tonks, N.K., Diltz, C.D., and Fischer, E.H. (1988) J. Biol. Chem. 263,   6722-6730
27. Charbonneau, H., Tonks, N.K., Kumar, S., Diltz, C.D., Harrylock, M.,   Cool, D.E., Krebs, E.G., Fischer, E.H., and Walsh, K.A. (1989) Proc. Natl. Acad. Sci. USA 86, 5252-5256
28. Charbonneau, H., Tonks, N.K., Walsh, K.A., and Fischer, E.H. (1988)   Proc. Natl. Acad. Sci. USA 85, 7182-7186
29. Thomas, M.L. (1989) Annu. Rev. Immunol. 7, 339-369
30. Barnea, G., Silvennoinen, O., Shaanan, B., Honegger, A.M., Canoll, P.D.,   D'Eustachio, P., Morse, B., Levy, J.B., LaForgia, S., Huebner, K., Musacchio, J.M., Sap, J., and Schlessinger, J. (1992) Mol. Cell. Biol. 13,   1497-1506
31. Krueger, N.X. and Saito, H. (1992) Proc. Natl. Acad. Sci. USA 89, 7417-7421
32. Walton, K.M. and Dixon, J.E. (1993) Annu. Rev. Biochem. 62, 101-120
33. Cool, D.E., Tonks, N.K., Charbonneau, H., Walsh, K.A., Fischer, E.H., and Krebs, E.G. (1989) Proc. Natl. Acad. Sci. USA 86, 5257-5261
34. Cool, D.E, Tonks, N.K., Charbonneau, H., Fischer, E.H., and Krebs, E.G.   (1990) Proc. Natl. Acad. Sci. USA 87, 7280-7284
35. Zander, N.F., Lorenzen, J.A., Cool, D.E., Tonks, N.K., Daum, G., Krebs, E.G., and Fischer, E.H. (1991) Biochemistry 30, 6964-6970
36. Cool, D.E., Andreassen, P.R., Tonks, N.K., Krebs, E.G., Fischer, E.H., and   Margolis, R.L. (1992) Proc. Natl. Acad. Sci. USA 89, 5422-5426
37. Fujiwara, K. and Pollard, T.D. (1976) J. Cell Biol. 71, 848-875
38. Sato, N., Yonemura, S., Obinata, T., and Tsukita, S. (1991) J. Cell Biol. 113, 321-330
39. Zander, N.F., Cool, D.E., Diltz, C.D., Rohrschneider, L.R., Krebs, E.G., and Fischer, E.H. (1993) Oncogene 8, 1175-1182
40. McDonough, S.K., Larsen, S., Brodey, R.S., Stock, N.D., and Hardy, W.D.Jr. (1971) Cancer Res. 31, 953-956
41. Sherr, C.J. (1988) Biochim. Biophys. Acta 948, 225-243

HYDROLYSIS OF PHOSPHOLIPIDS AND ACTIVATION OF PROTEIN KINASE C FOR INTRACELLULAR SIGNALLING

Shun-ichi Nakamura[1], Yoshinori Asaoka[2],
Kouji Ogita[1], and Yasutomi Nishizuka[1,2]

[1]Dept. of Biochemistry
Kobe University School of Medicine
Kobe 650, Japan
[2]Biosignal Research Center
Kobe University
Kobe 657, Japan

SUMMARY

Inositol phospholipid hydrolysis by phospholipase C was once thought to be the only mechanism to produce diacylglycerol that relays information of extracellular signals into intracellular events through activation of protein kinase C. It is becoming clear that agonist-induced hydrolysis of various membrane phospholipids, particularly choline phospholipid, by phospholipase A_2 and phospholipase D also takes part in cellular responses such as cell proliferation and differentiation. Possibly, the members of the protein kinase C family may be activated differently by various combinations of phospholipid degradation products, and play each distinct roles in signal transduction for the control of various cellular functions.

INTRODUCTION

It is well documented that protein kinase C (PKC) is activated by increased levels of diacylglycerol (DG) in the membrane that is produced as a result of hydrolysis of inositol phospholipids (PI) by agonist-induced activation of phospholipase C. Recently, attention has been paid to agonist-induced hydrolysis of other phospholipids, particularly phosphatidylcholine (PC), producing DG at a relatively later phase in cellular responses, and possible involvement of phospholipase D in PKC activation has been postulated (see for reviews, 1,2). Sustained activation of PKC is a prerequisite essential for long-term responses such as cell proliferation and differentiation (3-5). In addition, phospholipase A_2 is activated by most of the agonists which induce PI hydrolysis (6). Arachidonic acid regulates many physiological processes after its conversion to various eicosanoids, but other products of PC hydrolysis catalyzed by phospholipase A_2, greatly potentiate PKC activation, thereby contributing to

Cell Signal Transduction, Second Messengers, and Protein Phosphorylation in Health and Disease
Edited by A.M. Municio and M.T. Miras-Portugal, Plenum Press, New York, 1994

33

the signal transduction through the PKC pathway (3,4). Although the biochemical mechanism of activation of various phospholipases involved remains largely to be explored, this article will briefly summarize agonist-induced degradation of various membrane phospholipids which plays roles in transmitting information of extracellular signals across the membrane. Possibly, the members of the PKC family respond differently to various combinations of phospholipid degradation products.

PHOSPHOLIPASE A$_2$ AND PKC ACTIVATION

Phospholipase A$_2$, which hydrolyzes phospholipids to produce free fatty acids and lysophospholipids, is abundant in mammalian tissues, and the receptor-mediated activation of this enzyme has been proposed (6). The signals that provoke PI hydrolysis frequently release arachidonic acid, and hydrolysis of PC by phospholipase A$_2$ generates two additional molecules, free cis-unsaturated fatty acids and lysophosphatidylcholine (lysoPC), which are both effective to enhance subsequent cellular responses (7-9).

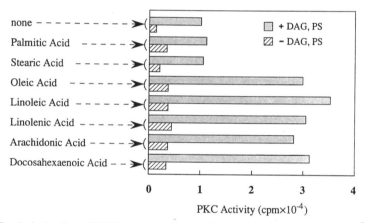

Fig. 1. Activation of PKC by various fatty acids in the presence of 1μM Ca^{2+} with phosphatidylserine (hatched bars) or with both phosphatidylserine and DG (stippled bars). The α-subspecies of PKC was assayed with H1 histone as a phosphate acceptor under the conditions described earlier (10). Various fatty acids (50 μM each) were added as indicated. Similar results were obtained with the β- and γ-subspecies of PKC. DAG, diacylglycerol; PS, phosphatidylserine.

Fig. 1 shows that several cis-unsaturated fatty acids such as oleic, linoleic, and linolenic, arachidonic, and docosahexaenoic acids, which are normally esterified to the 2 position of phospholipids, greatly enhance the DG-dependent activation of PKC in cell-free enzymatic systems. In the presence of both DG and fatty acid, the enzyme exhibits nearly full activity at the basal level of Ca^{2+} concentrations (10). Neither saturated fatty acid such as palmitic and stearic acids nor trans-unsaturated fatty acid such as elaidic acid was capable of enhancing the PKC activation.

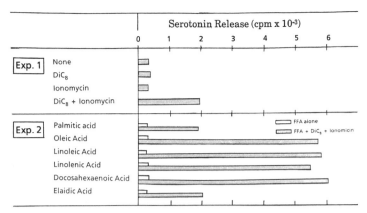

Fig. 2. Effect of various fatty acids on serotonin release from human platelets. [^{14}C]Serotonin-loaded platelets were stimulated for one min in the presence of 50μM various fatty acids with (shaded bars) or without (open bars) 25μM 1,2-dioctanoylglycerol plus 0.8μM ionomycin. Experiment 1 represents the serotonin release induced by 25μM 1,2-dioctanoylglycerol or 0.8μM ionomycin or both. Detailed experimental conditions were described elsewhere (7). FFA, fatty acid; DiC$_8$, 1,2-dioctanoylglycerol.

It is also shown that several cis-unsaturated fatty acids greatly enhance the DG-dependent activation of intact cells such as platelet release reaction (7). For such cellular responses, a membrane-permeant DG and Ca^{2+}-ionophore are both essential, and fatty acid alone is inactive. Consistent with the in vitro enzymatic reactions, the release reaction of platelets was enhanced by various cis-unsaturated fatty acids but not by trans-unsaturated and saturated fatty acid as shown in Fig. 2. Kinetic analysis with the Ca^{2+}-sensitive fluorescent dye fura-2 indicates that cis-unsaturated fatty acids markedly increase an apparent affinity of PKC activation to Ca^{2+}, thereby causing nearly full cellular responses at the basal level of Ca^{2+} concentration (7). It is possible, then, that PKC once activated initially by PI hydrolysis may intensify its enzymatic activity even after the Ca^{2+} concentration returns to the basal level, when DG and cis-unsaturated fatty acids become available as schematically illustrated in Fig. 3.

LysoPC, the other product of PC hydrolysis, shows a membranolytic activity, and is toxic to the cell. However, in the presence of a membrane-permeant DG or a phobol ester, lysoPC also greatly enhances cellular responses, particularly those in long-term such as cell proliferation and differentiation (8,9). For instance, lysoPC dramatically enhances the activation of human resting T-lymphocytes that is induced by a membrane-permeant DG and ionomycin, as determined by interleukin-2 receptor-α expression and thymidine incorporation as shown in Fig. 4. Similarly, lysoPC exerts its profound stimulatory effect on HL-60 cell differentiation to macrophage as measured by CD11b expression and appearance of phagocitic activity (9). In either case, lysoPC is active only when both DG and ionomycin are present, indicating that this lysophospholipid interacts with the PKC pathway. Other lysophospho-lipids are ineffective except for lysophosphatidylethanolamine

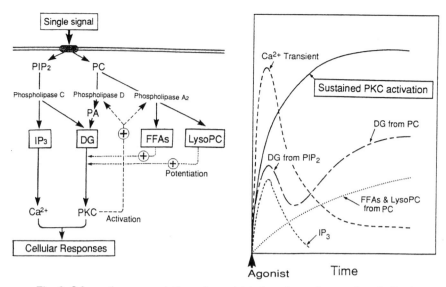

Fig. 3. Schematic representation of agonist-induced membrane phospholipid degradation cascade for sustained PKC activation. PIP_2, phosphatidylinositol 4,5-bisphosphate; FFAs, free <u>cis</u>-unsaturated fatty acids.

which is, however, far less active than lysoPC. The biochemical mechanism of this lysoPC action remains unclear, but it may be partly due to direct potentiation of the DG-dependent PKC activation (11). It is plausible, therefore, that the agonist-induced phospholipase A_2 activation may be directly involved in cell signalling through the PKC pathway.

The extracellular, secretory phospholipase A_2 (group II), when added directly to intact cells, is able to potentiate greatly the agonist-induced cellular responses such as T-lymphocyte activation (12). The phospholipase A_2, therefore, may take part in the propagation of inflammatory responses by degrading cell membrane phospholipids directly to produce <u>cis</u>-unsaturated fatty acids and lysoPC. The phospholipase A_2 (group II) has been long known to be secreted by a variety of cells such as platelets, neutrophils and mast cells at various sites of inflammation. The secretory phospholipase A_2 is also found in neuronal tissues (13), suggesting its role in synaptic processes, involving long-term potentiation.

POSSIBLE LINK BETWEEN PKC AND PHOSPHOLIPASE A_2

Phospholipase A_2 is known to be activated by some growth factors, such as epidermal growth factor (EGF) and platelet-derived growth factor. Studies with intact cell systems have suggested that PKC plays a role in the agonist-stimulation of phospholipase A_2, because phorbol esters and membrane-permeant DG provoke arachidonic acid release, sometimes in synergy with physiological agonists (see for a review, 4). In neutrophils, phorbol esters activate both arachidonic acid-selective and non-selective enzymes, thereby increasing the intracellular levels of various unsaturated fatty acids (14). An intracellular 85-kDa phospholipase A_2, that has been cloned (15, 16), is shown to be phosphorylated and activated by PKC and also by mitogen-activated protein kinases (MAP-kinases) (17). However, this type of phospholipase A_2 cleaves arachidonic acid selectively but not other fatty acids from the 2 position of phospholipids. Also note that PKC activation alone does not appear to be sufficient for the agonist-induced release of fatty acids. An analogue of GTP

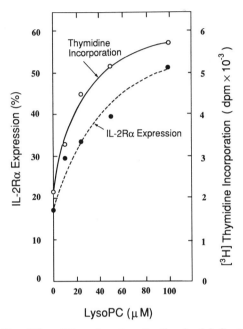

Fig. 4. Effect of lysoPC on T-lymphocyte activation by 1,2-dioctanoylglycerol and ionomycin. The resting T-lymphocytes (5×10^5 cells/ml) were stimulated with 1,2-dioctanoylglycerol (50μM) plus ionomycin (0.5μM) in the presence of various concentrations of lysoPC as indicated. T-lymphocyte activation was assayed by measuring the interleukin-2 receptor-α expression and also by thymidine incorporation. Detailed experimental conditions were described elsewhere (8). IL-2Ra, interleukin-2 receptor-α.

(GTP-γ-S) is shown to be capable of stimulating arachidonic acid release from permeabilized HL-60 cells in the presence of ATP (18). It has also been reported that EGF receptor tyrosine kinase may be involved in the phospholipase A_2 activation in fibroblast (19).

Although biochemical mechanism for the link between the activation of intracellular phospholipases A_2 and agonist-stimulation remains unclear, it seems likely that PKC once activated by DG plays a role for the PC hydrolysis by phospholipase A_2, thereby producing not only arachidonic acid but also other cis-unsaturated fatty acids as well as lysoPC to intensify various cellular responses through PKC activation (Fig. 3).

SUSTAINED ELEVATION OF DIACYLGLYCEROL

When cells are stimulated, DG is produced often in a biphasic manner during the subsequent cellular responses. Early peak of DG is transient, and temporally corresponds to the formation of inositol 1,4,5-trisphosphate (IP_3) and to the rise in intracellular Ca^{2+} concentration. The level of DG again increases with a slow onset but is more sustained, sometimes for several hours (Fig. 3). This second phase of DG is generally thought to be derived from the hydrolysis of PC by phospholipase D, yielding phosphatidic acid (PA) which is subsequently dephosphorylated to produce DG (1-5). This sustained elevation of DG is normally observed in response to various long-lasting agonists including some growth factors, phorbol esters, and oncogenic ras which are all capable of activating PKC and, as will be discussed below, PKC activation itself may be responsible, at least in part, for causing this sustained elevation of DG.

DG, either produced endogenously or added exogenously to intact cells, is metabolized very rapidly (20). On the other hand, tumour-promoting phorbol esters, that mimic DG for PKC activation, are metabolically stable, and thus the cellular responses caused by phorbol esters somewhat differ from those caused by a membrane-permeant DG. Experiments with multiple and repeated additions of a membrane-permeant DG, such as 1,2-dioctanoylglycerol, have confirmed that the prolonged activation of PKC is a prerequisite essential for causing long-term cellular responses, such as T-lymphocyte activation and HL-60 cell differentiation (20,21).

MECHANISM OF PHOSPHOLIPASE D ACTIVATION

Several mechanisms have been postulated for the agonist-induced formation of DG from PC. Cell-free preparations from various mammalian tissues reveal enzymatic activities that degrade PC, and phospholipase C that utilizes this phospholipid as substrate has been reported (see for reviews, 1-5). However, this is based on the observation that phosphorylcholine is produced from PC, and none of the phospholipase C-type enzymes which are reactive with PC has been unequivocally identified from mammalian tissues. It is worth noting that phosphorylcholine may be produced from PC by an alternative pathway involving phospholipase A_2 followed by the consecutive actions of lysoPC deacylase and phosphodiesterase.

On the other hand, several lines of evidence suggest that PC is hydrolyzed by phospholipase D in an agonist-dependent manner, resulting in the formation of PA, which is then converted to DG by the action of phosphomonoesterase (Fig. 3). Mammalian phospholipase D was reported first in rat brain, and subsequently in homogenetes and membranes from various tissues and cell types (see for reviews, 1-5). The enzyme is found primarily in particulate fractions, presumably membranes. However, kinetic properties of phospholipase D greatly vary from tissue to tissue, suggesting its extensive heterogeneity. The enzyme reacts preferentially with PC, and also phosphatidylethanolamine to lesser extents.

Phospholipase D from mammlian tissues also catalyzes transphosphatidylation reaction (base exchange reaction), and incorporates choline, ethanolamine, and serine into phospholipids (22). Apparently, multiple enzymes with different kinetics and distinct substrate specificities for base exchange reactions are present in various tissues with different intracellular localization. Although these enzymes have been extensively studied, the precise relationship between phospholipase D and base exchange enzymes remains to be clarified.

With several intact cell systems, phorbol esters are shown to stimulate PC breakdown to produce choline and phosphocholine (see for reviews, 1-5). In fact, phospholipase D is activated by a phorbol ester or a membrane-permeant DG, sometimes synergistically with Ca^{2+} ionophore, as assayed by measuring the phosphatidylethanol produced upon transphosphatidylation in which ethanol substitutes for water. With various membrane fractions and permeabilized cell preparations, both base exchange reactions and phospholipase D activities are shown to be activated by a GTP analogue (GTP-γ-S) and phorbol esters (see for reviews, 1-5). This activation by phorbol esters requires ATP and is inhibited by staurosporine as well as by PKC inhibitor peptides. However, these observations do not necessarily indicate the direct phosphorylation of phospholipase D by PKC.

It may be noted, however, that, analogous to phospholipase A_2 activation, PI hydrolysis per se may not be always essential nor sufficient for the agonist-induced phospholipase D activation. Presumably, several mechanisms may be responseble for the agonist-induced hydrolysis of PC. There are also several suggestions that protein tyrosine phosphrylation may be involved in phospholipase D activation (23-26). Nevertheless, it is attractive to surmise that PKC once activated by PI hydrolysis may take part to enhance PC hydrolysis and continuously provide DG by the action of phospholipase D, which is necessary for prolonged activation of PKC. Fig. 5 illustrates a hypothetical mechanism for the activation of phospholipases A_2 and D in analogous to that for the activation of phospholipase C. It is attractive to surmise that some isoforms of the enzymes may be regulated by reversible phosphorylation catalized by PKC or through activation of MAP-kinases. It has been recently hypothesized that sphingomyelin hydrolysis may play a role in the switch-off mechanism of this signal transduction pathway (5, 27).

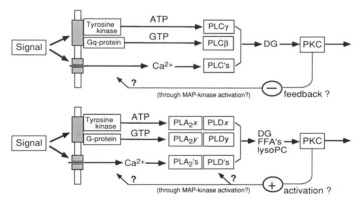

Fig. 5. A hypothetical mechanism of signal-induced activation of various phospholipases. PLA_{2x} and PLA_{2y} represent putative isoforms of phospholipase A_2. PLD_x and PLD_y represent putative isoforms of phospholipase D. FFA's, free unsaturated fatty acids; PLC, phospholipase C.

Ten subspecies of PKC have been identified in mammalian tissues, although some of these subspecies do not show typical characteristics of the classical PKC enzymes in their mode of activation. The enzymes may be devided tentatively into three groups depending on their requirements of Ca^{2+} and lipids for their activation, and the integrated nomenclature used herein for the PKC subspecies is described elsewhere (3,4). As schematically given in Fig. 6,

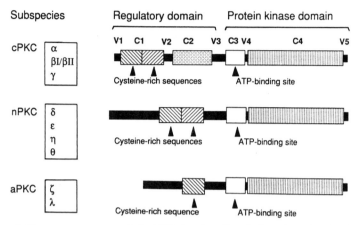

Fig. 6. Common structure of the PKC family. Detailed explanation is given elsewhere (4).

Table 1. Members of the PKC family from mammalian tissues
The activators for each subspecies are determined with calf thymus H1 histone and bovine myelin basic protein as model phosphate acceptors. The detailed enzymological properties of the η(L)-, θ- and λ-subspecies have not yet been clarified. PS, phosphatidylserine; FFA, cis-unsaturated fatty acid.

Group	Subspecies	Amino acid residues	Activators	Tissue expression
cPKC	α	672	Ca^{2+}, DG, PS, FFA, LysoPC	Universal
	βI	671	"	Some tissues
	βII	673	"	Many tissues
	γ	697	"	Brain only
nPKC	δ	673	DG, PS	Universal
	ε	737	DG, PS, FFA	Brain and others
	η (L)	683	?	Lung, skin, heart
	θ	707	?	Skeletal muscle (mainly)
aPKC	ζ	592	PS, FFA	Universal
	λ	596	?	Ovary, testis and others

cPKC consists of four classical or conventional PKC subspecies (α, βI, βII and γ) emerged from the initial screening. nPKC consists of four new or novel PKC subspecies (δ, ϵ, η(L) and θ) subsequently found. aPKC consists of two atypical PKC subspecies (ζ and λ) more recently characterized.

The members of the PKC family respond differently to various combinations of Ca^{2+}, DG, free fatty acids, lysoPC, and other phospholipid degradation products, and produce distinct activation patterns as indicated in Table 1. These subspecies show subtly distinct enzymological properties, differential tissue expression and specific intracellular localization (see for a review, 28). For instance, δ-PKC, which is present in the cell nucleus as well as in the cytoplasm, is insensitive to Ca^{2+} for its activtion. Treatment of CHO cells overexpressing δ-PKC with a phorbol ester prevents their mitosis possibly at G_2/M phase, suggesting that this PKC subspecies plays a role in the control of nuclear events. Presumably, multiple subspecies of the PKC family respond differently to various extracellular signals with respect to the extent, duration, and intracellular compartment. Such spatiotemporal aspects of the phospholipid degradation and the activation of PKC subspecies within a single cell are inevitable subjects to be explored for understanding further the molecular mechanism of intracellular signalling.

ACKNOWLEDGEMENTS

Skillful secretarial assistance of Miss Y. Nishimura and Mrs. S. Nishiyama is cordially acknowledged. This work was supported in part by research grants from the Special Research Fund of the Ministry of Education, Science, and Culture, Japan; Yamanouchi Foundation for Research on Metabolic Disorders; Sankyo Foundation of Life Sciences; New Lead Research Laboratories of Sankyo Company; and Terumo Life Science Foundation.

REFERENCES

1. Exton, J.H. (1990) J. Biol. Chem. 265, 1-4.
2. Billah, M.M., and Anthes, J.C. (1990) Biochem. J. 269, 281-291.
3. Asaoka, Y., Nakamura, S., Yoshida, K., and Nishizuka, Y. (1992) Trends Biochem. Sci. 17, 414-417.
4. Nishizuka, Y. (1992) Science 258, 607-614.
5. Zeisel, S.H. (1993) FASEB J. 7, 551-557.
6. Axelrod, J., Burch, R.M., and Jelsema, C.L. (1988) Trends Neurosci. 11, 117-123.
7. Yoshida, K., Asaoka, Y., and Nishizuka, Y. (1992) Proc. Natl. Acad. Sci. 89, 6443-6446.
8. Asaoka, Y., Oka, M., Yoshida, K., Sasaki, Y., and Nishizuka, Y. (1992) Proc. Natl. Acad. Sci. 89, 6447-6451.
9. Asaoka, Y., Yoshida, K., Sasaki, Y., and Nishizuka, Y. (1993) Proc. Natl. Acad. Sci. 90, 4917-4921.
10. Shinomura, T, Asaoka, Y., Oka, M., Yoshida, K., and Nishizuka, Y. (1991) Proc. Natl. Acad. Sci. 88, 5149-5153.
11. Sasaki, Y., Asaoka, Y., and Nishizuka, Y. (1993) FEBS Lett. 320, 47-51.
12. Asaoka, Y., Yoshida, K., Sasaki, Y., Nishizuka, Y., Nurakami, M., Kudo, I., and Inoue, K. (1993) Proc. Natl. Acad. Sci. 90, 716-719.
13. Oka, A., and Arita, H. (1991) J. Biol. Chem. 266, 9956-9960.
14. Conquer, J., and Mahadevappa, V.G. (1991) J. Lipid Mediators 3, 113-123.
15. Clark, J.D., Lin, L.-L., Kriz, R.W., Ramesha, C.S., Sultzman, L.A., Lin, A.Y., Milona, N., and Knopf, J.L. (1991) Cell 65, 1043-1051.

16. Sharp, J.D., White, D.L., Chiou, X.G., Goodson, T., Gamboa, G.C., McClure, D., Burgett, S., Hoskins, J., Skatrud, P.L., Sportsman, J.R., Becker, G.W., Kang, L.H., Roberts, E.F., and Kramer, R.M. (1991) J. Biol. Chem. 266, 14850-14853.

17. Nemenoff, R.A., Winitz, S., Qian, N.-X., Rutten, V.V., Johnson, G.L., and Heasley, L.E. (1993) J. Biol. Chem. 268, 1960-1964.

18. Xing, M., and Mattera, R. (1992) J. Biol. Chem. 267, 25966-25975.

19. Goldberg, H.J., Viegas, M.M., Margolis, B.L., Schlessinger, J., and Skorecki, K.L. (1990) Biochem. J. 267, 461-465.

20. Asaoka, Y., Oka, M., Yoshida, K., Nishizuka, Y. (1991) Proc. Natl. Acad. Sci. 88, 8681-8685.

21. Aihara, H., Asaoka, Y., Yoshida, K., and Nishizuka, Y. (1991) Proc. Natl. Acad. Sci. 88, 11062-11066.

22. Kanfer, J.N. (1980) Can. J. Biochem. 58, 1370-1380.

23. Uings, I.J., Thompson, N.T., Randall, R.W., Spacey, G.D., Bonser, R.W., Hudson, A.T., and Garland, L.G. (1992) Biochem. J. 281, 597-600.

24. Bourgoin, S., and Grinstein, S. (1992) J. Biol. Chem. 267, 11908-11916.

25. Dubyak, G.R., Schomisch, S.J., Kusner, D.J., and Xie, M. (1993) Biochem. J. 292, 121-128.

26. Kumada, T., Miyata, H., and Nozawa, Y. (1993) Biochem. Biophys. Res. Commun. 191, 1363-1368.

27. Dennis, E.A., Rhee, S.G., Billah, M., and Hannun, Y.F. (1991) FASEB J. 5, 2068-2077

28. Nishizuka, Y. (1988) Nature 334, 661-665.

PHOSPHOLIPID DEGRADATION AND ζPKC ACTIVATION DURING MITOGENIC SIGNAL TRANSDUCTION

Jorge Moscat and María T. Diaz-Meco

Centro de Biología Molecular,
CSIC-UAM, Canto Blanco
28049 Madrid, Spain

A number of recent studies suggest a role for phospholipid degradation in mitogenic signal transduction pathways. Particularly, the phosphodiesterase-mediated hydrolysis of phosphatidylcholine (PC) has been demonstrated in our and other laboratories to be activated in response to growth factors and oncogenes (1,2). In this regard, we have also produced evidence that activation of PLC-catalyzed hydrolysis of phosphatidylcholine (PC-PLC) is an important step in mitogenic signalling (2,3). In Xenopus laevis oocytes, microinjection of transforming ras p21 is a potent inducer of maturation, whereas microinjection of a neutralizing anti-ras p21 antibody specifically inhibits maturation induced by insulin but not by progesterone. Results from our laboratory demonstrate that microinjection of phosphatidylcholine-hydrolyzing phospholipase C (PC-PLC) is sufficient to induce maturation of Xenopus laevis oocytes (3).

Furthermore, microinjection of a neutralizing anti-PC-PLC antibody, specifically blocks the maturation program induced by ras p21/insulin but not by progesterone (3). These results suggest that PC-PLC activation is critical during mitogenic signalling in oocytes. We also have shown that this is also true in mammalian cells. Thus, we have demonstrated that expression of a permanently activated form of PC-PLC, by-passes the inhibitory actions of a dominant negative mutant of the ras oncogene (2,4). Furthermore, very recent data from our laboratory establishes that uncontrolled hydrolysis of PC is sufficient to confer a transforming phenotype to mouse fibroblasts (5).

Cell Signal Transduction, Second Messengers, and Protein Phosphorylation in Health and Disease
Edited by A.M. Municio and M.T. Miras-Portugal, Plenum Press, New York, 1994

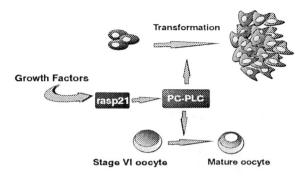

Fig. 1. Microinyeccion of a permanently activated form of PC-PLC into Xenopus oocytes induces maturation in the absence of any other stimuli. Also transfection of the PC-PLC gene into fibroblasts induces a transforming phenotype

However, cell growth is a finely tuned balance between positive and negative signals. One important example of the latter are those brought about by the TGF-β family. We have obtained results that strongly suggest that the mechanism whereby TGF-β blocks cell proliferation most probably involves the inhibition by this cytokine of the ability of ras p21 to activate PC-PLC (6). Taken together all these results indicate that PC degradation by PLC is a decisive step in mitogenic signal transduction.

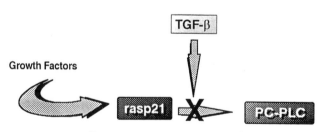

Fig. 2. TGF-β inhibits activation of PC-PLC by ras p21

The mechanism whereby PC-PLC activates downstream signals is another important issue. Since PLC-mediated PC hydrolysis produces diacylglycerol (DAG), and this is an important activator of protein kinase C (PKC), the involvement of this kinase in PC-PLC-triggered pathways is a conceivable possibility. Thus, we have demonstrated that exposure of fibroblasts to a permanently activated PC-PLC leads to the translocation of αPKC to the plasma membrane, a well established parameter of PKC activation (7). However, this translocation does not correlate with an increase in the phosphorylation of p80, a known substrate of the classical PKC isotypes (c-PKC) whose phosphorylation is associated with mitogenic signalling by PKC (7,8). Taken together, these results imply that although αPKC is somehow activated by

PC-PLC, it does not appear to be involved in the stimulation of cell proliferation by PC hydrolysis. Further evidence for this assumption was obtained from down-regulation experiments (2).

Cell Membrane

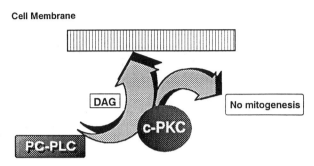

Fig. 3. PC-PLC activation promotes the translocation of αPKC to the membrane but this is not involved in the control of the mitogenic activity

Therefore, although PMA-sensitive PKCs (c-PKCs) could be critically involved in mitogenic signal transduction in some cell systems, this does not appear to be a general phenomenon. Interestingly, we have recently demonstrated that the specific inhibition of the isotype ζPKC by pseudosubstrate peptides, or its depletion with antisense RNA or oligonucleotides, leads to the blockade of the maturation pathway activated by insulin/ras/PC-PLC in Xenopus laevis oocytes with little or no effect on the progesterone route (9). Therefore, ζPKC appears to be necessary for mitogenic signalling by ras p21/PC-PLC at least in that system. More recent data from our laboratory show that this is also true in mammalian cells (10). Thus, we have presented experimental evidence demonstrating that activation of ζPKC is not only necessary but also sufficient by itself to activate maturation in oocytes and to produce deregulation of growth control in mouse fibroblasts (10). Furthermore, by using a dominant kinase-defective mutant of ζPKC, we confirm that this kinase is required for mitogenic activation in oocytes and fibroblasts (10). These results permit one to propose ζPKC as a critical step downstream of rasp21 in mitogenic signal transduction.

On the other hand, great effort is being invested in elucidating the mechanisms responsible for the transmission to the nucleus of signals generated by cytokine-receptor interactions at the plasma membrane. Activation of transcription factors is a pivotal step in these pathways. Thus, Nuclear Factor κB (NF-κB) is involved in the regulation of a large number of genes, including the enhancer of the immunodeficiency virus (HIV). NF-κB is a heterodimer of 50 kDa (p50) and 65 kDa (p65) subunits located in the cytoplasm in an inactive form, bound to the inhibitory protein IκB through the p65 molecule. Upon stimulation of cells, IκB dissociates from

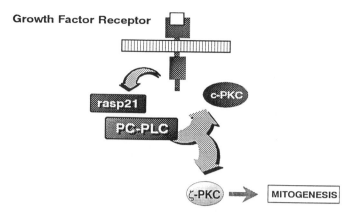

Fig. 4. The activation of PC-PLC by ras p21 generates two signals, one induces translocation of cPKC to the plasma membrane, and the other activates ζPKC which is critical for cell mitogenic activation

NF-κB allowing its translocation to the nucleus where it carries out the transactivation function. The pathways that regulate NF-κB translocation to the nucleus are presently unclear. The fact that NF-κB activation takes place following the stimulation with cytokines that stimulate progressively better characterized transmembrane signalling mechanisms, prompted the investigation of the precise role played by these signal transduction pathways in NF-κB activation. Thus, for example, the contribution of PMA-sensitive PKC isotypes to the stimulation of this parameter has previously been addressed (11). From those results, it is clear that although phorbol esters may be

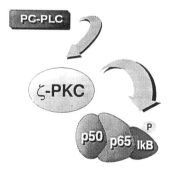

Fig. 5. ζPKC activates NF-κB probably trough inactivation of IκB by phosphorylation

able to activate NF-κB, the role of PMA-sensitive PKCs in vivo is not so clear. Interestingly, we have recently demonstrated in the model system of Xenopus laevis oocytes that the PMA-insensitive PKC isotype, ζPKC, is a required step in the activation of NF-κB in response to ras p21 (11). This is particularly challenging in light of the role of ras in lymphocyte stimulation. We have also demonstrated that the overexpression of ζPKC is by itself sufficient to activate NF-κB in NIH-3T3 fibroblasts. Furthermore, expression of a dominant negative mutant of ζPKC dramatically inhibits the κB-dependent transactivation of a CAT reporter plasmid (12).

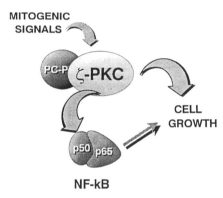

Fig. 6. ζPKC is a key step in the signalling pathways that control cell proliferation and NF-κB activation

These findings support the notion that ℋPKC is critically involved in cell cycle control and NF-κB activation and therefore appears to play a pivotal role in the signalling cascades of several cytokines and growth factors. The mechanism whereby ζPKC funnels signals coming from either ras or TNFα is presently unclear. An interesting link between both findings could be provided by PC-PLC. Thus, both ras p21 and TNFα are potent stimulants of this phospholipid degradative pathway, whose function has been established to be critical in mitogenic signalling. Therefore, all these results could fit in a model according to which the tandem PC-PLC/ζPKC transmit signals originated from different sources and that lead to NF-κB activation and cell growth.

REFERENCES

1. Lopez-Barahona, M., Kaplan, Paul L., Cornet, E., Larrodera, P., Diaz-Meco, M.T., Moscat, J. (1990) Kinetic evidence of a rapid activation of phosphatidylcholine hydrolysis by Ki-ras oncogene: Possible involvement in late steps of the mitogenic cascade. J.Biol.Chem. 265, 9022-9026

2. Larrodera, P., Cornet, M.E., Diaz-Meco, M.T., Lopez-Barahona, M., Diaz-Laviada, I., Guddal, P.H., Johansen, T., Moscat, J. (1990) Phospholipase C-mediated hydrolysis of phosphatidylcholine is an important step in PDGF-stimulated DNA synthesis. Cell 61, 1113-1120

3. García de Herreros, Dominguez, I., Diaz-Meco, M.T., Graziani, G., Guddal, P.H., Johansen, T., Moscat, J. (1991) Requirement of phospholipase C-hydrolysis of phosphatidylcholine for maturation of Xenopus laevis oocytes in response to insulin and ras p21. J.Biol.Chem. 266, 6825-6829

4. Cai, H., Erhart, P., Szeberenyi, J., Diaz-Meco, M.T., Moscat, J., Cooper, G.M. (1992). Hidrolysis of phosphatidylcholine is stimulated by Ras proteins during mitogenic signal transduction. Mol.Cell.Biol. 12, 5329-5335

5. Johansen, T., Bjorkoy, G., Overvatn, A., Diaz-Meco, M.T., Traavik, T., Moscat, J. (1993). NIH 3T3 cells stably transfected with the gene encoding phosphatidylcholine-hydrolysing phospholipase C from Bacillus cereus acquire a transformed phenotype Mol.Cell. Biol. 14, (In Press)

6. Diaz-Meco, M.T., Dominguez, I., Sanz, L., Municio, M.M., Berra, E., Cornet, M.E., Moscat, J. (1992) Phospholipase C-mediated hydrolysis of phosphatidylcholine is a target of TGF-ß1 inhibitory signals. Mol.Cell.Biol. 12, 302-308

7. Diaz-Laviada, I., Larrodera, P., Diaz-Meco, M.T., Cornet, M.E., Guddal, P.H., Johansen, T., Moscat, J. (1990) Phospholipase C-catalyzed hydrolysis of phosphatidylcholine is responsible for protein kinase C regulation in fibroblasts transformed by Ras and Src oncogenes. EMBO J. 9, 3907-3912

8. Diaz-Laviada, I., Larrodera, P., Cornet, M.E., Diaz-Meco, M.T., Sanchez, M.J., Guddal, P.H., Johansen, T., Moscat, J. (1991) Mechanism of inhibition of adenylate cyclase by phospholipase C-catalyzed hydrolysis of phosphatidylcholine. J.Biol.Chem. 266, 1170-1176

9. Dominguez, I., Diaz-Meco, M.T., Municio, M.M., Berra, E., Garcia de Herreros, A., Cornet, M.E., Sanz, L., Moscat, J. (1992) Evidence for a role of Protein Kinase

C ζ subspecies in maturation of <u>Xenopus laevis</u> oocytes. <u>Mol.Cell.Biol.</u> 12, 3776-3783

10.Berra, E., Diaz-Meco, M.T., Dominguez, I., Municio, M.M., Sanz, L., Lozano, J., Chapkin, R.S., Moscat, J. (1993). Protein kinase C ζ isoform is critical for mitogenic signal transduction. <u>Cell</u> 74, 555-563

11.Dominguez, I., Sanz, L., Arenzana-Seisdedos, F., Diaz-Meco, M.T., Virelizier, J.L., Moscat, J. (1993). Inhibition of Protein Kinase ζ subspecies blocks the activation of a NF-κB-like activity in <u>Xenopus laevis</u> oocytes. <u>Mol.Cell.Biol.</u> 13, 1290-1295

12.Diaz-Meco, M.T., Berra, E., Municio, M.M., Sanz, L., Lozano, J., Dominguez, I., Diaz-Golpe, V., Lain de Lera, M.T., Alcamí, J.,Arenzana, F., Payá, C.V., Virelizier, J.L., Moscat, J. (1993). Dominant negative protein kinase C ζ subspecies blocks NF-κB activation. <u>Mol.Cell.Biol.</u> 13, 4770-4775

INDUCTION OF NITRIC OXIDE SYNTHASE AFTER PROTEIN KINASE C ACTIVATION BY PHORBOL ESTERS

Sonsoles Hortelano, Ana M. Genaro, María J.M. Díaz-Guerra
and Lisardo Boscá

Instituto de Bioquímica (CSIC-UCM)
Facultad de Farmacia
Universidad Complutense
Madrid, E 28040
Spain

SUMMARY

Activation of protein kinase C by biologically active phorbol esters (4β-phorbol 12,13-dibutyrate or 4β-phorbol 12-myristate,13-acetate) promotes the expression of the inducible form of nitric oxide synthase in primary cultures of peritoneal macrophages and hepatocytes. The induction process is clearly observed after short (hours) incubation periods with phorbol esters, and is antagonized by endotoxins and some cytokines, recognized by their ability to induce nitric oxide synthase, such as the lipopolysaccharides from the outer wall of Gram negative bacteria. The co-stimulatory effect among cytokines, growth factors, tumor promoters and endotoxins reveal a complex regulatory mechanism in the control of the nitric oxide synthase expression.

INTRODUCTION

Since the unexpected discovery of nitric oxide (NO) as the molecule responsible for the vasodilation produced by acetylcholine as well as by other well known substances that require the presence of endothelial cells to express their effect (Furchgott 1988; Moncada

Cell Signal Transduction, Second Messengers, and Protein Phosphorylation in Health and Disease
Edited by A.M. Municio and M.T. Miras-Portugal, Plenum Press, New York, 1994

51

1992; Moncada *et al.* 1991), a major effort has been made to identify the NO generating system and to determine the physiological and biological role of this molecule (for a review see Madison 1993; Knowles and Moncada 1992). Pharmacological studies previous to 1987 established the existence of an elusive molecule that once released by stimulated vascular endothelium was able to produce vascular relaxation even in unstimulated systems. This substance was recognized by its lability and was termed Endothelial Derived Relaxing Factor (EDRF). In 1987 the nature of this molecule was identified as nitric oxide (Palmer *et al.* 1987, 1988; Sakuma *et al.* 1988). As a result of these pioneer studies, the scope of the biological effects mediated through NO, especially in the vascular system, increased enormously (Palmer 1993). At present, nitric oxide has emerged as an important intra- and intercellular regulatory molecule with functions as diverse as vasodilation, neural communication and host defense. Paralleling this variety of functions, NO is produced by many different cell types such as neurons, endothelial cells, monocytes and muscular cells, although the regulatory mechanisms involved in its synthesis vary among tissues (Lowenstein and Snyder 1992; Bredt *et al.* 1990a; Moncada *et al.* 1991; Moncada 1992).

Because of its chemical nature, a gaseous substance, NO is a molecule that not only acts on the agonist stimulated producing cell (i.e. endothelial cells), but may also exert its physiological action over other neighboring cells, thus defining a new type of cellular communication in which the synthesis of second messengers is achieved in a particular environment in the absence of additional transmembrane signalling events required to perceive extracellular messages by the responding cell (Bredt and Snyder 1989, 1992).

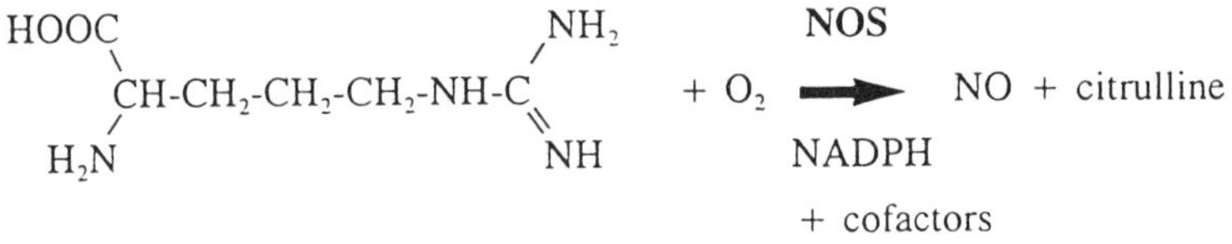

Figure 1. Nitric oxide is produced by nitric oxide synthase from arginine and molecular oxygen. The enzyme shares important structural homologies with cytochrome P-450 reductase and is virtually identical to the NADPH-dependent diaphorase. This collateral activity allows an easy localization of the enzyme using immunohistochemistry techniques.

Nitric oxide synthase (NOS) is the enzyme involved in the production of nitric oxide from molecular O_2 and arginine (Fig. 1), and appears to be a growing family of isoenzymes. At least four isoforms have been identified in mammalian tissues after cDNA analysis. These isoenzymes are encoded by at least three distinct genes (Lowenstein and Snyder 1992; Knowles and Moncada 1992): two genes encode for the "constitutively" expressed enzymes, and one other gene is responsible for the expression of the "inducible" form of NOS. Neural and endothelial cells are representative of the two forms of constitutive enzyme that requires Ca^{2+} and calmodulin to be active. The cytokine-inducible isoenzyme is expressed in macrophages, hepatocytes and monocytes after stimulation by a wide array of cytokines (interferon γ, tumor necrosis factor α) and endotoxins (lipopolysaccharide), and is Ca^{2+} and calmodulin independent (Billiar *et al.* 1990; Hauschildt *et al.* 1990; Hevel *et al.* 1991). However, despite this diversity in origin, all these isoenzymes share an important degree of structural and functional homology (Lowenstein and Snyder 1992).

The expression of the Ca^{2+}-independent cytokine-inducible NOS has been described in various cell types, in addition to macrophages (Stuehr and Marletta 1985; Stuehr *et al.* 1991). The established view that the enzyme induced by cytokines is the same in all responding tissues is doubtful because the iNOS induced in interleukin-1-stimulated human hepatocytes exhibits an important degree of calcium/calmodulin dependence in contrast to the absolute independence exhibited by the enzyme expressed in macrophages (Geller *et al.* 1993b). However, macrophage iNOS is by far the best characterized isoenzyme among the cytokine-inducible forms, both from the biological and chemical points of view, and the presence of a complex regulatory mechanism in the control of NOS expression by cytokines and endotoxins has been well established (Marletta *et al.* 1988; Ding *et al.* 1990; Geller *et al.* 1993).

Nitric oxide synthase forms

The constitutive form of NOS expressed in neural and endothelial tissues has been studied in detail (Bredt and Snyder 1990b; Dawson *et al.* 1991; Hope *et al.* 1991). The main difference between both forms is the ability of the endothelial enzyme to be acylated in the amino terminal domain of the protein with myristoyl residues in such a way that the enzyme is recovered as a membrane-bound form in contrast to the soluble localization of the neural constitutive enzyme (Michell *et al.* 1993; Lamas *et al.* 1992).

Regarding the role of NOS in neural tissues, it is clear that NO is a potent neurotransmitter and neuromodulator in the central nervous system (see for details Böhme *et al.* 1991; Bredt and Snyder 1991, 1992; Madison 1993). The enzyme has consensus sites of phosphorylation by the cyclic AMP-dependent protein kinase, as the endothelial enzyme, by calcium/calmodulin dependent protein kinase and by protein kinase C. Although phosphorylation of the endothelial enzyme seems to affect its subcellular distribution, the role of phosphorylation on enzyme activity remains to be established (Michell *et al.* 1993).

Regarding the signals generated after NO transient in neural tissues, the phosphorylation of DARPP-32, a protein substrate of the cyclic AMP-dependent protein kinase which is phosphorylated *in vivo* upon stimulation of dopamine receptors plays an important role in inhibiting protein phosphatase 1 (Tsou *et al.* 1993). These effects on DARPP-32 can be obtained with permeant cyclic GMP analogues, since the increase in the concentration of cyclic GMP is one of the early events elicited after NO generation. This process is achieved after binding of NO to the heme group of the soluble form of guanylate cyclase, and many of the physiological actions of NO could be obtained *via* the use of cyclic GMP derivatives (Bredt and Snyder 1989).

The archetypic form of inducible NOS (iNOS) is the enzyme expressed in macrophages in response to various cytokines and endotoxins (Nathan 1992). Convergent lines of evidence accumulated during the period previous to the identification of EDRF as NO showed that under pathological situations related to bacterial infection (septic shock, diarrhoea) the urinary excretion of nitrites and the plasma concentration of nitrites and nitrates were increased (Wagner *et al.* 1983). The ability of different macrophage-like cell lines such as RAW 264.7 to release NO after activation made possible a detailed study on the mechanisms of response of these cells to combinations of cytokines and endotoxins. These types of studies revealed the existence of an extensive and complex relationship between individual factors acting synergistically in most cases. Specifically, combinations of interferon γ, tumor necrosis factor α and lipopolysaccharide produced one of the highest inductions of the enzyme in macrophages, in keeping with its role in response to the host-defense (Liew *et al.* 1991; Lyons *et al.* 1992).

The main difference between the constitutive and inducible enzyme activities, in addition to its short-term regulation, is the amount of NO released, quantitatively more important in cells expressing the inducible enzyme.

Regarding the biological role of NO, its effects on the neural and endothelial systems are well documented both from the physiological and pharmacological points of view (Knowles and Moncada 1992; Palmer 1993). However, the role of NO released by the inducible enzyme is more difficult to be understood. Nitric oxide, in addition to promoting the activation of guanylate cyclase, inhibits enzymes (aconitase, ribonucleotide reductase, ADP-ribosylation of proteins; Drapier and Hibbs 1986; Lepoivre *et al.* 1990; Brüne and Lapetina 1989), metabolic pathways (mitochondrial respiration, DNA synthesis in some types of cells; Granger *et al.* 1980; Garg and Hassid 1989), and, presumably through these actions, participates in different processes, such as host defense and rejection of engrafted tissues (McCartney *et al.* 1993; Hoffman *et al.* 1990).

Induction of iNOS by phorbol esters

As indicated above, iNOS expression can be induced by a wide array of cytokines, growth factors and endotoxins, for most of which the activation of specific transcriptional factors has been detected in the analysis of the promoter region of this gene (Xie *et al.* 1993). Specifically, for many of these biomolecules, the targets of their biological actions are the plasma membrane receptors located in the extracellular face of the cells. However, for other important iNOS inducers, such as bacterial lipopolysaccharide, the nature of a putative receptor from the classical point of view is elusive despite the efforts made regarding its identification (Knowles and Moncada 1992; Glasser *et al.* 1990). The release of moieties of lipopolysaccharide with a potential messenger role has been reported (lipid A and lipid X; Wightman and Raetz 1984). Therefore, for all these extracellular factors, specific transmembrane signalling pathways are required to internalize the conformational change induced after ligand binding to its receptor (or the lipid moiety in the case of lipopolysaccharide). Several different types of internalization mechanisms have been characterized throughout the last decade; they have been classified into three major groups according to their chemical structure and mechanism of function (Miyajima *et al.* 1992). The receptors involved in iNOS induction mainly belong to the group of one-domain transmembrane proteins, except for the lipopolysaccharide receptor. The second messengers released after binding of the factors involved in iNOS induction have not yet been fully characterized.

Despite the different chemical structures of the substances involved in iNOS expression, they may partially share common transduction steps; therefore the identification of individual messengers (i.e. calcium mobilization, diacylglycerol, cyclic AMP, activation of protein kinase cascades) contributes to the understanding of the mechanism of action of these stimuli. Moreover, the release of second messengers can be achieved through mechanistically different signalling pathways resulting in the activation of enzymes placed in a common branch, for example, the activation of protein kinase C can be achieved by diacylglycerol, *cis*-unsaturated fatty acids, and in some cases calcium. These effectors could be released by pathways as different as those involving several forms of phospholipase C, phospholipase D, and calcium mobilization *via* different mechanisms (including the release from intracellular stores, and the efflux through the plasma membrane channels).

The assessment of the relevance of individual steps of these signalling pathways in the control of iNOS expression is complex. One example is the ascertainment of the role of protein kinase C in the process of iNOS induction by lipopolysaccharide, a factor for which the subcellular signalling involves protein kinase C activation. Because some effectors of protein kinase C are also potential activators of other transduction pathways such as various systems of protein kinases (i.e. calcium and calmodulin dependent protein kinase), the use of specific activators of the enzyme is suitable in order to bypass the requirement of other polyvalent second messengers (calcium and diacylglycerols). The use of phorbol esters is one of the best means to solve this problem since they specifically activate protein kinase C in the absence of calcium mobilization or production of lipid molecules (diacylglycerol or unsaturated fatty acids). Phorbol esters are able to activate protein kinase C *via* their structural homology with the 1,2-diacylglycerols through the esterification present in the 12 and 13 positions of the phorbol ring. Moreover, this system has an additional advantage over other lipidic activators of protein kinase C (mezerein, terpenoids). It is possible to use biologically inactive stereoisomers to address the specificity of the effect elicited by the phorbols (Nishizuka 1984, 1988). The 4β-phorbol esters however, promote protein kinase C activation, and the 4α-stereoisomers fail to activate protein kinase C both *in vivo* and *in vitro*. In addition, the use of phorbol esters as activators of protein kinase C offers two additional advantages over the physiological stimulators. (1) They allow a long-sustained activation of the kinase, whereas under physiological conditions the calcium increase is a transient process, and the 1,2-diacylglycerols are short-lived molecules due to the rapid degradation elicited by several diacylglycerol-consuming enzymes such as diacylglycerol kinase. (2) Phorbol esters are sufficiently hydrophobic to be prepared in aqueous solution, whereas other permeant synthetic diacylglcyerols exhibit serious problems of solubility and disposal. It was therefore important to test whether the incubation of cells lacking any form of NOS could respond to phorbol esters stimulation expressing the inducible NOS enzyme. Two cell types were used to determine the ability of phorbol esters to promote the expression of the iNOS: Peritoneal macrophages and hepatocytes were isolated from rat and after *in vitro* culture, were incubated with phorbol esters (Hortelano *et al.* 1992, 1993). We observed that protein kinase C activation by phorbol esters is sufficient to promote the expression of iNOS in systems as different as macrophages and hepatocytes (Fig. 2). These results were unexpected, since the complex regulatory mechanism elicited by various cytokines better suited a model in which phorbol esters would modulate (potentiate or inhibit) the action of other iNOS inducers (lipopolysaccharide, tumor necrosis factor α) rather than to promote iNOS expression by itself. The involvement of protein kinase C in the process was clear, since the inactive 4α-phorbol ester failed to promote iNOS induction, and in cells treated with calphostin C, staurosporine or the isoquinoline H7, all potent inhibitors of protein kinase C, the induction was significantly blocked (Fig. 3).

In view of the complex regulation of iNOS expression reported by several groups, it was supposed that other cooperative signals might potentiate the effect of phorbol esters in the expression of this enzyme (Geller *et al.* 1993a). However, our experimental evidence was the opposite: the positive effect of phorbol esters on iNOS expression was antagonized by lipopolysaccharide, a well known inducer of nitric oxide synthase in macrophages, and the same was observed with those cytokines that synergize with lipopolysaccharide in macrophages, specifically the combination of tumor necrosis factor α and lipopolysaccharide (Fig. 2). In a systematic search of hormones, cytokines and growth factors with a potential role in the positive modulation of iNOS induction by phorbol esters, insulin and especially epidermal growth factor, potentiate the effect of phorbol 12,13-dibutyrate (Fig. 3). The molecular basis of this regulation remains to be established.

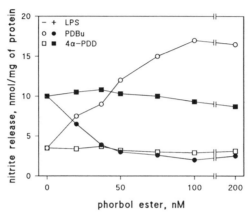

Figure 2. Expression of iNOS in macrophages incubated with lipopolysaccharide and phorbol esters. Resident peritoneal macrophages were incubated for 15 hours in the presence of active (PDBu) or inactive (4α-PDD) phorbol ester, and in the absence or presence of 10 μg/ml of lipopolysaccharide. The nitrite released was measured following the method of Griess (Hortelano *et al. et al.* 1992).

The specificity of iNOS induction by phorbol esters has been confirmed by the detection of mRNA levels by Northern blot analysis, as well as by the measurement of its enzymatic activity either following the formation of a nitrosyl-hemoglobin complex producing a specific shift in the absorbance of hemoglobin at 401/411 nm or following the release of [^{14}C]citrulline from its substrate [^{14}C]arginine (Hortelano *et al.* 1992,1993). Furthermore, a less specific reaction as the NADPH-dependent diaphorase activity, is associated with NOS as a collateral reaction (Dawson *et al.* 1991), evidenced *in situ* the expression of iNOS in hepatocyte cultures incubated in the presence of phorbol esters.

The increase in intracellular cyclic GMP, a feature dependent on NO release also confirmed the expression of iNOS by phorbol esters in macrophages (Knowles *et al.* 1989). Although this cellular response is clearly observed in macrophage cultures, this situation is not so clear in hepatocytes. Several possibilities may explain the low "yield" in cyclic GMP in the hepatic cell: The form expressed in hepatocytes when assayed *in vitro* seems to exhibit a certain degree of calcium-dependence, and the activity *in vivo* may be more calcium dependent than previously supposed. Therefore, the amount of NO released in the absence of calcium mobilization would be lower. Because of the suggestion that the hepatocyte plasma membrane has a special permeability to cyclic GMP that easily diffuses to the extracellular milieu (Billiar *et al.* 1992), the low levels of cyclic GMP could be explained. In agreement with this, exposure of hepatocytes to "authentic" gaseous NO also failed to trigger a drastic increase in the intracellular cyclic GMP concentration (the maximal increase is only a four-fold the basal value whereas in other cells treated under similar conditions this increase is of at least one order of magnitude).

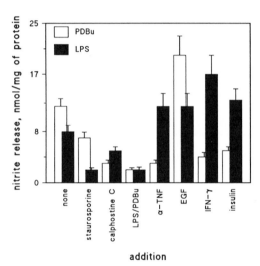

Figure 3. Effect of phorbol esters and lipopolysaccharide on iNOS expression by different cytokines in primary cultures of resident peritoneal macrophages. Cells were incubated for 15 hours in the presence of 10% heat-inactivated calf serum and the production of nitrites was measured. All cytokines and factors were used at saturant concentrations.

The induction of iNOS in macrophages by phorbol esters is accompanied by a stimulation of arginine influx within the cell (table 1). This observation agrees with other data suggesting that expression of iNOS is accompanied by an improvement of the NOS effector system: the activity of enzymes involved in tetrahydrobiopterine synthesis also increases as result of NOS induction since this cofactor appears to be essential for the expression of a functional enzyme (Kwon *et al.* 1989).

Table 1. Apparent Km for arginine transport and maximal arginine transport activity in intact macrophages incubated for five hours with saturant concentrations of PDBu or lipopolysaccharide. The transporter activities were calculated after extrapolation at 0 time of the arginine uptake assayed in fractions of 15 seconds during one minute.

Addition	apparent K_m (μM)	V_{max} (nmolxmin^{-1}/mg of protein
none	4.2 ± 0.4	0.43
PDBu	2.5 ± 0.4	2.09
lipopolysaccharide	2.9 ± 0.3	1.92

Mechanism of action of phorbol esters on iNOS expression

Since protein kinase C constitutes a family of isoenzymes (also known as subspecies) from which at least 10 different molecular forms have been so far characterized by cDNA analysis, it is possible that not all of them will be involved in the process of iNOS expression. Three structurally different protein kinase C groups have been characterized: the so called *classic* subspecies (α, βI, βII and γ) are dependent on calcium and diacylglycerol to be active. The *new* subspecies (δ, ϵ, θ, η) only require diacylglycerol or phorbol esters for activation and are widely distributed among tissues including hepatocytes and macrophages, whereas the subspecies of the *atypic* group (ζ, λ) are activated by diacylglycerols in the absence of calcium and are the only forms not stimulated by phorbol esters (Parker *et al.* 1989). According to this functional classification, different experimental approaches could address the nature of the protein kinase C subspecies involved in iNOS expression. Activation of hepatocytes with a combination of a calcium ionophore and a permeant diacylglycerol analogue is sufficient to induce iNOS expression. Therefore, it is likely that all protein kinase C forms may be involved in iNOS induction, at least under pharmacological conditions. An alternative way to approach the contribution of the different subspecies to the process of iNOS induction is by using cell lines with low endogenous protein kinase C basal levels and by expressing monospecifically isoenzymes upon transfection.

To understand the mechanism by which phorbol esters induce iNOS expression, several possibilities exist when the structure of the promoter region of the iNOS gene is considered (Xie *et al.* 1993). Expression of inducible genes in eukaryotic cells is mainly controlled through the activity of transcriptional factors that once activated by the second messengers released in transmembrane signalling promote its binding to the DNA and after alteration of the DNA structure offer new sites of binding for the transcriptional machinery allowing mRNA synthesis to proceed. The number of identified transcriptional factors is continuously increasing, and for some of them, the more ubiquitous (Roeder 1991) the

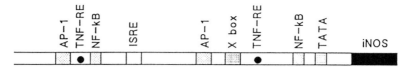

Figure 4. Structure of the promoter region of iNOS from murine macrophages. Only some of the consensus sequences for transcriptional activators is represented (adapted from Xie *et al.* 1993).

mechanism of action is well known. As figure 4 shows, the promoter region of of the murine macrophage iNOS gene has two AP-1 and NF-κB consensus sequences.

This structure could be well conserved among species in view of the parallelism in the mechanism of iNOS induction in equivalent cell types. Both AP-1 and NF-κB have been reported to be activated by the action alone of phorbol ester incubation of the cells (Fig. 5). Regarding NF-κB it is noteworthy that this transcriptional factor has been clearly implicated in the mechanism of action of lipopolysaccharide, tumor necrosis factor α, and interleukin 1β as well as in the pharmacological cell activation with phorbol esters (Grimm and Baeuerle 1993). NF-κB is a complex composed by three subunits. The trimer is inactive due to the inhibition shared by the I-κB subunit. When NF-κB is activated, the affinity for I-κB decreases and the remaining dimer is translocated to the nucleus where it exerts its transcriptional activation. However, according to the proposed mechanisms of NF-κB and AP-1 activation, it is difficult to understand the negative cooperation between signals transmitted through the tumor necrosis factor α/interferon γ, and phorbol esters/protein kinase C pathways. Work is in progress in order to follow up the early steps in the activation

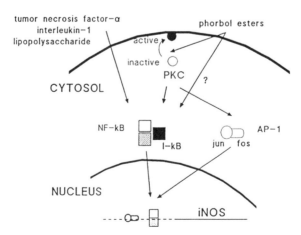

Figure 5. Proposed mechanism for iNOS induction in cells responding to lipopolysaccharide, tumor necrosis α, interleukin 1 and phorbol esters. The black box refers to the I-κB subunit. Cross-talk between the phorbol ester pathway and the lipopolysaccharide signalling pathways prevents iNOS expression.

of these transcriptional factors, and preliminary results suggest that activation through the lipopolysaccharide/tumor necrosis α pathway alter signalling through protein kinase C. In conclusion, analysis of the initial steps of activation of transcription factors may provide definitive clues to understand the complex regulatory network that operates in the control of iNOS expression in response to activatory and inhibitory cytokines.

REFERENCES

Billiar, T.R., Curran, R.D., Stuehr, D.J., Stadler, J., Simmons, R.L. and Murray, S.R., 1990, Inducible cytosolic enzyme activity for the production of nitrogen oxides from L-arginine in hepatocytes, *Biochem. Biophys. Res. Commun.* 168:1034.

Billiar, T.R., Curran, R.D., Harbrecht, B.G., Stadler, J., Williams, D.L., Ochoa, J.B., DiSilvio, M., Simmons, R.L. and Murray, S.R., 1992, Association between synthesis and release of cGMP and nitric oxide biosynthesis by hepatocytes, *Am. J. Physiol.* 262:C1077.

Böhme, G.A., Bon, C., Stutzman, J.M., Doble, A. and Blanchard, J.C., 1991, Possible involvement of nitric oxide in long-term potentiation, *Eur. J. Pharmacol.* 199:379.

Bredt, D.S. and Snyder, S.H., 1989, Nitric oxide mediates glutamate-linked enhancement of cGMP levels in the cerebellum, *Proc. Natl. Acad. Sci. USA* 86:9030.

Bredt, D.S., Hwang, P.M. and Snyder, S.H., 1990a, Localization of nitric oxide synthase indicating a neural role for nitric oxide. *Nature* 347:768.

Bredt, D.S. and Snyder, S.H., 1990b, Isolation of nitric oxide synthase, a calmodulin-requiring enzyme, *Proc. Natl. Acad. Sci. USA* 87:682.

Bredt, S.D., Hwang, P.M., Glatt, C.E., Lowenstein, C., Reed, R.R., and Snyder, S.H., 1991, Cloned and expressed nitric oxide synthase structurally resembles cytochrome P-450 reductase, *Nature* 351:714.

Bredt, D.S. and Snyder, S.H., 1992, Nitric oxide, a novel neuronal messenger, *Neuron* 8:3.

Brüne, B. and Lapetina, E.G., 1989, Activation of a cytosolic ADP-ribosyl-transferase by nitric oxide-generating agents, *J. Biol. Chem.* 264:8455.

Dawson, T.M., Bredt, D., Fotuki, M., Hwang, P.M. and Snyder, S.H., 1991, Nitric oxide synthase and neuronal NADPH diaphorase are identical in brain and peripheral tissues, *Proc. Natl. Acad. Sci. USA* 88:7797.

Ding, A.H., Nathan, C.F., Graycar, J., Derynck, R., Stuehr, D.J. and Srimal, S., 1990, Macrophage deactivating factor and transforming growth factors β1, β2 and β3 inhibit induction on macrophage nitrogen oxide synthesis by IFN-γ, *J. Immunol.* 145:940.

Drapier, J.C. and Hibbs, J.B.Jr., 1986, Murine cytotoxic activated macrophages inhibit aconitase in tumor cells. Inhibition involves the iron-sulfur prosthetic group and is reversible, *J. Clin. Invest.* 78:790.

Drapier, J.C., Wietzerbin, J. and Hibbs, J.B.Jr., 1988, Interferon-γ and tumor necrosis factor induce the L-arginine-dependent cytotoxic effector mechanism in murine macrophages, *Eur. J. Immunol.* 18:1587.

Furchgott, R.F., 1988, Studies on relaxation of rabbit aorta by sodium nitrite: the basis for the proposal that the acid activatable inhibitory factor from retractor penis is inorganic nitrite and the endothelium-derived relaxing factor is nitric oxide. In Mechanism of Vasodilation, ed. Vanhoutte, P.M., pp 401. Raven Press, New York.

Garg, U.C. and Hassid, A., 1989, Nitric oxide-generating vasodilators and 8-bromo-cyclic guanosine monophosphate inhibit mitogenesis and proliferation of cultured rat vascular smooth muscle cells, *J. Clin. Invest.* 83:1774.

Geller, D.A., Lowenstein, C.J., Shapiro, R.A., Nussler, A.K., Di Silvio, M., Wang, S.C., Nakayama, D.K., Simmons, R.L., Snyder, S.H. and Billiar, T.R., 1993a, Molecular cloning and expression of inducible nitric oxide synthase from human hepatocytes, *Proc. Natl. Acad. Sci. USA* 90:3491.

Geller, D.A., Nussler, A.K., Di Silvio, M., Lowenstein, C.J., Shapiro, R.A., Wang, S.C., Simmons, R.L. and Billiar, T.R., 1993b, Cytokines, endotoxin, and glucocorticoids regulate the expression of inducible nitric oxide synthase in hepatocytes, *Proc. Natl. Acad. Sci. USA* 90:522.

Glaser, K., Asmis, R., Denis, E.A., 1990, Bacterial lipopolysaccharide priming of P388D1 macrophage-like cells for enhanced arachidonic acid metabolism, *J. Biol. Chem.* 265:8658.

Granger, D.L., Taintor, R.R., Cook, J.L and Hibbs, J.B., 1980, Injury in neoplastic cells by murine macrophages leads to inhibition of mitochondrial respiration, *J. Clin. Invest.* 65:357.

Grimm, S. and Baeuerle, P.A., 1993, The inducible transcription factor NF-kB: structure-function relationship of its protein subunits, *Biochem. J.* 290:297.

Hauschildt, S., Luckhoff, A., Mulsch, A., Kohler, J., Bessler, W. and Busse, R., 1990, Induction and activity of NO synthase in bone-marrow derived macrophages are independent of Ca^{2+}, *Biochem. J.* 270:351.

Hevel, J.M. White, K.A. and Marletta, M.A., 1991, Purification of the inducible murine macrophage nitric oxide synthase. Identification as a flavoprotein, *J. Biol. Chem.* 266:22789.

Hoffman, R.A., Langrehr, J.M., Billiar, T.R., Curran, R.D. and Simmons, R.L., 1990, Alloantigen-induced activation of rat splenocytes is regulated by oxidative metabolism of L-arginine, *J. Immunol.* 145:2220.

Hope, B.T., Michael, G.J., Knigge, K.M. and Vincent, S.R., 1991, Neuronal NADPH diaphorase is a nitric oxide synthase, *Proc. Natl. Acad. Sci. USA* 88:2811.

Hortelano, S., Genaro, A.M. and Boscá, L., 1992, Phorbol esters induce nitric oxide synthase activity in rat hepatocytes. Antagonism with the induction elicited by lipopolysaccharide, *J. Biol. Chem.* 267:24937.

Hortelano, S., Genaro, A.M. and Boscá, L., 1993, Phorbol esters induce nitric oxide synthase and increase arginine influx in cultured peritoneal macrophages, *FEBS Lett.* 320:135.

Knowles, R.G., Palacios, M., Palmer, R.M.J. and Moncada, S., 1989, Formation of nitric oxide from L-arginine in the central nervous system: A transduction mechanism for stimulation of the soluble guanylate cyclase, *Proc. Natl. Acad. Sci. USA* 86:5159.

Knowles, R.G. and Moncada, S., 1992, Nitric oxide as a signal in blood vessels, *Trends Biochem. Sci.* 17:399.

Kwon, N.S., Nathan, C.F. and Stuehr, D.J., 1989, Reduced biopterin as a cofactor in the generation of nitrogen oxides by murine macrophages, *J. Biol. Chem.* 264:20496.

Lamas, S., Marsden, P.A., Li, G.K., Tempst, P. and Michell, T., 1992, *Proc. Natl. Acad. Sci. USA* 89:6348.

Lepoivre, M., Chenais, B., Yapo, A., Lemaire, G., Thelander, L. and Tenu, J-P., 1990, Alterations of ribonucleotide reductase activity following induction of the nitrite-generating pathway in adenocarcinoma cells, *J. Biol. Chem.* 265;14143.

Liew, F.Y., Li, Y., Severn, A., Millott, S., Schmidt, J., Salter, M. and Moncada, S., 1991, A possible novel pathway of regulation by murine T helper type-2 (Th2) cells of a Th1 cell activity via the modulation of the induction of nitric oxide synthase on macrophages, *Eur. J. Pharmacol.* 21:2489.

Lowenstein, C.J. and Snyder, S.H., 1992, Nitric oxide, a novel biological messenger, *Cell* 70:705.

Lyons, C.R., Orloff, G.J. and Cunningham, J.M., 1992, Molecular cloning and functional expression of an inducible NOS from a murine macrophage cell line, *J. Biol. Chem.* 267:6370.

Madison, D.V., 1993, Pass the nitric oxide, *Biochemistry* 90:4329.

Marletta, M.A., Yoon, P.S., Iyengar, R., Leaf, C.D. and Wishnok, J.S., 1988, Macrophage oxidation of L-arginine to nitrite and nitrate: Nitric oxide is an intermediate. *Biochemistry* 27:8706.

McCartney-F., N., Allen, J.B., Mizel, D.E., Albina, J.E., Xie, Q-W., Nathan, C.F. and Wahl, S.M., 1993, Supression of arthritis by an inhibitor of nitric oxide synthase, *J. Exp. Med.* 178:749.

Michell, T., Li, G.K. and Busconi, L., 1993, Phosphorylation and subcellular translocation of endothelial nitric oxide synthase, *Proc. Natl. Acad. Sci. USA* 90:6252.

Miyajima, A., Hara, T. and Kitamura, T., 1992, Common subunits of cytokine receptors and the functional redundancy of cytokines, *Trends Biochem. Sci.* 17:378.

Moncada, S., 1992, The L-arginine:nitric oxide pathway, *Acta Physiol. Scand.* 145:201.

Moncada, S., Palmer, R.M.J. and Higgs, E.A., 1991, Nitric oxide: physiology, pathophysiology and pharmacology, *Pharmacol. Rev.* 43:109.

Nathan, C., 1992, Nitric oxide as a secretory product of mammalian cells, *FASEB J.* 6:3051.

Nishizuka, Y., 1984, The role of protein kinase C in cell surface signal transduction and tumour promotion, *Nature* 308:693.

Nishizuka, Y., 1988, The molecular heterogeneity of protein kinase C and its implications for cellular regulation, *Nature* 334:661.

Palmer, R.M.J., 1993, The discovery of nitric oxide in the vessel wall, *Arch. Surg.* 128:396.

Palmer, R.M.J., Ferridge, A.G. and Moncada, S., 1987, Nitric oxide accounts for the biological activity of endothelium-derived relaxing factor, *Nature* 327:524.

Palmer, R.M.J., Asthton, D.S. and Moncada, S., 1988, Vascular endothelial cells synthesize nitric oxide from L-arginine, *Nature* 333:664.

Parker, P.J., Kour, G., Marais, R.M., Mitchell, F., Pears, C., Schaap, D., Stabel, S. and Webster, C., 1989, Protein kinase C- a family affair, *Mol. Cell. Biol.* 65:1.

Roeder, R.G., 1991, The complexities of eukaryotic transcription initiation: regulation of preinitiation complex assembly, *Trends Biochem. Sci.* 16:402.

Sakuma, J., Stuehr, D.J., Gross, S.S., Nathan, C. and Levi, R., 1988, Identification of arginine as a precursor of endothelium-derived relaxing factor (EDRF), *Proc. Natl. Acad. Sci. USA* 85:8664.

Stuehr, D.J. and Marletta, M.A., 1985, Mammaliam nitrate biosynthesis: Mouse macrophages produce nitrite and nitrate in response to Escherichia coli lipopolysaccharide, *Proc. Natl. Acad. Sci. USA.* 82:7738.

Stuehr, D.J., Cho, H.J., Kwon, N.S. and Nathan, C.F., 1991, Purification and characterization of the cytokine-induced macrophage nitric oxide synthase: a FAD- and FMN-containing flavoprotein, *Proc. Natl. Acad. Sci. USA* 88:7773.

Tsou, K., Snyder, G.L. and Greengard, P., 1993, Nitric oxide/cGMP pathway stimulates phosphorylation of DARPP-32, a dopamine- and cAMP-regulated phosphoprotein, in the substantia nigra, *Proc. Natl. Acad. Sci. USA* 90:3462.

Wagner, D.A., Young, V.R., and Tannenbaum, S.R:, 1983, Mammalian nitrate biosynthesis: Incorporation of $^{15}NH_3$ into nitrate is enhanced by endotoxin treatment, *Proc. Natl. Acad. Sci. USA* 80:4518.

Xie, Q-W., Whisnant, R and Nathan, C., 1993, Promoter of the mouse gene encoding calcium-independent nitric oxide synthase confers inducibility by interferon γ and bacterial lipopolysaccharide, *J. Exp. Med.* 177:1779.

METABOTROPIC GLUTAMATE RECEPTORS AND THE ACTIVATION OF PROTEIN KINASE C IN THE CONTROL OF GLUTAMATE RELEASE

José Sánchez-Prieto, Elena Vázquez,
and Inmaculada Herrero

Departamento de Bioquímica
Facultad de Veterinaria
Universidad Complutense
Madrid-28040, Spain

SUMMARY

The activation of protein kinase C, (PKC), with phorbol esters in nerve terminals enhances the release of neurotransmitter glutamate. The presynaptic locus of PKC action is at the K^+-channels controlling the duration or frequency of the action potentials; for this reason, the facilitation of glutamate release is observed only in synaptosomes under the repetitive depolarizations induced by the K^+-channel blocker 4-aminopyridine, (4-AP), but not by the clamped depolarization induced by high KCl. The PKC isoform responsible for this increase in release is unknown but it is sensitive to both phorbol esters and cis-unsaturated fatty acids. A metabotropic receptor for glutamate generating diacylglycerol is involved in the activation of this PKC-dependent pathway. However, the potentiation of glutamate release is observed only in the presence of arachidonic acid. Arachidonic acid does not modify the generation of diacylglycerol induced by metabotropic agonists, but it is necessary for the phosphorylation of the major PKC-substrate in nerve terminals, the myristoylated-alanine-rich-C-Kinase substrate, (MARCKS), suggesting that the enhancement of release is based on the synergistic activation of PKC by diacylglycerol (DG) and arachidonic acid.

INTRODUCTION

Glutamate is the most abundant excitatory amino acid in the mammalian nervous system and is involved in several forms of synaptic plasticity such as long term potentiation as well as in neuronal cell degeneration. Transmitter glutamate is accumulated in small synaptic vesicles and it is now well established that these vesicles dock to the plasma membrane at the active zone, a specialized region of the nerve terminal containing clusters of voltage-dependent Ca^{2+} channels[1]. Glutamate exocytosis is triggered by the transient elevation of $[Ca^{2+}]_c$ that follows the opening of voltage dependent Ca^{2+}-channels after depolarization of the plasma membrane. The preparation of isolated nerve terminals (synaptosomes) has provided a valuable model to investigate the events that mediate between the depolarization of the plasma membrane and the release of neurotransmitters. In this preparation, it has been shown that the vesicular release of glutamate is triggered by a localized increase in the $[Ca^{2+}]_c$ because Ca^{2+} entering the nerve terminal through voltage dependent Ca^{2+}-channels is far more efficient to initiate release than the de-localized entry of Ca^{2+} provided by a Ca^{2+}-ionophore[2]. The Ca^{2+}-channel linked to the Ca^{2+}-dependent release of glutamate is different from the classic L,N or T-types but inhibited by toxins from the funnel web spider venom, ω-Aga IVA[3] and Aga GI[4]. In addition, tetanus and botulinum toxins partially attenuated the Ca^{2+}-dependent evoked release of glutamate[5, 6] due to the ability of some clostridial toxins to cleave vesicle proteins[7] and to interfere with the vesicle docking and fusion processes.

The synaptosomal preparation is also a good model to study the modulation of glutamate release. Recently, it has been shown that the K^+-channel blocker 4-aminopyridine, (4AP), is able to induce transient depolarizations mimicking the invasion of the nerve ending by action potentials[8]. The depolarization, the elevation in $[Ca^{2+}]c$ and the release of glutamate induced by 4AP are sensitive to the Na^+-channael blocker tetradotoxin, (TTX),[8]. Thus, with 4AP the modulation of glutamate release based on the control of transient K^+ or Na^+ channels can be analyzed. In contrast, the clamped depolarization caused by KCl abolishes the effects due to Na^+ or K^+ channel modulation, but it allows the investigation of those modulatory mechanisms for glutamate release where Ca^{2+}-channels or the Ca^{2+}-secretion coupling are involved.

In this paper we review recent data regarding the facilitatory role of protein kinase C activation on the exocytotic release of glutamate. We also describe how a metabotropic receptor for glutamate coupled to the generation of diacylglycerol, and located at the glutamatergic nerve terminals is responsible for the activation of presynaptic PKC. The enhancement of release by the metabotropic agonist occurs only in the presence of the cis-unsaturated fatty acid arachidonic acid further indicating that the synergistic activation of PKC is the trigger of this positive feedback control of glutamate release.

PKC IN NERVE TERMINALS: POTENTIATION OF GLUTAMATE RELEASE

Protein kinase C is present at high concentrations in neuronal tissues, where it has been implicated in several neuronal functions such as the modulation of ionic channels activity (9) the enhancement of neurotransmitter release [10,11] and the increase in the efficacy of synaptic transmission at the hippocampus known as Long Term Potentiation (LTP) [12].

Biochemical and inmunochemical studies have demonstrated the presence in nerve terminals of several isoforms of PKC including α, β, γ and ϵ[13,14]. With the exception of ϵ that does not require Ca^{2+}, all the isoforms are activated by Ca^{2+}, phosphatidylserine, diacylglycerol; and this activation is further enhanced by cis unsaturated fatty acids[15]. Nerve terminals also contain the ζ isoform[16] activated by phosphatidylserine and fatty acids but not affected by diacylglycerol, phorbol esters or Ca^{2+}[15]. Phorbol esters mimic the action of diacylglycerols and also activate PKC in vitro. The activation of presynaptic PKC with phorbol dibutyrate, 4ß-PDBu, leads to an increase in glutamate release[11] prevented by the PKC-inhibitor Ro- 31-8220[17]. The PKC isoform involved in the facilitation of glutamate release is not known but has to be one of the subspecies with high sensitivity for cis-unsaturated fatty acids, since arachidonic acid in combination with either phorbol esters[18] or a metabotropic glutamate receptor agonist producing DG[19,20] is required for the optimal facilitation of glutamate release.

The mechanism of PKC activation for glutamate release has been elucidated by comparing the effects of phorbol esters on 4AP and KCl induced release to determine whether the locus of PKC is in the ionic channels that control the duration or frequency of the action potentials induced by 4AP or whether the kinase affects the Ca^{2+}-channels or the Ca^{2+}-secretion coupling itself. Using this experimental approach it has been shown that PKC activation increases the depolarization, the cytosolic free Ca^{2+} concentration, $[Ca^{2+}]_c$ and glutamate release evoked by 4AP, but has no effect in KCl-depolarized synaptosomes[11]. These results suggests that the likely action of PKC is to inhibit a K^+-channel that controls the duration of the action potentials initiated by 4AP.

Experiments with the selective PKC inhibitor Ro 31-82200, capable of inhibiting the α, ß, γ and ϵ-PKC isoforms[17], have indicated that there is no PKC involvement in Ca^{2+}-exocytosis coupling itself or in any direct modulation of the non-inactivating Ca^{2+}-channel coupled to glutamate release[2] because this PKC inhibitor, while preventing both the depolarization-evoked and the phorbol esters-evoked phosphorylation of the PKC-substrate MARCKS, is without effect on KCl-evoked Ca^{2+}-dependent release of glutamate[17]. However, evidence for an extensive modulatory role for a Ro 31-8220-sensitive PKC isoform in glutamate exocytosis is available. The PKC inhibitor not only prevents the phorbol esters-induced enhancement of glutamate release in 4AP-stimulated synaptosomes but almost completely blocks 4AP-evoked exocytosis[17] due to the inhibition of an "intrinsic" PKC activity (Coffey et al., unpublished results) maintained by external glutamate through a presynaptic metabotropic receptor linked to diacylglycerol generation[20]. Thus, these experiments indicate a wide modulation by PKC of the ion channel activity controlling glutamate release from almost complete inhibition to maximal release.

SYNERGISTIC ACTIVATION OF PKC: ROLE OF CIS UNSATURATED FATTY ACIDS

Previous studies have shown that the facilitation of the Ca^{2+}-dependent release of glutamate by PKC activation requires relatively high concentrations of phorbol esters $(0.1-1\ \mu M)$[11] indicating either that the PKC isoform coupled to glutamate exocytosis has a low sensitivity to phorbol esters or that these compounds do not fully mimic "in vivo" PKC activation. Recently, it has been shown that several cis-unsaturated fatty acids including oleic, linoleic, linolenic and arachidonic acid greatly enhance the DG-dependent activation of PKC[21,22]. In nerve terminals from the cerebral cortex an enhanced sensitivity of glutamate exocytosis to both phorbol esters or diolein is observed in the presence of low concentrations of arachidonic acid (Figure 1). These results suggest that a protein kinase C synergistically activated by diacylglycerol and arachidonic acid control the facilitatory pathway for glutamate release as has been demonstrated in other presynaptic events such as the phosphorylation of the presynaptic PKC substrates MARCKS (Coffey et al., unpublished results) and B-50[23].

mGluRs LINKED TO PKC ACTIVATION

The agonist-induced activation of PLC and hydrolysis of inositol phospholipids produce a rapid and transient increase in the DG. A late but more sustained production of DG is also observed by signal-dependent induced hydrolysis of phosphatidylcholine. Metabotropic glutamate receptors, mGluRs, are coupled to both

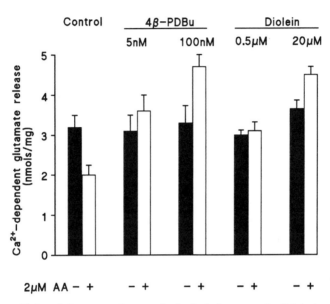

Fig. 1.- The sensitivity of glutamate release to both phorbol esters and diolein is enhanced in the presence of arachidonic acid. Different concentrations of 4ß-PDBu or diolein were added 60 sec before depolarization of rat cerebrocortical synaptosomes with 1 mM 4-aminopyridine. 2 μM arachidonic acid (AA) was added 30 sec before depolarization. The Ca^{2+}-dependent glutamate release was calculated by substracting release obtained after 5 minutes of stimulation in medium with a free Ca^{2+} concentration of 200 nM from release at 1.33 mM Cl_2Ca.

PLC activation and to the modulation in the intracellular levels of cyclic AMP. A presynaptic receptor for glutamate of the metabotropic type linked to a rapid (within seconds) and transient generation of diacylglycerol[20] has been recently found to be involved in the facilitation of glutamate release from nerve terminals[19]. However, the metabotropic receptor agonist (1S,3R-ACPD) increases release only if arachidonic acid is co-applied[19]. The fact that only cis-unsaturated fatty acids (arachidonic, oleic) facilitate release (Figure 2) suggests the involvement of a synergistic activation of presynaptic PKC by the DG generated by the mGluR and the free fatty acid. In agreement with this suggestion is the finding that the phosphorylation of the protein kinase C substrate MARCKS is enhanced only the combination of (1S,3R)-ACPD/AA (Coffey et al., unpublished results).

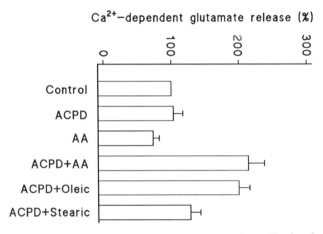

Fig. 2.- Effects of free fatty acids on the facilitation of glutamate release. The free fatty acids ($2\mu M$) were added 30 sec prior depolarization of synaptosomes with 50 μM 4-aminopyridine, in the presence of the metabotropic receptor agonist (1S,3R)-ACPD(100 μM) which was added 60 sec prior depolarization.

Further evidence of the involvement of PKC in this transduction pathway was obtained from the experiments presented in Figure 3 where the potentiation by ACPD+AA was prevented both in synaptosomes preincubated with the protein kinases inhibitor staurosporine and in synaptosomes in which the PKC activity was partially "downregulated" by pretreatment of the synaptosomes with 1 μM of phorbol esters[24].

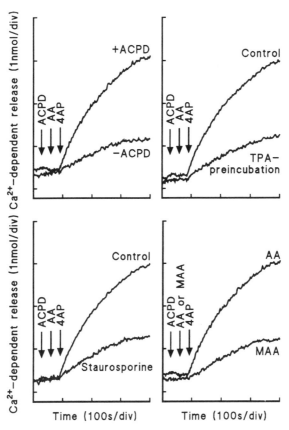

Fig. 3.- PKC-involvement in the facilitatory action of glutamate release induced by (1S,3R)-ACPD and arachidonic acid. In control experiments, synaptosomes were depolarized with 50 μM 4-aminopyridine in the presence of 100 μM (1S-3R)-ACPD and 2μM arachidonic acid (AA) which were added 60 and 30 sec, respectively, before depolarization. In experiments with staurosporine (St), the synaptosomes were incubated with protein kinase inhibitor at 100 nM for 30 min. TPA-preincubated synaptosomes were treated with 1 μM 12-O-tetradecanoylphorbol-13-acetate (TPA) for 30 min. After preincubation the synaptosomes were pelleted and assayed for glutamate release. Experiments with methyl arachidonate (MAA) were similar to the control but methyl arachidonate was used instead of arachidonic acid.

In addition the activation of presynaptic PKC either by phorbol esters or by the agonist-induced formation of diacylglycerol leads to the desensitization of the metabotropic receptor[20], in agreement with the well documented inhibitory action of PKC on agonist-induced hydrolysis of phosphoinositides in many neuronal preparations[25,26]. As expected, the desensitization of the metabotropic receptor prevented the facilitation of glutamate release[20].

All these data provide evidence on the involvement of PKC from glutamatergic nerve terminals in a positive feedback control for glutamate exocytosis. The PKC isoform that trigger the facilitation of release has not been identified but respond only to the coincident presence of DG and AA, (Figure 4). Thus, the activation of the mGluR with the agonist ACPD in the absence of arachidonic acid, although generating DG, does not activate PKC to phosphorylate MARCKS, and also fails to increase release. Arachidonic acid alone inhibits glutamate exocytosis [27] by a mechanism independent of PKC activation [28]. However, the ACPD-induced increase in DG in the presence of AA activates PKC inhibits the delayed rectifier K$^+$-

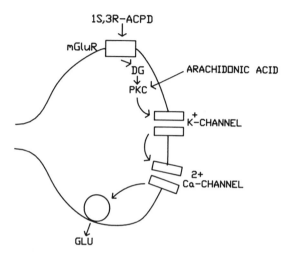

Fig. 4.- Scheme for the PKC-dependent facilitation of glutamate release by activation of presynaptic mGluR in the presence of arachidonic acid. (1S,3R)-ACPD activates mGluR generating DG and synergisticaly activating PKC in the presence of arachidonic acid. PKC-mediated inhibition of a K$^+$-channel (probably the delayed rectifier) results in prolonged action potential and in an increased entry of Ca^{2+} and glutamate release. Arachidonic acid does not alter the DG production by mGluR indicating that the locus for AA is down stream of PLC at the level of PKC because it is needed for the phosphorylation of the PKC-substrate.

channels, potentiates presynaptic action potentials and Ca^{2+} entry and further enhances the neurotransmitter release. The action of AA is localized at PKC since the fatty acid does not potentiate the DG production but is required for the phosphorylation of the PKC substrate MARKCS (Coffey et al unpublished results). According to the role of arachidonic acid as a possible retrograde messenger during synaptic plasticity, coupling glutamate release to the postsynaptically generated arachidonic acid, this positive feedback control of glutamate exocytosis could play a role in presynaptic aspects of synaptic plasticity.

ACKNOWLEDGEMENTS

This work was supported by a grant from DGICYT (Spain)(PM 92/0029). I.Herrero was supported by a fellowship from Universidad Complutense de Madrid (Spain). We also thank Erik Lundin for his help in the preparation of the manuscript.

REFERENCES

1. R. Llinas, M. Sugimori, and R.B., Silver R.B. Microdomains of high calcium concentration in a presynaptic terminal, Science. 256: 677 (1992).

2. H.T. McMahon, and D.G. Nicholls, Transmitter release from isolated nerve terminals: evidence for biphasic release and triggering by localized Ca^{2+}, J. Neurochem. 56: 86 (1991).

3. T.J. Turner, M.E. Adams, and K. Dunlap, Calcium channels coupled to glutamate release identified byω-Aga-IVA, Science. 258: 310 (1992).

4. J.M. Pocock, and D.G. Nicholls. A toxin (Aga-GI) from the venom of the spider Agelenopssis aperta inhibits the mammalian presynaptic Ca^{2+}-channel coupled to glutamate exocytosis, Eur. J. Pharmacol. 226: 343 (1992).

5. J. Sánchez-Prieto, T.S. Sihra, D. Evans, A. Ashton, J.O. Dolly, and D.G. Nicholls, Botulinum toxin A blocks glutamate exocytosis from guinea-pig cerebral cortical synaptosomes, Eur.J. Biochem. 165: 675 (1987).

6. H. T. McMahon, P. Foran, J.O. Dolly, M. Verhage, V. M. Wiegant and D.G. Nicholls, Tetanus toxin and botulinum toxins type A and B inhibit glutamate, γ-aminobutyric acid, aspartate , and Met-enkephalin release from synaptosomes, J. biol.Chem. 267: 21338 (1992).

7. E. Link, L. Edelmann, J.H. Chou, T. Binz, S. Yamasaki, U. Eisel, M. Baumert, T.C. Südhof, H. Niemann, R. Jahn, Tetanus toxin action: inhibition of neurotransmitter release linked to synaptobrevin proteolysis, Biochem. Biophys. Res. Commun 189: 1017 (1992).

8. G.R. Tibbs, A.P. Barrie, F. Van-Mieghem, H.T. McMahon, and D.G. Nicholls, Repetitive action potentials in isolated nerve terminals in the presence of 4-aminopyridine: effects on cytosolic free Ca^{2+} and glutamate release, J. Neurochem. 53: 1693 (1989).

9. M. S. Shearman, K. Sekiguchi, Y. Nishizuka, Modulation of ion channel activity: a key function of the protein kinase C family, Pharmacol. Rev. 41: 211 (1989).

10. R.A. Nichols, J.W. Haycock, J.K.T. Wang, and P. Greengard, Phorbol ester enhancement of neurotransmitter release from rat brain synaptosomes, J. Neurochem. 48, 615 (1987).

11. A.P. Barrie, D.G. Nicholls, J. Sánchez-Prieto, and T.S. Sihra, An ion channel locus for the protein kinase C potentiation of transmitter glutamate release from guinea pig cerebrocortical synaptosomes, J.Neurochem. 57: 1398 (1991).

12. P.A. Colley, and A. Routtenberg, Long-term potentiation as synaptic dialogue, Brain Res. Rev. 18: 115 (1993).

13. M.S. Shearman, T. Shinomura, T. Oda, Y. Nishizuka, Synaptosomal protein kinase C subspecies: Dinamic changes in the hippocampus and cerebellar cortex concomitant with synaptogenesis, J. Neurochem. 56: 1255 (1991).

14. N. Saito, A. Itoji, Y. Totani, I., Osawa, H. Koide, N. Fujisawa, K. Ogita, C. Tanaka, Cellular and intracellular localization of ϵ-subspecies of protein kinase C in the brain; presynaptic localization of the ϵ-subspecies, Brain Res. 607: 241 (1993).

15. Y. Nishizuka, Intracellular signaling by hydrolysis of phospholipids and activation of PKC, Science 258: 607 (1992).

16. D.M. Terrian, D. Kirk Ways, R.L. Gannon,. and D.A. Zetts, Transduction of a protein kinase C-generated signal into a long-lasting facilitation of glutamate release, Hippocampus, 3: 205 (1993).

17. E.T. Coffey, T.S. Sihra, and D.G. Nicholls, Protein kinase C and the regulation of glutamate exocytosis from cerebrocortical synaptosomes, J. Biol. Chem. 268: 21060 (1993).

18. I. Herrero, M.T. Miras-Portugal, and J. Sánchez-Prieto, Activation of PKC by phorbol esters and arachidonic acid required for the optimal potentiation of glutamate exocytosis, J. Neurochem. 59: 1574 (1992).

19. I. Herrero, M.T. Miras-Portugal, and J. Sánchez-Prieto, Positive feed back of glutamate exocytosis by metabotropic presynaptic receptor stimulation, Nature. 360: 163 (1992).

20. I. Herrero, M.T. Miras-Portugal, and J. Sánchez-Prieto, Rapid desensitization of the metabotropic receptor that facilitates glutamate release in rat cerebrocortical nerve terminals, Eur. J. Neurosci. 6: (1994). (in the press).

21. M.S. Shearman, T. Shinomura, T. Oda, Y. Nishizuka, Protein Kinase C subspecies in adult rat hippocampal synaptosomes. Activation by diacylglycerol and arachidonic acid, FEBS Letters 279: 261 (1991).

22. H. Koide, K. Ogita, U. Kikkawa, Y. Nishizuka, Isolation and characterization of the ε-subspecies of protein kinase C from rat brain, Proc. Natl. Acad. Sci. USA 89: 1149 (1992).

23. J.D. Schaechter, and L.I. Benowitz, Activation of protein kinase C by arachidonic acid selectively enhances the phosphorylation of GAP-43 in nerve terminal membranes, J. Neurosci. 13: 4361 (1993).

24. T. Oda, M.S. Shearman, and Y. Nishizuka, Synaptosomal protein kinase C subspecies: B. Down-regulation promoted by phorbol ester and its effect on evoked norepinephrine release, J. Neurochem. 56: 1263 (1991).

25. O.J.J. Monzoni, F. Finiels-Marlier, I. Sassetti, J. Bockaert, and F.A.J. Sladeczeck, The glutamate receptor of the Qp-type activates PKC and is regulated by protein kinase C. Neurosci Lett. 109: 146 (1990).

26. M. V. Catania, E. Aronica, M.A. Sortino, P.L. Canonico, and F. Nicoletti, Desensitization of metabotropic glutamate receptors, J. Neurochem. 56: 1329 (1991).

27. I. Herrero, M.T. Miras-Portugal, and J. Sánchez-Prieto, Inhibition of glutamate release by arachidonic acid in rat cerebrocortical synaptosomes, J. Neurochem. 57: 718 (1991).

28. I. Herrero, M.T. Miras-Portugal, and J. Sánchez-Prieto, PKC-independent inhibition of glutamate exocytosis by arachidonic acid in rat cerebrocortical synaptosomes, FEBS Letters, 296: 317 (1992).

MARCKS, A MAJOR IN VIVO SUBSTRATE OF PROTEIN KINASE C PURIFICATION, INTERACTION WITH MODEL MEMBRANE, AND DEMYRISTOYLATION

S. Manenti[1], O. Sorokine[2], A. Van Dorsselaer[2] and H. Taniguchi[3]

[1]Biocenter of the University, Basel, Switzerland
[2]Université Louis Pasteur, Strasbourg, France
[3]Fujita Health University, Toyoake, Japan

INTRODUCTION

The protein kinase C (PKC) family has been implicated in a large number of cellular responses. In order to better understand the cellular function of these enzymes, the identification of physiologically important substrates has received a peculiar attention during the last decade. The myristoylated alanine-rich C kinase substrate (MARCKS) is a widely distributed in vivo and in vitro substrate that has been often used as a marker of the activation of the kinase. (see Aderem, 1992 and Blackshear , 1993). Its phosphorylation by PKC in the cells leads to its translocation from the plasma membrane to the cytoplasmic compartment. The process is reversible, the dephosphorylation by cellular phosphatases leading to its reassociation with the plasma membrane (Thelen et al., 1991).

The protein is very acidic (pI 4.12-4.42) and presents an aberrant migration at 85 kDa in SDS gels for a real molecular mass of 32 kDa. It is heat stable, a characteristic which facilitates its purification. No hydrophobic segment suceptible to cross the bilayer can be deduced from the sequence. The primary sequence from different species (Blackshear, 1993) indicates two functional highly conserved domains. The N-terminus, which contains the myristoylation site, and the phosphorylation site domain (PSD) a very basic domain located in the middle of the sequence containing three sites of phosphorylation (Ser_{159}, Ser_{163} and Ser_{170}) by PKC. The myristoylation of proteins consists in the covalent attachment through an amide bond of myristic acid (C14:0) to the N-terminal glycine. The reaction is catalysed by N-myristoyl transferase, an enzyme very specific for myristic acid. It is generaly accepted that this process is cotranslational and irreversible (Rudnick et al., 1993). The function of this modification is not clear, but it seems to be involved in protein-lipid and protein-protein interactions. In the case of MARCKS, the myristoylation is necessary for the interaction with the membrane but not for the phosphorylation by PKC (Graff et al., 1989).

The cellular function of MARCKS was not established untill now. Two main hypothesis have been derived from in vitro experiments. The first one concerns the

Cell Signal Transduction, Second Messengers, and Protein Phosphorylation in Health and Disease
Edited by A.M. Municio and M.T. Miras-Portugal, Plenum Press, New York, 1994

77

interaction with calmodulin (McIlroy et al., 1991). This interaction is calcium-dependent, and occurs with a very high affinity (Kd: 2-5nM). It is inhibited when MARCKS is phosphorylated by PKC. The site of interaction was localized in the PSD, which explains the inhibition by PKC. This interaction opens the possibility that MARCKS acts as a regulator of the calmodulin concentration in the cells, the phosphorylation by PKC leading to the release of this protein from the plasma membrane where nonphosphorylated MARCKS is localized. Recently, a possible role of MARCKS in the regulation of the adenylate cyclase activity through the release of calmodulin was proposed (Sawai et al., 1993). MARCKS was also found to cross-link actin filaments in vitro (Hartwig et al., 1992). This interaction is inhibited by the presence of calmodulin and calcium, as well as by PKC phosphorylation. The site of interaction was again identified in the phosphorylation site domain. Based on these data, the role of MARCKS in the regulation of the actin microfilament network was proposed. In macrophages, the protein was colocalized in focal contacts,with vinculin and talin, and its release from these regions was observed when PKC was activated in the cells (Rosen et al., 1990), a process known to involve actin filaments rearrangement.

These possible functions of MARCKS must be considered in the view of recent data concerning the expression of the protein in the cells. It was recently shown that MARCKS is expressed when the cells are in G_0 phase, and is down regulated when they enter in mitosis through the effect of mitogenic agents (Herget et al., 1993). It is clear that these data must be taken into consideration to interpret the possible function of MARCKS as an actin microfilament regulator. Furthermore, down regulation of MARCKS in various transformed cells was also described (Wolfman et al., 1987; Joseph et al., 1992).

In this work, we describe the purification of MARCKS under native conditions, the interaction of the purified protein with artificial membranes (liposomes) and the regulation of this interaction by PKC. We also describe the purification of a non-myristoylated population of the protein, and demonstrate the existence of a demyristoylation activity in synaptosomes which may explain the presence of the non-myristoylated population. The implication of these new data for the regulation of MARCKS in the cells is discussed.

RESULTS

Purification of MARCKS in native conditions

Since MARCKS is translocated from the plasma membrane to the cytoplasmic compartment of the cells when it is phosphorylated by PKC, we purified the protein from the soluble and the particulate fractions of bovine brain to compare these two populations (Manenti et al., 1992). We adapted already described methods of purifications, by improving the separation at various steps with HPLC, supressing the heating or acidic precipitation steps, and introducing at the end of the process a calmodulin affinity chromatography. We finally obtained the cytoplasmic and membrane pure fractions of the protein and compared them by various methods. The electrospray mass spectrometry analysis revealed a higher degree of phosphorylation of the cytoplasmic protein compared with the membrane population (Fig 1). However we could not detect, even in the cytoplasmic fraction, the stoechyometry described in vivo (three moles of phosphate incorporated per mole of protein). This may come either from a loss of this population during the purification process, or from a nonspecific dephosphorylation occuring during the purification. Interestingly, the mass measured for the nonphosphorylated population $(31,750 \pm 4$ Da) is different from the mass calculated from the sequence (31,970 Da). This discrepancy may reflect either post-translational unknown modifications of the purified protein, or mistakes in the published sequence of the bovine protein. We are now investigating these two possibilities. We also found that the N-terminus of both populations is blocked, probably indicating that they are myristoylated, as was previously described in the literature. These data obtained on the purified protein partially confirm the results described by other groups at the cell level, and they clearly demonstrate that mass spectrometry analysis of proteins can be very helpfull, specially for the analysis of covalent modifications.

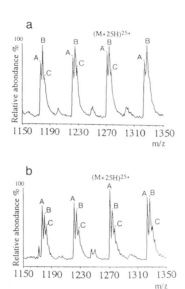

Figure 1: Electrospray mass spectrum of the cytoplasmic (a) and membrane (b) purified fractions of MARCKS. The peaks A, B, and C correspond to 0, 1 and 2 phosphorylations of the protein. The major peak is B (one phosphorylation) for the cytoplasmic fraction, while A (no phosphorylation) is more represented in the membrane fraction. The calculated mass for each of these populations are 31,750±3 Da (A), 31,825±6 Da (B), and 31,895±10 Da (C).

The interaction of MARCKS with artificial membranes

In order to elucidate the type of interaction of MARCKS with the membrane we used a model system of artificial membranes (liposomes) with different lipidic compositions. The incubation of the purified protein (membrane fraction) was performed with the vesicles, and the free and bound populations of the protein were separated by ultracentrifugation on sucrose gradient (Taniguchi and Manenti, 1993). The results of these experiments are summarized in Fig. 2. MARCKS binds to pure phosphatidylcholine (PC: neutral) vesicles and the subsequent phosphorylation of the protein-liposome complex by PKM, the catalytic fragment of PKC, does not affect the binding of the protein to the liposomes. When vesicles containing 20% of phosphatidylserine (PS: negatively charged) were used, the protein was found bound to the membranes, with higher affinity than with the PC liposomes. In this case the phosphorylation by PKM induced a release of the protein to the soluble fraction. Similarly, when MARCKS was phosphorylated by PKM previously to the incubation with PC-PS liposomes, the protein did not bind to the vesicles. The same results were obtained with synaptic vesicles membranes from bovine brain. These data indicate that MARCKS directly interacts with artificial membranes, and that the presence of acidic phospholipids is necessary for the regulation of this binding by PKC phosphorylation. In order to test whether the phosphorylation site (PSD) was directly involved in this interaction, we performed the same experiment as above with a synthetic peptide of 25 residues corresponding to the sequence of the PSD. This peptide does not bind at all to the pure PC vesicles, but shows a high affinity for the PC-PS liposomes. The phosphorylation by PKM completly inhibits the interaction. These data indicate that the PSD is responsible for the interaction of MARCKS with acidic phospholipids, and contains the site of regulation of this interaction by PKC.

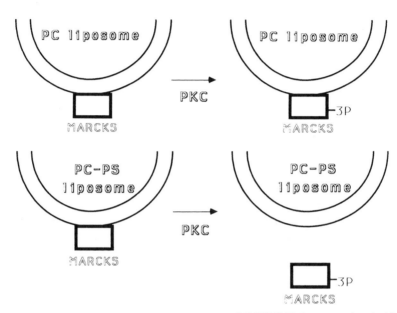

Figure 2: Interaction of MARCKS with artificial membranes. MARCKS binds to pure phosphatidylcholine (PC) and phosphatidylcholine-phosphatidylserine (PC-PS) liposomes. The phosphorylation by PKC induces the the release of the protein from the membrane only when acidic lipids (PS) are present in this membrane.

Purification of a non-myristoylated population of MARCKS from bovine brain

The myristoylation of MARCKS in myocytes and other cell types was described as a cotranslational and irreversible process (James and Olson, 1989), as it is generally the case for other myristoyl proteins. This myristoylation is clearly necessary for the interaction with the plasma membrane, since a mutant protein lacking the myristoylation N-terminal site and expressed in COS cells was found in the cytoplasmic compartment of these cells (Graff et al., 1989). During the purification of MARCKS described above, we also purified a product of 70 kDa (in SDS gel) with similar characteristics to MARCKS (Fig. 3 and Manenti et al., 1993). Surprisingly, by mass spectrometry analysis and N-terminal sequencing, we could identify this protein as the non-myristoylated form of MARCKS This population was found mainly in the cytoplasmic fraction of the cells. Because of the irreversible character of myristoylation, this finding was quite unexpected.

Interestingly, the release of the myristic acid, which corresponds to a decrease of the molecular mass of 210 Da, leads to a large increase of the mobility of the protein in SDS gel (15 kDa), probably reflecting an important change of conformation of the molecule. This hypothesis of a conformational change was confirmed by the observation that the interaction of MARCKS with calmodulin is different for the myristoylated and non-myristoylated populations. Since the site of this interaction was localized in the phosphorylation site domain, which is 150 residues far away from the N-terminus in the primary sequence, this probably means that these two domains are interlinked. Whether this reflects a direct intramolecular interaction between the two domains or a change of conformation of the protein can not be deduced from our data.

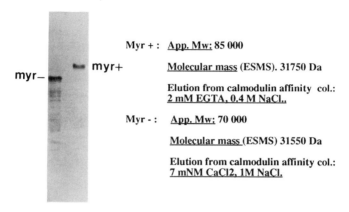

Myr + : App. Mw: 85 000

Molecular mass (ESMS). 31750 Da

Elution from calmodulin affinity col.:
2 mM EGTA, 0.4 M NaCl.,

Myr - : App. Mw: 70 000

Molecular mass (ESMS) 31550 Da

Elution from calmodulin affinity col.:
7 mNM CaCl2, 1M NaCl.

Figure 3: Comparison of the myristoylated (myr+) and non-myristoylated (myr-) forms of MARCKS. SDS gel of the two populations of the protein is shown, and the app. Mw, real mass (measured by mass spectrometry) and conditions of elution from the calmodulin affinity column are indicated on the right. For more details, see Manenti et al., 1993.

A demyristoylation activity is present in synaptosomes

In order to identify the origin of this non-myristoylated population of MARCKS, we tested the possible existence of a demyristoylation activity capable to remove the myristoyl moiety from purified MARCKS. For this purpose, we used synaptosomes from bovine brain for two reasons. First because MARCKS was purified from bovine brain, and it was more relevant to use the same biological source for our study. Second, because the existence of a non-myristoylated pool of MARCKS was reported in the cytoplasmic fraction of synaptosomes, as well as the post-translational myristoylation of this pool in vitro (McIlhinney and McGlone, 1990). We prepared cytoplasmic and membrane synaptosomal fractions, and tested the possible presence of a demyrisotylation activity by incubating aliquots of these fractions with purified MARCKS under various conditions (Manenti et al., in press). We found that with the cytoplasmic fraction, and in the presence of ATP in the reaction, a product of 70 kDa in SDS gel appeared after 24 h of incubation at 37°C, concomitantly with the disappearance of MARCKS (Fig. 4A). This product did not appear when the incubation was done with the membrane fraction of synaptosomes, or when ATP was omitted. Further purification of this product from the incubation medium and analysis by mass spectrometry and N-terminal sequencing confirmed that it was indeed the demyristoylated form of MARCKS. The enzymatic nature of this process was confirmed by heat inactivation and concentration-dependence of the activity. This is to our knowledge the first time that such an enzymatic activity is reported in the litterature.

The molecular mass (31630 ± 6 Da) of this demyristoylated form when compared with the mass of the non-myristoylated purified fraction (31550± 6 Da) suggested that a phosphorylation occured during the incubation with synaptosomes. This was confirmed by performing the demyristoylation in the presence [^{32}P] ATP and showing by autoradiography that both the myristoylated and demyristoylated forms were labelled after the incubation (Fig. 4B). We did not establish yet wether this phosphorylation is necessary for the demyristoylation, and wether it is due to synaptosomal endogenous PKC or to an other kinase present in the system. The identification of this phosphorylation site, which is now under progress in our laboratory, will probably give the answer to this problem.

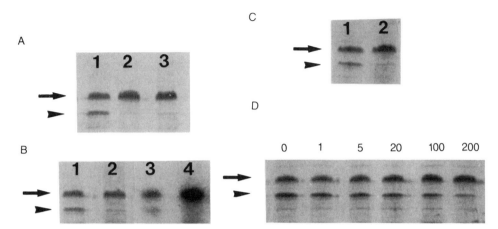

Figure 4:A. <u>Demyristoylation of MARCKS by a synaptosomal fraction</u> . MARCKS was incubated with the cytoplasmic fraction of synaptosomes in the presence (lane 1) or the absence (lane 2) of ATP, for 24 hours at 37°C. SDS gel of the incubation mixture was then performed. A control at time 0 is shown (lane 3). B. <u>MARCKS is phosphorylated during the demyristoylation.</u> The same experiment as in A was performed both with the cytoplasmic (lanes 1 and 3) and the membrane (lanes 2 and 4) fractions of synaptosomes, except that ^{32}P ATP was included. SDS gel of these fractions (lanes 1 and 2) and corresponding autoradiogram (lanes 3 and 4) are shown. The autoradiogram clearly shows that both myristoylated and demyristoylated MARCKS are phosphorylated. C. <u>Phosphatidylcholine liposomes inhibit the demyristoylation</u> The same experiment as in A was performed, but PC liposomes were added to the incubation medium (lane 2). This clearly inhibits the demyristoylation. D. <u>Calmodulin inhibits the demyristoylation.</u> Increasing concentrations of calmodulin were added to the incubation medium (the molar ratio of cam versus MARCKS is indicated fro each line). This was performed in the presence of calcium. When calcium was omitted, this inhibition was completely suppressed (not shown). In all the figures the 85 kDa and 70 kDa bands are indicated by the arrow and the arrowhead respectively.

Since we did not find any significant activity in the membrane fraction of synaptosomes, we tested the effect of artificial membranes on the demyristoylation activity. The incubation was performed with the cytoplasmic synaptosomal fraction in the presence of phosphatidylcholine liposomes. A complete inhibition of the demyristoylation was observed in these conditions as well as with liposomes containing 20% phosphatidylserine (Fig. 4C). These results suggest that in the presence of membranes, the myristoyl moiety of MARCKS is inserted into the bilayer. In this configuration of the protein, the acces of the enzyme to its recognition site is blocked.

Since we observed a modified affinity for calmodulin of the non-myristoylated population compared with the myristoylated one, it was of interested to investigate an eventual effect of calmodulin on the demyristoylation process. We found that increasing concentrations of calmodulin in the incubation medium in the presence of calcium resulted to increased inhibition of the demyristoylation (Fig. 4D). This effect was completely abolished when calcium was omitted and replaced by EGTA. This probably indicates that interaction of MARCKS with calmodulin occuring at the PSD, blocks the access of the enzyme to its recognition site. However, we cannot rule out that calmodulin interacts with another factor involved in the demyrisotylation process, for example the enzyme itself. Further experiments are needed to resolve this question.

DISCUSSION

In this work we investigated the interaction of purified native MARCKS with model membranes and tried to establish the importance of the two functional domains (the myristoylated N-terminus and the phosphorylation site domain) for this interaction.

The developpement of a new method of purification of MARCKS by introducing calmodulin affinity chromatography was justified to eliminate the heat and acidic precipitations as well as reversed phase chromatography used by others (Patel and Kligman, 1987; Brooks et al., 1990). Whether the native structural properties of the protein were absolutely conserved after these drastic separation steps was not clearly investigated. In fact, we have unpublished evidences that the heated protein, as well as the one purified by reversed phase have a modified susceptibility to the interaction with membranes and to demyristoylation. The relative amounts of protein that we purified from the cytoplasmic and membrane fractions are in good agrement with previous data showing that 70-80 % of MARCKS is bound to the plasma membrane in the cells. However, we could not identify any population of the cytoplasmic protein phosphorylated on the three sites described for PKC. This maybe implicates that the pool of cytoplasmic MARCKS described in resting cells is not as much phosphorylated as the protein translocated from the membrane to the cytoplasm when PKC is activated by mitogens or other extracellular factors. However, we cannot rule out that the population of MARCKS phosphorylated on the three sites is lost during our purification, or that endogenous specific or non-specific phosphatases dephosphorylate the protein during the purification.

The experiments of interaction of the purified MARCKS with the liposomes clearly indicate that the protein can directly interact with the phospholipids of the membrane. The difference of comportment with the PC and PC-PS vesicles of the phosphorylated molecule strongly suggests that two types of interaction are involved one of them being dependent on the presence of negatively charged phospholipids. We propose the following model (Fig 5): first MARCKS binds to the bilayer by the insertion of the myristoyl moiety into the

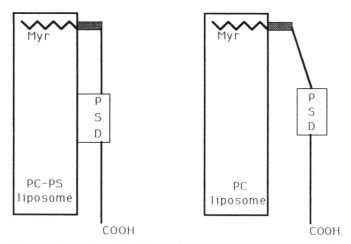

Figure 5: A model for the interaction of MARCKS with the membrane. In the presence of acidic phospholipids (PC-PS), MARCKS interacts with the membrane through hydrophobic interactions of its myristoyl moiety, and electrostatic interaction of the phosphorylation domain (PSD) with the surface of the bilayer. With neutral lipids (PC), the interaction happens only through the myristoyl moiety.

hydrophobic core. This type of insertion was not published in the literature, but we have recent evidences obtained by photolabeling method, that the myristic acid is indeed inserted into the membrane. Second, the basic phosphorylation site domain has electrostatic interactions with the surface of the membrane when negatively charged head groups are present. The phosphorylation by PKC of this domain induces a great change of charge (9 negative additional charges for 3 phosphorylations) which implicates repulsive interactions with the acidic lipids and leads to the release of the protein from the membrane. These in vitro data are in good correlation with the comportment of the protein in the cell (Thelen et al., 1991). However they do not take into account the possible interaction of MARCKS with other proteins like calmodulin, actin, or an eventual membrane receptor. For example they can not explain the specific immunolocalization of MARCKS in focal contacts of macrophages. We are now interested to use this in vitro system to study more complex interactions including for example actin or calmodulin together with MARCKS.

Since the myristoyl moiety is involved in the interaction of MARCKS with membranes, our purification of a natural non-myristoylated form of the protein may have important implications at the cell level. The localization of this population in the cytoplasmic fraction confirms previous data describing the cytoplasmic localization of a non-myristoylated mutant expressed in COS cells (Graff et al., 1989). Myristoylation of proteins is considered as cotranslational and irreversible, but there are few reports in the litterature of the existence of non-myristoylated populations of myristoyl proteins (Da Silva and Klein, 1990). In the case of MARCKS, McIlhinney and McGlone (1990) recently described myristoylation occuring in a cytoplasmic fraction of synaptosomes in the absence of protein synthesis, implicating the existence of a pool of non-myristoylated MARCKS in this fraction. This is probably the same pool that we purified from bovine brain.

The main question opened by these data concerns the origin of this pool. Does it reflect a loose coupling between MARCKS synthesis and the myristoylation, or is it the result of the demyristoylation of previously myristoylated MARCKS? We partialy answered this question by identifying a demyristoylation activity in the cytoplasmic fraction of synaptosomes. Our results are confirmed by the observation of Mc Ilhinney and McGlone that a product of 70 kDa recognized by a specific antibody against MARCKS appears in synaptosomes concomitantly with the release of labelled myristic acid from the protein and disapearance of MARCKS itself. These authors explained their results by the N-terminal degradation of MARCKS, but they probably visualized by this method the demyristoylation of the protein. The existence of this activity is in contradiction with previous work describing the cotranslational and irreversible myristoylation of MARCKS in various cell types (James and Olson, 1989). However, the pulse chase experiment used by these authors to detect the eventual demyristoylation was performed during 30 min., while the demyristoylation observed by Mc Ilhinney and McGlone could be detected after 4 hours of chase. This difference in the protocols may explain why demyristoylation could not be detected in myocytes. Furthermore, the cell systems are completely different (synaptosomes and myocytes), and nothing is known untill now about the mechanisms of activation of the demyristoylase activity. A more precise characterization of this enzymatic activity will be necessary to investigate its presence and its function in various cell types.

The presence of a demyristoylation activity in the cells suggests the existence of cycles of myristoylation-demyristoylation of proteins, as it was described for other types of acylation, like palmitoylation (Magee, 1990). The existence of such cycles may be of a very great importance for the function of other myristoyl proteins. For example, a non-myristoylated mutant of pp60^{v-src} from Roux Sarcoma Virus looses its transforming activity and its localization at the plasma membrane (Jove and Hanafusa, 1987). Interestingly, its tyrosine kinase activity is conserved. Nef 27 from HIV also looses its negative activity on the virus replication when the N-terminus is not myristoylated (Yu and Felsted, 1992). The non-myristoylated α-subunit of various small trimeric G proteins loose their ability to associate with the β and γ chain (Gallego et al., 1992). As can be seen from these examples the biological implication of the demyristoylation activity may be of a crucial importance in the next years. If this activity is confirmed for other proteins than MARCKS, it may become one of the molecular events to be considered for the elucidation of the signal transduction pathways.

ACKNOWLEDGEMENTS

We thank Prof. G. Schwarz for helpful discussions and Dr. P. Jenö for the synthesis of the PSD peptide. This work was supportedin part by Swiss National foundation for Promotion of Science Grants 31-25230.88 and 31-32188.91

REFERENCES

Aderem, A., 1992, The MARCKS brothers: a family of protein kinase C substrates. *Cell* 71:713.

Blackshear, P.J., 1993, The MARCKS family of cellular protein kinase C substrates *J. Biol. Chem.* 268:1501.

Brooks, S. F., Erusalimsky, J. D., Totty, N. F., and Rozengurt, E., 1990, Purification and internal amino acid sequence of the 80 kD protein kinase C substrate from Swiss 3T3 fibroblasts. *FEBS Letters.* 268:291.

Da Silva, A.M.,and Klein, C., 1990, A rapid postranslational myristylation of a 68-kD protein in D. discoideum. *J. Cell Biol.* 111: 401.

Gallego, C., Gupta, S. K., Winitz, S., Eisfelder, B. J., and Johnson, G. L., 1992, Myristoylation of the G_{ai2} polypeptide, a G protein a subunit, is required for its signaling and transformation functions. *Proc. Natl. Acad. Sci. USA.* 89: 9695.

Graff, J.M., Gordon, J.I., and Blackshear, P.J.,1989, Myristoylated and non-myristoylated forms of protein are phosphorylated by protein kinase C. *Science* 246:503.

Hartwig, J.H., Thelen, M., Rosen, A., Janmey, P.A., Nairn, A.C., and Aderem, A. 1992, MARCKS is an actin filament crosslinking protein regulated by protein kinase C and calcium-calmodulin. *Nature* 356:618.

Herget, T., Brooks, S. F., Broad, S., and Rozengurt, E., 1993, Expression of the major protein kinase C substrate, the acidic 80-kilodalton myristoylated alanine-rich C kinase substrate, increases sharply when Swiss 3T3 cells move out of cycle and enter G0 *Proc. Natl. Acad. Sci. USA* 90:2945.

James, G., and Olson, E.N., 1989, Myristoylation, phosphorylation and subcellular distribution of the 80-kDa protein kinase C substrate in BC3H1 myocytes *J. Biol. Chem.* 264:20928.

Joseph, C., Qureshi, S. A., Wallace, D. J., and Foster D. A., 1992, MARCKS protein is transcriptionally down-regulated in v-src-transformed BALB/c 3T3 cells *J. Biol. Chem.* 267:1327.

Jove, R., and Hanafusa, H., 1987, Cell transformation by the viral src oncogene. *Annu. Rev. Cell Biol.* 3:31.

Magee, A. I., 1990, Lipid modification of proteins and its relevance to protein targeting. *J. Cell Sci.* 97:581.

Manenti, S., Sorokine, O., Van Dorsselaer, A., and Taniguchi, H., 1992, Affinity purification   and characterization of myristoylated alanine rich C kinase substrate (MARCKS) from bovine brain. *J. Biol. Chem.* 267: 22310.

Manenti, S., Sorokine, O., VanDorsselaer, A., and Taniguchi, H., 1993, Isolation of the non-myristoylated form of a major substrate of protein kinase C (MARCKS) from bovine brain. *J. Biol. Chem.* 268:6878.

Manenti, S., Sorokine, O., VanDorsselaer, A., and Taniguchi, H. Demyristoylation of the major substrate of protein kinase C (MARCKS) by a cytoplasmic fraction of bovine brain synaptosomes. *J. Biol. Chem.* In press.

McIlhinney, R.A.J. and McGlone, K., 1990, Evidence for a non-myristoylated pool of the 80-kDa prtotein kinase C substrate of rat brain. *Biochem. J.* 271:681.

Patel, J., and Kligman, D., 1987, Purification and characterization of an Mr 87,000 protein kinase C substrate from rat brain. *J. Biol. Chem.* 262:16686.

Rudnick D. A., McWherter C. A., Gokel G. W., and Gordon., J. I., 1993, Myristoyl CoA:protein N-myristoyltransferase. *Adv. Enzym.* 67:375.

Rosen, A., Keenan, K. F., Thelen, M., Nairn, A. C., and A. Aderem., 1990, Activation of protein kinase C results in the displacement of its myristoylated, alanine-rich substrate from punctate structures in macrophage filipodia. *J. Exp. Med.* 172:1211.

Sawai, T., Negishi, M., Nishigaki, N., Ohno, T., and Ichikawa, A., 1993, Enhancement by protein kinase C of prostacyclin receptor-mediated activation of adenylate cyclase through a calmodulin/myristoylated alanine-rich C kinase substrate (MARCKS) system in IC2 mast cells *J. Biol. Chem.* 268:1995.

Taniguchi, H. and Manenti, S., 1993, Interaction of myristoylated alanine-rich C kinase substrate (MARCKS) with membrane phospholipids. *J. Biol. Chem.* 268:9960.

Thelen, M., Rosen, A., Nairn, A.C., and Aderem, A., 1991, Regulation by phosphorylation of reversible association of a myristoylated protein kinase C substrate with the plasma membrane. *Nature* 351:320.

Wolfman, A., Wingrove T. G., Blackshear, P. J., and Macara I. G., 1987, Down-regulation of protein kinase C and of an endogenous 80-kDa substrate in transformed fibroblasts *J. Biol. Chem.* 262:1546.

Yu, G., and Felsted, R. L., 1992, Effect of myristoylation on p27nef subcellular distribution and suppression of HIV-LTR transcription. *Virology* 187:46.

BIOCHEMICAL AND IMMUNOLOGICAL CHARACTERIZATION OF PROTEIN KINASE C ISOZYMES IN RAT KIDNEY PROXIMAL TUBULE: EFFECT OF ANGIOTENSIN II

J. Poggioli, N. Defontaine, L. Micheli, and M. Paillard

INSERM U356 and Université Paris VI
15, Rue de L'école de médecine
75270 Paris Cédex 06, France

INTRODUCTION

The rat proximal tubule is the site for reabsorption of 75% of the normal filtered load of bicarbonate. This luminal acidification results from the following cascade of events : filtered HCO_3^- binds a H^+ to give H_2CO_3. Proton issued from hydration of cellular CO_2 is secreted into the tubular fluid (mainly via the apical Na/H antiporter) and combines to filtered HCO_3^- to give H_2CO_3. H_2CO_3 is converted into CO_2 and H_2O as rapidly as it is formed, because of the presence of apical carbonic anhydrase in contact with the tubular fluid. Bicarbonate generated in the cell also from hydration of CO_2 is transported into fluid via the basolateral Na^+-HCO_3^- cotransporter. Peptide hormones are potent regulators of bicarbonate reabsorption in this nephron segment. AngiotensinII (AII) stimulates it by increasing both Na^+/H^+ antiporter and Na^+-HCO_3^- cotransporter activities. We have recently shown that AII receptors were coupled to the inositol phosphate-Ca^{++} signaling pathway (Poggioli et al., 1992) and to well known cAMP pathway (Woodcock and Johnson 1982). Thus, our results indicated that AII was able to activate the Ca^{++}-phospholipid -dependent protein kinase C (PKC, Nishizuka, 1988). Both functional and biochemical approaches suggested that the effect of AII on transepithelial transport was mediated, at least in part, through PKC activation. In vivo microperfusion experiments have shown that this AII effect was reproduced by phorbol esters and inhibited by PKC inhibitor H7 (Liu and Cogan,1990 ; Wang and Chan,1991). Weinman et al. (1989) demonstrated on brush border membrane vesicles that an increase in the activity of the luminal Na^+/H^+ antiporter resulted from a PKC-mediated phosphorylation. Ruiz and Arruda (1992) reported on basolateral membrane vesicles that an increase in Na^+-HCO_3^- cotransporter activity resulted from a PKC-mediated phosphorylation too.

The aim of the present work was to determine what PKC isotype could be involved in the regulation of H^+/HCO_3^- transporters. Thus we attempted to establish the PKC isoform profile and to make a pharmacological characterization of the isoforms present in the tissue. Then we investigated the effect of phorbol esters and of AII on PKC activity.

Cell Signal Transduction, Second Messengers, and Protein Phosphorylation in Health and Disease
Edited by A.M. Municio and M.T. Miras-Portugal, Plenum Press, New York, 1994

MATERIALS AND METHODS

Partial Purification of PKC

A suspension of proximal tubules free of glomeruli was prepared as previously described (Poggioli et al. 1992). 0.5ml samples of tubule suspension in a Ringer medium (mM:137 NaCl, 3 KCl, 1 MgSO$_4$, 0.2 KH$_2$PO$_4$, 0.8 K$_2$HPO$_4$, 10 Hepes, 1 CaCl$_2$, 25 NaHCO$_3$, 5 glucose, 5 leucine and 0.1% bovine serum albumin) were equilibrated at 37°C under an atmosphere of O$_2$/CO$_2$ (95/5) for 4 minutes before the addition of the agent that had to be tested. The incubation was stopped by adding 2.5ml of ice cold buffer A (mM: 250 sucrose, 20 Tris/HCl pH 7.5, 2 EDTA, 10 EGTA, 2 dithiothreitol (DTT) ,10 benzamidine, 0.25 phenyl methylsulfonyl fluoride (PMSF) and 10µg/ml leupeptin, 100µg/ml aprotinin). The suspension was homogeneized by sonication (Branson sonifier, setting 1 for 15sec. at 4°C). The samples were centrifuged at 100,000g for 1 hour. The supernatants were collected and used as soluble crude fractions. The pellets were sonicated in buffer A containing 0.5% (v/v) TritonX-100 and used as particulate crude fractions after an additional centrifugation. The crude extracts were loaded onto a column of DE52 equilibrated in buffer B (mM: 20 Tris/HCl pH 7.5, 2 EGTA, 1 EGTA, 2 DTT, O.5 PMSF, 10 benzamidine and 0.1µg/ml aprotinin). After loading the samples, the columns were washed with 5ml of buffer B and 5ml of 20mM NaCl in buffer B. The PKC was eluted with 4ml of 90mM NaCl in buffer B containing 10% glycerol as described by Dong et al. (1991) ; appropriate fractions were stored at -80°C. Protein concentration was determined by the method of Lowry after precipitation with HCLO$_4$ 5%.

PKC Activity Assay

Partially purified PKC was assayed in the presence of Ca^{++}, phosphatidylserine (PS) and diglycerides (DG) by measuring the incorporation of ^{32}P into histone type III-S as described by Kikkawa et al.(1982). 35µl aliquots corresponding to 0.5-2µg protein were incubated in a 100µl reaction mixture consisting of 20mM Tris/HCl, pH7.5, 10mM MgCl$_2$, 0.2mg/ml histone, 75µg/ml PS, 12.5µg/ml DG, 0.5mM EGTA and 16µM γ^{32}P ATP (Specific activity 250-500cpm/pmol) in the presence or absence of 0.55mM CaCl$_2$. Samples were incubated at 30°C for 2 min. An aliquot of each tube was spotted onto phosphocellulose papers washed into 25% trichloracetic acid at 4°C. Under the conditions employed the activity was linear with respect to time and protein concentration. The non-phospholipid-dependent phosphorylation was substracted from the total amount of ^{32}P incorporated in order to determine the total specific Ca-dependent PKC activity. The radioactivity on the phosphocellulose papers was measured by liquid scintillation counting.

Immunoblot Analysis of PKC Isozymes

Aliquots of 0.15ml of crude fractions were mixed with 0.05ml of Laemmli buffer and boiled for 10 min. For whole cell extracts, an aliquot equivalent to 15µg (to detect the presence of ζ and δ PKC isoforms) or to 30µg of total protein (to detect the presence of α, β, γ and ϵ) were subjected to SDS-PAGE (10%) as described by Laemmli (1970). For cytosolic and particulate fractions equal aliquots were loaded to allow direct comparison of PKC content (10 µg for ζ , 20µg for δ and 40 µg for α and ϵ). Proteins on the gel were electrophoretically transferred onto nitrocellulose membranes (Schleicher & Schuell, 0.45µm) by using a Tyler ET 10 apparatus. The blotting buffer used was 50mM Tris, 380mM Glycine pH 8.3 , SDS 0.1% in 20% (v/v) ethanol. The membranes were blocked in 5% (w/v) non fat milk in phosphate buffer saline (PBS:137 mM NaCl, 2 mM KCl, 8 mM Na$_2$HPO$_4$,1.5 mM KH$_2$PO$_4$) containing 0.2% (v/v) tween overnight at 4°C. They were rinsed with PBS containing 2.5% non fat milk and 0.1% tween and incubated with specific anti-PKC isozyme antibodies (GIBCO) diluted in the same buffer. After overnight incubation with anti-PKC antibodies at 4°C, the nitrocellulose membranes were washed 3 times with PBS containing 2.5% milk and 0.1% tween and incubated with alkaline phosphatase conjugate anti-IgG antibody (Sigma Chemical Co) for 1 hour at room temperature. After washing (5 x10 min) in Tris buffer saline (TBS; 50mM Tris, 150mM NaCl, pH 7.4) the immunoreactive bands were visualized by color development using

0.4mM nitro blue tetrazolium and 4.5mM 5-bromochloro-3-indoyl phosphate dissolved in 100mM Tris, 100mM NaCl, 5 mM MgCl$_2$ pH 9.5. In some experiments the secondary antibody was ^{125}I-labelled Ig (Amersham). The membranes loaded with anti-PKC antibodies, were incubated for 1 hour with 0.5 to 2 µCi/ml depending on the PKC isoform that had to be detected. After washes (5x10 min) in PBS the bands were dried and detection was done by autoradiography by exposing the membranes to Kodak- X-Omat film for 60 hours at -80°C . The radioactive bands identified as PKC were sliced from the nitrocellulose and the radioactivity determined by liquid scintillation.

RESULTS AND DISCUSSION

A Western analysis of the isoforms of PKC expressed in proximal tubule is shown in Figure 1. It is evident from the immunoreactivity with specific antibodies that PKC α, δ, ε and ζ are present. PKC C β and γ could not be detected in the proximal tubule but were detected in rat brain extracts used as positive controls to test the β and γ subtype specific antibodies.

As shown in Figure 2 (lane A row 1), PKC α specific immunoreactivity is almost completely restricted to the cytosolic compartment. PKC δ (lane A row2) and protein kinase C ε (lane A row3) are preferentially located in the particulate fraction with little cytosolic location. PKC ζ (lane A row4) is equally well distributed between cytosolic and particulate fraction. PKC α and δ corresponded to about 80KD, PKC ε corresponded to 90KD and PKC ζ corresponded to a 72-80KD doublet. Data obtained on tubules treated with 10^{-7}M PdBu are presented in Figure 2, lane B. Protein kinase C α, δ, ε are both retained in the particulate fraction following a 4 min treatment. This translocation of protein kinase C α, δ and ε is consistent with the primary structure of these isozymes. There are two cysteine rich sequences in the first conserved region which are responsible for phorbol ester binding (for a review see Azzi et al.,1992). As expected from the primary structure of PKC ζ which contains only one set of cysteine rich sequence (Ono et al. ,1989) no translocation was observed after PdBu treatment. In parallel experiments the Ca^{++} sensitive membrane association of the different isoforms was also tested. The results are presented in Figure 2, lane C. Only PKC α was associated with the particulate fraction when extracted in the presence of Ca^{++} whereas δ, ε and ζ remained in the cytosol. The failure of Ca^{++} to affect PKC- δ, ε and ζ is due to the absence of the second conserved domain in their primary sequence. This domain seems to confer the Ca dependence of α, β and γ isoenzymes of PKC (for a review see Azzi et al.,1992). Additional experiments were performed to quantify the distribution of the different PKC isoforms using a radiolabelled secondary antibody.

Table 1: Distribution of PKC isoforms

Controls		PdBu		Ca^{++}		
soluble	particulate	soluble	particulate	soluble	particulate	
81± 8	19±8	24±13 *	76±13*	52±7 $	48±7$	(1)
35± 7	65±7	20±12	80±12	27±2	73±2	(2)
18±2	82±2	10±3	90±3	22±6	78±6	(3)
62±6	38±6	61±13	39±13	60±1	40±1	(4)

Each isoform is expressed as % of total (soluble+particulate). The total counts were the same in the three conditions . Row 1 is for α, row 2 for δ, row 3 for ε and row 4 for ζ. The data shown are means±sem of 3 experiments. .*p<0.02 vs controls;$ p<0.05vs controls.

As shown in Table 1, the distribution of PKC isotypes in unstimulated tubules as well as their pharmacological characterization by means of PdBu and Ca^{++} were confirmed and extended by these experiments using another way of detection of PKC. The above results have clearly shown the presence of 1 Ca^{++}-dependent PKC, α and 3 Ca^{++}-independent PKC, δ, ε and ζ in the rat renal proximal tubule.

To further characterize PKC, in the second set of experiments, the effect of PdBu, Ca^{++} and AII on PKC activity was investigated.

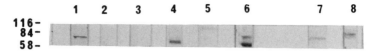

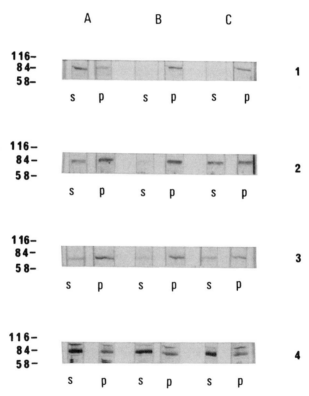

Figure1: PKC isoforms present in rat proximal tubules.

An aliquot of total protein from proximal tubules (left) or brain (right) was subjected to SDS-PAGE and Western blotting as described in Materials and Methods. Blots were probed with antisera specific to PKC-α(1), PKC-β(2), PKC-γ(3), PKC-δ (4), PKC-ε (5) and PKC-ζ(6). PKC β (7) and γ (8) were detected in brain extracts.

Figure 2: Cytosol and particulate distribution of PKCs

Proximal tubule fragments were either untreated (A and C) or treated with 10^{-7}M PdBu(B). The cells were homogenized in the presence of 10mM EGTA (A and B) or in the absence of EGTA (C) and fractionated into soluble (S) and particulate fractions (P). Equal volume of each fraction were subject to SDS-PAGE and immunoblotting with specific antisera as stated in Materials and Methods. (row 1, PKC-α, row 2, PKC-δ , row 3, PKC-ε and row 4 , PKC-ζ).

Table 2: Effect of phorbol dibutyrate, ionomycin and angiotensin II on the specific activity of soluble and particulate protein kinase C.

	SOLUBLE	PARTICULATE
PHORBOL DIBUTYRATE	pmol/μgP/2min.	pmol/μgP/2min
0	0.736±0.060	0.847±0.107
10^{-7}M	0.265±0.053***	1.557±0.178**
IONOMYCIN		
0	0.726±0.107	0.711±0.043
10^{-7}M	0.638±0.096	0.964±0.044**
ANGIOTENSIN II		
0	0.940±0.189	0.675±0.091
10^{-7}M	0.819±0.158	0.976±0.081*

Results are means±sem of 6 experiments for phorbol esters and ionomycin and of 12 determinations in 9 experiments for angiotensin II;***$p<0.001$; **$p<0.01$;*$p<0.05$.

As illustrated in Table 2, 10^{-7}M PdBu added for 4min induced a decrease in soluble enzyme activity and an increase in the particulate one. 10^{-7}M of the inactive phorbol ester 4α phorbol didecanoate was without effect (not shown). PdBu did not change the total PKC activity. The time course of the PdBu-induced PKC activation is presented in Figure 3. A rapid and large increase in particulate enzyme contribution was observed. The effect reached a plateau after 2min. A corresponding decrease in the contribution of the soluble enzyme was observed, indicating that PdBu induced a translocation.

The effect of Ca^{++} was tested by means of the Ca^{++} ionophore ionomycin. 10^{-6}M ionomycin elicited a stable rise in ionized free Ca^{++}, determined with fura 2, from 135 to 250nM. This was of the same magnitude as the Ca^{++} signal elicited by 10^{-7}M AII. 10^{-6}M ionomycin added during 4min. increased particulate activity as shown in Table 2. It did not change the total enzyme activity which allowed us to estimate its effect on PKC distribution. Increasing internal free Ca^{++} in a physiological range induced a statistically significant increase in particulate enzyme from 35.71±1.37 to 43.94±1.40 ($p<0.01$) and a decrease in the soluble enzyme from 64.27±1.37 to 55.58±1.35% ($p<0.01$) of the total respectively.

These findings were in agreement with those obtained by Western blot analysis and clearly indicated that PKC is activated by translocation from a soluble to a bound form when PdBu was present or when Ca^{++} was elevated.

This pharmacological characterization was a prerequisite to the study of the effect of AII on PKC activity. A 4 min. incubation with 10^{-7}M results in an increase in particulate PKC activity (Table2). AII did not modify the total enzyme activity. The time course of AII effect on PKC repartition is shown in Figure 4. An increase in bound-form of the enzyme associated with a decrease in soluble enzyme was observed as soon as 30sec. after hormone addition. This dual effect reached a plateau after 4min. The nature of the receptor activated by AII was determined by means of losartan, a nonpeptide specific antagonist of the AT_1 subtype receptor. These receptors are coupled positively to phospholipase C and negatively to adenylyl cyclase in the proximal tubule (Poggioli et al., 1992). A 4min. preincubation in the presence of 10^{-5}M losartan totally inhibited the AII-induced increase in particulate PKC activity (from 117.72±4.47 to 94.49±3.32% of controls, n=9) as well as AII-induced PKC translocation (not shown). In summary of that second part, we have shown that AII induced a time-dependent activation of PKC which corresponded to a translocation of a soluble to a bound form of the enzyme. The receptors activated are of AT_1 subtype since the effect was inhibited by losartan.

In conclusion, immunoblotting procedures demonstrated that kidney proximal tubular cells express PKC: α, δ, ε and ζ isozymes but not PKC β and γ. The presence of PKC β has been reported in the whole kidney by Wetsel et al.(1992) and in the kidney cortex by Laporta and Comolli (1993). Since PKC β has not been detected in our preparation of proximal tubules, it is likely associated with the glomeruli. PKC: α, δ, ε were translocated by phorbol

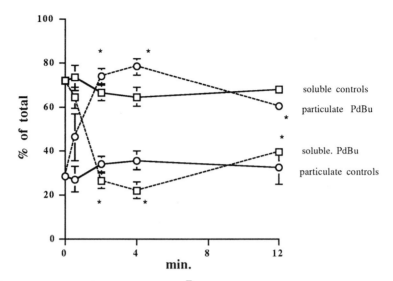

Figure 3: Time course of the effect of 10^{-7}M PdBu on PKC distribution.
Results are means ± sem of 3 to 6 determinations in 3 separate experiments. Each determination was performed in duplicate. *p<0.001

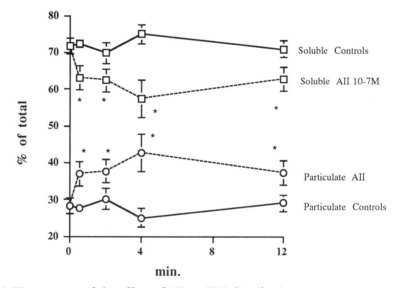

Figure 4: Time course of the effect of AII on PKC distribution.
Results are means ± sem of 5 to 11 different experiments. Each determination was performed in duplicate. *p<0.05

esters and thus are good candidates to regulate Na^+/H^+ and $Na^+-HCO_3^-$ transporter activities. Several questions arise then as to : 1)the PKC isotypes activated by AII ? PKC activity determined by histone phosphorylation and increased by the hormone probably represented for a large part the activity of PKC-α . However PKC-ζ has been reported to phosphorylate histone too (Ways et al.,1992). 2)the subcellular compartment in which the translocation occurs ? 3)the direct or indirect effect of PKC on the transporters? Four Na^+/H^+ isotypes have been isolated presently. NHE$_1$ is the ubiquitous house-keeping form located at the basolateral side of the epithelial cells. There is growing evidence that it is stimulated by activation of PKC (for review see Weinman and Shenolikar 1993). NHE$_2$ and NHE$_3$ are supposedly located at the luminal side of the epithelial cells and possess the consensus sequence for PKC (Tse et al.1993). $Na^+-HCO_3^-$ transporter has not been sequenced yet. This left open the possibility of either a direct regulation of this cotransporter by a PKC-mediated phosphorylation or by a phosphorylation cascade.

REFERENCES

Azzi,A., Boscoboinik,D., and Hensey, C., 1992, "The protein kinase C family", Eur J Biochem 208:547-557.

Dong,L., Stevens,J.,L., and Jaken, S., 1991, "Biochemical and immunological characterization of renal protein kinase C", Am J Physiol, 261:F679-F687.

Kikkawa,U., Takai,Y., Minakuchi, R., Inohara, S., and Nishizuka, Y., 1982 , "Calcium-activated, phospholipid-dependent protein kinase from rat brain", J Biol Chem, 257:13341-13348.

Laemmli,U., 1970, "Cleavage of structural proteins during the assembly of the head of bacteriophage T4", Nature, 227:680-685.

LaPorta , C.,Å.,M., and Comolli, R., 1993, "Biochemical and immunological characterization of calcium-dependent and -independent PKC isoenzymes in renal ischemia", Biochem Biophys Res Commun, 191:1124-1130.

Liu, F.,Y., and Cogan, M., G., 1990, "Effects of intracellular calcium on proximal bicarbonate absorption", Am J Physiol, 259:F451-F457.

Nishizuka, Y., 1988, "The molecular heterogneity of protein kinase C and its implications for cellular regulation", Nature, 334:661-665.

Ono, Y., Fujii,T., Ogita,K., Kikkawa,U.,Igarashi,K., and Nishizuka,Y., 1989, "Protein kinase C ζ subspecies from rat brain : its structure, expression, and properties", Proc Natl Acad Sci U.S.A. 86:3099-3103.

Poggioli, J., Lazar, G., Houiller, P., Gardin, J.,P., Achard, J.,M., and Paillard, M., 1992, "Effects of angiotensin II and nonpeptide receptor antagonists on transduction pathways in rat proximal tubule", Am J Physiol, 263: C750-C758.

Ruiz,O.,S., and Arruda, J.,A.,L., 1992, "Regulation of the renal Na-HCO$_3$ cotransporter by cAMP and Ca-dependent protein kinases", Am J Physiol, 262:F560-F565.

Tse, C-M., Levine,S.,A.,Yun,C.,H.,C., Montrose,M.,H., Little, P.,J., Pouyssegur,J., and Donowitz, M., 1993, "Cloning and expression of a rabbit cDNA encoding a serum-activated ethylisopropylamiloride-resistant epithelial Na^+/H^+ exchanger isoform (NHE-2)", J Biol Chem, 268:11917-11924.

Wang, T., and Chan, Y.,L., 1991, "The role of phosphoinositide turnover in mediating the biphasic effect of angiotensin II on renal tubular transport", J Pharmacol,Exp Ther, 256:309-317.

Ways,D.,K., Cook, P.,P., Webster,C., and Parker,P.,J., 1992, "Effect of phorbol esters on protein kinase C-ζ*", J Biol Chem, 267:4799-4805.

Weinman,E.,J., Dubinsky,W., and Shenolikar, S., 1989, "Regulation of the renal Na^+-H^+ exchanger by protein phosphorylation", Kidney Int, 36:519-525.

Weinman, E., J., and Shenolikar, S., 1993, "Regulation of the renal brush border membrane Na^+-H^+ exchanger", Ann Rev Physiol, 55:289-304.

Wetsel,W.,C., Khan,W.,A., Merchenthaler, I., Rivera, H., Halpern, A.,E., Phung, H.,M., Negro-Vilar, A. and Hannun, Y, 1992, "Tissue and cellular distribution of the extended family of protein kinase C isoenzymes", J Cell Biol, 117:121-133.

Woodcock, E.,A., and Johnson,C.,I., 1982, "Inhibition of adenylate cyclase by angiotensin II in rat renal cortex*", Endocrinology, 111:1687-1691.

PHOSPHOLIPID DEGRADATION, SECOND MESSENGERS AND ACTIVATION OF CELL SIGNALING GENES

Nicolas G. Bazan and Geoffrey Allan

LSU Eye Center and LSU Neuroscience Center
Louisiana State University Medical Center School of Medicine
2020 Gravier Street, Suite B
New Orleans, Louisiana 70112
USA

INTRODUCTION

Platelet-activating factor (PAF) is one of the most potent biologically active lipids known. It is a mediator of inflammatory and immunological events at concentrations down to the picomolar range in a wide variety of cells and tissues (for a review see Braquet et al., 1987). A role for PAF in the nervous system has been suggested by the ability of seizures (Kumar et al., 1988) and certain neurotransmitters (Bussolino et al., 1986; Sogos et al., 1990) to trigger the formation of PAF in the brain and in neuronal cells. Moreover, the finding that PAF antagonists confer neuroprotection during ischemia-reperfusion injury (Panetta et al., 1987) highlights a potential physiological effect of PAF in the nervous system. PAF is undetectable in resting cells. However, seizures or cerebral ischemia rapidly activate phospholipases A_2 (Bazan, 1970) and C, resulting in the release of PAF and other short-lived second messengers. The study of the signal transduction events that control PAF synthesis is only now beginning to evolve. However, two new clues about the biological activity of PAF in the brain have been found. In presynaptic nerve terminals, PAF enhances excitatory neurotransmitter release in hippocampal neurons (Clark et al., 1991; Clark et al., 1992 a, b); and, through a pharmacologically distinct intracellular site, PAF activates transcription of immediate early genes in the brain (Marcheselli et al., 1990 a, b; Bazan et al., 1991). These novel effects could have both physiological and pathophysiological significance.

PHARMACOLOGICAL CHARACTERIZATION OF DISTINCT PAF RECEPTORS IN BRAIN

Specific PAF binding sites have been pharmacologically characterized in synaptic and intracellular membranes of rat cerebral cortex (Marcheselli et al., 1990 b; Marcheselli and Bazan, 1993). Three sites have been identified, one synaptic and two microsomal. These sites differ in their affinities for PAF and PAF analogues, and also in their sensitivity to distinct PAF antagonists. The microsomal sites have a higher affinity for PAF than the

Cell Signal Transduction, Second Messengers, and Protein Phosphorylation in Health and Disease
Edited by A.M. Municio and M.T. Miras-Portugal, Plenum Press, New York, 1994

95

synaptosomal sites, with Kd values of 22.5 pM and 25 nM in ligand displacement studies. The hetrazepine PAF antagonist BN-50730 (Braquet and Esanu, 1991) competes with PAF binding only at the microsomal sites (Marcheselli and Bazan, 1993). The antagonist displays very specific binding to each microsomal site, with an IC_{50} of 585 pM (Figure 1). The ginkgolide BN-52021 has a much stronger affinity for the synaptosomal site (Marcheselli et al., 1990 b). Thus, the two drugs are useful tools for determining the modes of PAF action which elicit particular cellular responses.

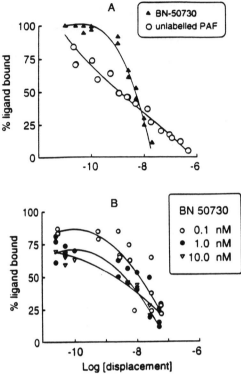

Figure 1 [^{3}H] PAF binding displacement by the hetrazepine BN-50730 on brain microsomal membranes. A: Comparison of [^{3}H] PAF ligand displacement curve with increasing concentrations of unlabelled PAF and BN-50730. B: Displacement curves of [^{3}H] PAF with increasing concentrations of unlabelled PAF in the presence of BN-50730. (From Marcheselli and Bazan, 1994)

THE PRESYNAPTIC SITE OF PAF ACTION AND ITS ROLE IN LONG-TERM POTENTIATION

PAF enhances glutamate release at excitatory synapses from rat hippocampal neuronswithout affecting GABA release at inhibitory synapses. This is a presynaptic effect and is inhibited by BN-52021 (Clark et al., 1992 a, b; Bazan et al., 1993b). This phenomenon may contribute to the neurological protection afforded by PAF antagonists in experimental models of cerebral ischemia and against ischemia-reperfusion damage (Panetta et al., 1987). Also, PAF antagonists inhibit the accumulation of free polyunsaturated fatty acids in cerebral ischemia. It is possible that the synaptic site is involved in the onset of brain damage through an enhancement of excitotoxicity-mediated events in cerebral ischemia, trauma and seizures The onset of cerebral ischemia triggers the hydrolysis of phospholipids in excitable membranes through the activation of phospholipases. Activation occurs mainly through ischemia-induced neurotransmitter release and impairment of calcium homeostasis, that, in turn, enhance receptor-mediated activation of phospholipases A_2 and C. The result is the accumulation of cytokine precursors, such as arachidonic acid, and second messengers, including diacylglycerol, and PAF. Additionally, by regulating excitatory synaptic transmission, PAF may play a role in long-term potentiation (Arai and Lynch, 1992; Clark et al., 1992a; Wieraszko et al., 1993)

PAF AS A MODULATOR OF NEURONAL GENE EXPRESSION

PAF Induction of Neuronal Gene Expression *In Vitro*

PAF is known to transiently enhance immediate-early gene (IEG) expression in some neuronal cell lines. These include the human neuroblastoma SH-SY5Y (Squinto et al., 1989) and the rodent neuroblastoma x glioma hybrid cell line NG108-15 (Doucet and Bazan, submitted). Induction of IEG expression has also been described in some non-neuronal cell lines (Mazer et al., 1991; Schulam et al., 1991; Tripathi et al., 1991); and there is evidence that the induction of metalloproteinase gene expression in corneal epithelium by PAF may involve a gene cascade mechanism (Bazan et al., 1993a).

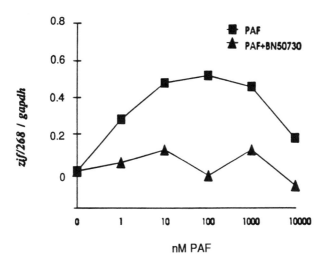

Figure 2. Concentration-dependent stimulation by PAF of *zif/268* expression in NG108-cells. Cells were treated for 30 minutes with the appropriate concentration of PAF. Where indicated, cells were incubated with 100 nM BN-50730 prior to addition of PAF. mRNA expression was assessed by Northern analysis and the results expressed as *zif/268* signal divided by the signal intensity for the control gene glyceraldehyde-3-phosphate dehydrogenase (gapdh). (From Doucet and Bazan, submitted)

PAF induces *c-fos* and *c-jun* expression in SH-SY5Y cells. The *c-fos* and *c-jun* gene products can combine to form the AP-1 transcription factor complex. Induction can be blocked using the ginkgolide BN-52021. Studies using nuclear run-on and promoter deletion mutants indicate that the effect on *c-fos* expression is exerted at the level of transcription and requires a region of the promoter which coincides with the calcium responsive element (Squinto et al., 1989).

A transient induction of transcription factor gene expression by PAF has also been observed in NG108-15 cells. In this instance, the target gene is that which encodes the zinc-finger protein *zif/268* (also called TIS 8, egr-1, NGFIA). In contrast to the SH-SY5Y line, *c-fos* and *c-jun* expression are unaffected in NG108-15 cells. PAF produces a broad, concentration-dependent induction of *zif/268* expression. Concentrations of PAF as low as 1 pM are effective, although peak induction is achieved at 50-100 nM (Figure 2). This induction is completely blocked by pretreatment of the cells with nanomolar concentrations of BN-50730, which implicates a cognate of the cerebral cortical intracellular receptor in this effect. Other classes of PAF antagonists, including BN-52021 and the PAF analogue WEB2086, are largely ineffective. Furthermore, BN-50730 also blocks acetylcholine-induced *zif/268* expression. Given that atropine also completely inhibits this induction, PAF is implicated as in intracellular signal generated by muscarinic acetylcholine receptor activation.

The Intracellular PAF Receptor and Induction of the Primary Genomic Response in Brain

Many of the stimuli which induce acute biochemical effects in the brain, such as PAF synthesis and accumulation, also result in the transcriptional activation of some immediate-early genes. This primary genomic response is thought to couple some short-term biochemical responses with longer term adjustments in neuronal phenotype. The consequences could be either in the homeostatic restoration of normal brain functions or a pathophysiological alteration of neuronal circuitry.

Electrically induced seizure activity induces profound changes in IEG expression in specific brain regions, with the hippocampus being particularly sensitive (Morgan et al.,

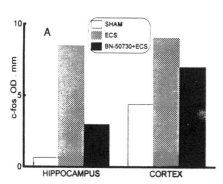

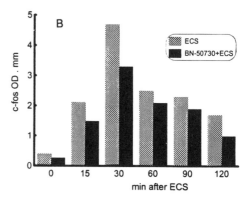

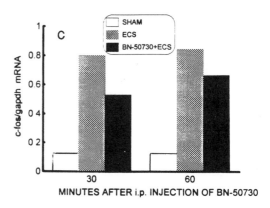

Figure 3. Effect of i.p. administration of BN-50730 on ECS-induced *c-fos* mRNA expression in rat brain. A: Comparison between hippocampus and cortex. Drug or vehicle were injected 30 min before ECS. Animals were decapitated 30 min after ECS. Shams received drug or vehicle, but no ECS. B: Time course of *c-fos* mRNA expression after ECS in the hippocampus and effect of BN-50730 pretreatment. C: Inhibition of *c-fos* mRNA expression at different times of pretreatment with BN-50730 in the hippocampus. Twenty mg per kg of the drug were delivered in 150μl of dimethylsulfoxide. Data are the average of two to four animals per condition. (From Marcheselli and Bazan, 1994)

1987). A role for PAF, and specifically the intracellular PAF receptor in this induction has been implied. A single electroconvulsive shock (ECS) transiently elevates levels of the mRNAs coding for *c-fos* and *zif/268* in rat hippocampus and cerebral cortex, while pretreatment of animals with BN-50730 inhibits this accumulation (Marcheselli and Bazan, 1993). Some inhibition is observed when the drug is delivered intra-peritoneally (i.p.) (Figure 3), but stronger inhibition is seen when the drug is injected directly into the lateral cerebral ventricles (intracerebroventricularly, i.c.v.) (Figure 4). This suggests that the drug exerts its effects directly in the central nervous system. When i.c.v delivery is employed, there is a dose-related inhibition of *zif/268* mRNA expression. Three micrograms administered to 175-250 gram rats 30 minutes before ECS results in a 75% inhibition of induced *zif/268* expression in hippocampus. By contrast, the same dose gives only a 50% inhibition of *c-fos* expression. Thirty microgrammes of the drug, which inhibits ECS-induced *zif/268* expression in the hippocampus by 80%, only marginally inhibits *c-fos* expression. These data suggest that there are multiple pathways by which *c-fos* and *zif/268* may be induced as a result of ECS, and that molecular events channelled through the intracellular PAF receptor result in both zinc-finger and AP-1 transcription factor activation. Current studies are concerned with the consequences of this inhibition on trauma-induced changes in neuronal function.

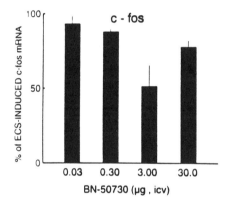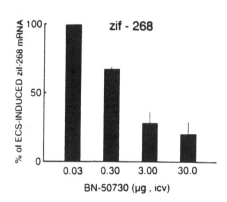

Figure 4. Dose-dependent effect of i.c.v. injection of BN-50730 on ECS-induced *zif/268* and *c-fos* mRNA expression in the rat hippocampus. Data are normalized to the control gene gapdh and are the average of at least six individuals in each condition ± one standard deviation. (From Marcheselli and Bazan, 1994)

ACKNOWLEDGEMENTS

This work was supported by the National Institutes of Health, National Eye Institute R01 EY05121.

REFERENCES

Arai, A., and Lynch, G. (1992) Antagonists of the platelet-activating factor receptor block long-term potentiation in hippocampal slices. *European J. of Neurosci.* 4, 411-419.

Bazan, H.E.P., Tao, Y., Bazan, N.G. (1993a) Platelet-activating factor induces collagenase expression in corneal epithelial cells. *Proc. Natl. Acad. Sci. U.S.A.* 90, 8678-8682.

Bazan, N.G. (1970) Effects of ischemia and electroconvulsive shock on free fatty acid pool in the brain. *Biochim. Biophys. Acta* 218, 1-10.

Bazan, N.G., Squinto, S.P., Braquet, P., Panetta, T., Marcheselli, V.L. (1991) Platelet-activating factor and polyunsaturated fatty acids in cerebral ischemia or convulsions: Intracellular PAF-binding sites and activation of a Fos/Jun/Ap-1 transcriptional signaling system. *Lipids* 26, 1236-1242.

Bazan, N.G., Zorumski, C.F., and Clark, G.D. (1993b) The activation of phospholipase A_2 and release of arachidonic acid and other lipid mediators at the synapse: The role of platelet-activating factor. *J. Lipid Mediators* 6: 421-427

Braquet, P., Touqui, L., Shen, T.Y., and Vargaftig, B.B. (1987) Perspectives in platelet-activating factor research. *Pharmacol Rev* 39, 97-145.

Braquet, P. and Esanu A. (1991) New trends in PAF antagonist research: a new series of potent hetrazepine-derived PAF antagonists. *Agents. Actions.* 32, 34-36

Bussolino, F., Gremo, F., Tetta, C., Pescarmona, G.P., and Camussi, G. (1986) Production of platelet-activating factor by chick retina. *J. Biol. Chem.* 261, 16502-16508.

Clark, G.D., Happel, L.T., Zorumski, C.F., and Bazan, N.G. (1991) Platelet-activating factor augments excitatory synaptic transmission in cultured rat hippocampal neurons. *Soc. Neurosci.* 17, 951.

Clark, G.D., Happel, L.T., Zorumski, C.F., and Bazan, N.G. (1992a) Enhancement of hippocampal excitatory synaptic transmission by platelet-activating factor. *Neuron* 9, 1211-1216.

Clark, G.D., Happel, L.T., Zorumski, C.F., and Bazan, N.G. (1992b) A novel presynaptic receptor that modulates the release of excitotoxic neurotransmitters. In: J. Krieglstein, H. Oberpichler-Schwenk (eds), Pharmacology in Cerebral Ischemia,, Wissenschaftliche Verlagsgesellschaft mbH, Stuttgart, pp 65-71.

Doucet, J., and Bazan, N.G. (submitted) Muscarinic induction of *zif/268* immediate-early gene in NG108-15 neurohybrid cells is mediated by platelet-activating factor.

Kumar, R., Harvey, S., Kester, N., Hanahan, D., and Olson, M. (1988) Production and effects of platelet-activating factor in the rat brain. *Biochim. Biophys. Acta* 963, 375-383.

Lynch, J.M. and Henson, P.M. (1986) The intracellular retention of newly synthesized platelet-activating factor. *J. Immunol.* 137, 2653-2661.

Marcheselli, V.L. and Bazan, N.G. (1994) Platelet-activating factor is a messenger in the electroconvulsive shock-induced transcriptional activation of c-*fos* and *zif*-268 in hippocampus. *J.Neurosci. Res.* 37, 54-61.

Marcheselli, V.L., Doucet, J.P., and Bazan, N.G. (1990a) PAF antagonist decreases *fos* expression in rat hippocampus induced by single seizure. *Soc. Neurosci.* 16, 629.

Marcheselli, V.L., Rossowska, M., Domingo, M.T., Braquet, P., and Bazan, N.G. (1990b) Distinct platelet-activating factor binding sites in synaptic endings and in intracellular membranes of rat cerebral cortex. *J. Biol. Chem.* 265, 9140-9145.

Mazer, N., Domenico, J., Sawami, H., and Gelfand E. W. (1991) Platelet-activating factor induces an increase in intracellular calcium and expression of regulatory genes in human B lymphoblastoid cells. *J. Immunol.* 146, 1914-1920.

Morgan, J.L., Cohen D.R., Hempstead J.L., and Curran, T. (1987) Mapping patterns of c-*fos* expression in the central nervous system after seizure. *Science* 237, 192-197.

Panetta, T., Marcheselli, V. L., Braquet, P., Spinnewyn, B. and Bazan, N. G. (1987). Effects of a platelet-activating factor antagonist (BN-52021) on free fatty acids, diacylglycerols, polyphosphoinositides and blood flow in the gerbil brain: inhibition of ischemia-reperfusion induced cerebral injury. *Biochem. Biophys. Res. Commun.* 149, 580-587.

Schulam, P.G., Kuruvilla, A., Putcha, G., Mangus, L., Franklin-Johnson, J. , and Schearer W.T. (1991) Platelet-activating factor induces phospholipid turnover, calcium flux, arachidonic acid liberation, eicosanoid generation, and oncogene expression in a human B cell line. *J. Immunol.* 146, 1642-1648.

Sogos V., Bussolino E., Pilia E., Torelli S., and Gremo F. (1990) Acetylcholine-induced production of platelet-activating factor by human fetal brain cells in culture. *J. Neurosci. Res.* 27, 706-711

Squinto, S.P., Block, A.L., Braquet, P., and Bazan, N.G. (1989) Platelet-activating factor stimulates a Fos/Jun/AP-1 transcriptional signaling system in human neuroblastoma cells. *J. Neurosci. Res.* 24, 558-566.

Tripathi, Y.B., Kandala, J.C., Guntaka, R.V., Lim R.W., and Shukla, S.D. (1991) Platelet-activating factor induces expression of early response genes c-fos and TIS-1 in human epidermoid carcinoma A-431 cells. *Life Sci.* 49, 1761-1767.

Wieraszko, A.. Li, G.. Kornecki, E.. Hogan, M.V., Ehrlich, Y.H. (1993) Long-term potentiation in the hippocampus induced by platelet-activating factor. *Neuron* 10, 553-557.

PLATELET-ACTIVATING FACTOR AND CALCIUM SIGNALING: ITS IMPLICATION IN CELLULAR RESPONSES

Matyas Koltai, David Hosford, and Pierre G. Braquet

Institut Henri Beaufour, Le Plessis Robinson, France

SUMMARY

Platelet-activating factor (PAF), a potent phospholipid inflammatory mediator, known to affect cellular phosphoinositide metabolism, through formation of inositol triphosphate, transiently releases intracellular calcium. This highly sensitive cellular signaling process is involved in a great variety of cellular responses in many organs, and it may play a pivotal role in regulating meditor release in the cell. Through activating phospholipase A_2, leading to arachidonic acid release and eicosanoid generation, PAF may induce downregulation of mediator release, an important process to maintain normal cell metabolism. On the other hand, via interacting with various cytokines or bacterial lipopolysaccharides, PAF may up-regulate mediator release, initiating an important role for the phospholipid noy only under normal conditions, but also in pathophysiological alterations of cellular responses and metabolism.

INTRODUCTION

Transduction of important signals brought about by endogenous mediators and agonists from cell membrane receptors to the site of specific cellular activation has a pivotal role in cellular function, and it is therefore is continuously in focus of interest. A number of these pathways, e.g. signal transduction through inositol phosphates and cAMP has now been well defined (Hall, 1993). Phosphoinositides play an important role in the release of intracellular calcium $[Ca^{2+}]_i$ which appears to be involved in a great variety of cellular responses. (Berridge, 1993a).

Cell Signal Transduction, Second Messengers, and Protein Phosphorylation in Health and Disease
Edited by A.M. Municio and M.T. Miras-Portugal, Plenum Press, New York, 1994

Cells have two intracellular channels for regulating the release of Ca^{2+} from internal stores (Berridge, 1993b), these are inositol 1,4,5-triphosphate (IP_3) receptors and ryanodine receptors. Although the messenger responsible for opening the first pathway was recognized as ITP_3 long ago, the latter was unknown until recent mounting evidence that ryanodine receptors might be controlled by cyclic ADP ribose.

Platelet-activating factor (PAF), identified as 1-O-alkyl-2-acetyl-sn-glyceryl-3-phosphorylcholine (Hanahan et al., 1980), is a potent phospholipid autacoid generated rapidly from its inactive precursor lyso-PAF by an acyltransferase and released from the cells, however, its considerable part remains cell-associated, indicating that this phospholipid induces fundamental intracellular changes (Prescott et al., 1990). PAF has been implicated as mediator in inflammation, ischemic disorders, and shock, furthermore, it is produced by a variety of cells, including endothelial cells, polymorphonuclear leukocytes, platelets, monocytes, basophils, eosinophils, mast cells, and lymphocytes (Braquet et al., 1987, Koltai et al., 1991a,b).

Early studies clearly showed that PAF affects cellular phosphoinositide metabolism in platelets of various species. This was characterized by an early rapid and transient decrease in phosphatidylinositol 4,5-biphosphate in washed rabbit platelets labeled with [^{3}H]inositol when stimulated with low concentrations of PAF (Shukla & Hanahan, 1983). Lyso-PAF had no such an effect and the response was clearly observed within 5 s after addition of PAF, and proved to be concentration-dependent. Interestingly, within 15 s an increase in [^{3}H]lyso-phosphatidylinositol was also seen, pointing to a close relationship of PAF to changes of phosphoinosite metabolism. The stimulation by PAF of phosphoinositide metabolism was first suggested to be linked to the mobilization of membrane-bound [$Ca^{2+}]_i$ in horse platelets (Billah & Lapetina, 1983). Alternatively, trifluoperazine and N-(6-aminohexyl)-5-chloro-l-naphthalene sulfonamide, two calmodulin antagonists, were shown to inhibit PAF biosynthesis in neutrophil leukocytes stimulated by Ca^{2+} ionophore A23187 (Billah & Siegel, 1984). Later on, when the fluorescent probe fura-2 has become frequently used, PAF-induced release of [$Ca^{2+}]_i$ was confirmed by exact measurements in U937 human monocytic cell line, and its response was antagonized by a variety of specific PAF receptor antagonists, again emphasizing the role of PAF in [$Ca^{2+}]_i$ signaling (Ward & Westnick, 1988). This signal transduction involves activation of GTPase (Homma & Hanahan, 1988, Hwang & Lim 1986), and linked to G-protein coupled receptors (Braquet et al., 1990, Collins, 1993, Lee & Kerlavage, 1993, Paubert-Braquet et al., 1991), and it directly affects membrane ion pumps (Pedemonte, 1993). There is also convincing evidence that, through activation of phospholipase A_2 (PLA_2), PAF releases arachidonic acid (AA), resulting in eicosanoid formation both on the cyclooxygenase and lipoxygenase pathways (Braquet et al., 1987, Prescott et al., 1990).

As reviewed recently (Rink & Sage, 1990), these biochemical alterations are clearly related to platelet functions. The present review is an attempt to summarize functional changes induced by PAF in various cells through the activation of phosphoinisitide metabolism and release of [$Ca^{2+}]_i$.

PAF-INDUCED CALCIUM SIGNALING IN VARIOUS CELLS

This chapter describes characteristic examples, supporting the view that Ca^{2+} signaling contributes to diverse biological responses induced by PAF. More recently, intensive research has been devoted to clarify the mechanisms involved in cell adhesion. Platelet aggregation is mediated via fibrinogen binding to membrane glycoprotein IIB-IIIA complexes which become exposed when the cells are stimulated with PAF (van Willigen & Akkerman, 1991). Stimulation is enhanced by agents that raise cAMP levels, inhibit protein kinase C (PKC), or disrupt the energy supply, and has been shown to be accompanied by an increase in $[Ca^{2+}]_i$. In another study, the role of PAF in neutrophil adherence to interleukin-1 (IL-1) prestimulated endothelial cells (EC) has been examined (Kuijpers et al., 1991). Cultured human umbilical cord vein EC were prestimulated for 4 to 6 hours with rIL-1β. Freshly isolated human blood neutrophils bound very avidly to these cells. Adherence was inhibited for more than 80% by anti-ELAM-1 F(ab)$_2$ fragments and for 40 to 50% by anti-CD18 F(ab)$_2$ fragments, whereas combined addition of these antibodies caused a complete inhibition. Neutrophil adherence was prevented when the ATP content of the cells was depleted, furthermore, neutrophils responded with a transient rise in $[Ca^{2+}]_i$ upon binding to rIL-1β-pretreated EC which ran parallel with PAF production and release. Although the removal of PAF from EC or pretreatment with EC with WEB 2086 had no effect on the extent a cell adhesion, the initial trigger to activate this response appears to be strongly dependent of a transient PAF-induced Ca^{2+} signaling.

As mentioned above, the human monocytic cell line U937 responded to stimulation with the naturally occurring stereoisomer (R)-PAF with a transient increase in $[Ca^{2+}]_i$ (Ward & Westwick, 1988). Interestingly, the $[Ca^{2+}]_i$ content of U937 cells was shown to be increased not only by PAF but also by the chemotactic peptide fMLP and leukotriene B$_4$, but it was unaffected by stimulation with interferon and lipopolysaccharide (LPS) (Maudsley & Morris, 1987). These findings are in accord with that found by Conrad and Rink (1986) in guinea-pig peritoneal macrophages, and obviusly related to phagocytosis, maturation, and other cellular functions.

In human B-lymphocytes PAF has also been shown to increase $[Ca^{2+}]_i$ (Mazer et al., 1991, 1992, Schulam et al., 1990), a process which involves G-proteins (Harnett & Klauss, 1988), and related to immunoglobulin secretion.

PAF AND ENDOTHELIUM-DEPENDENT VASORELAXATION

The nitric oxide synthase (NOS)-nitric oxide (NO) system, producing endothelium-dependent vasorelaxation, has been suggested to maintain a constant vasodilator tone. These enzymes exist in two forms: (i) a constitutive, Ca^{2+}-dependent enzyme located predominantly in EC is active under normal conditions and induces transient vasodilation, whereas (ii) an inducible, apparently Ca^{2+}-independent enzyme, which has however been shown to nclude calmodulin as a subunit (Forstermann et al., 1991), is synthesized de novo in a variety of inflammatory cells, and has been thought to have a great pathophysiological significance in septic shock. These enzymes exhibit similar properties, for example, both synthesize NO and activate guanylate cyclase, however, the constitutive enzyme produces NO at only picomolar concentrations that induce transient vasodilation, whereas the inducible enzyme longlastingly synthesizes nanomolar concentrations of NO, thus inducing a powerful and sustained drop of blood pressure. The inducible enzyme is produced in macrophages, PMN (Rees et al., 1990) and freshly isolated human hepatocytes (Nussler et al., 1992) exposed to IL-1, TNF, γ-interferon, or LPS.

Since PAF releases $[Ca^{2+}]_i$, it may induce NO formation through activation of the constitutive NOS. Indeed, Kamata et al. (1989) have demonstrated that PAF produces dose-dependent, long-lasting, powerful vasodilation ($IC_{50} = 0.04$ nM) in the perfused mesenteric vascular bed of the rat, and this response can be abolished after detergent infusion and CV-6209, but not by indomethacin, suggesting an endothelium-dependent mechanism. These results were confirmed in the rat perfused mesenteric bed and isolated mesenteric artery (Chiba et al., 1990), where CV-3988 inhibited the response. Both PAF- and ACh-induced endothelium-dependent vasodilation was blocked by hydroquinone and methylene blue, or canavanine. Two PLA_2 inhibitors, chinacrine and ONO-RS-082 abolished the effect of ACh but did not affect that induced by PAF. More recently, Moritoki et al. (1992) have reported the involvement of NO pathway in the PAF-induced relaxation of isolated rat thoracic aorta. Concentrations of PAF causing relaxation induced a marked increase in cGMP concentration, and mechanical removal of endothelium, or treatment with CV-3988, CV-6209 and FR-900,452 as well as with N^G-nitroarginine, an inhibitor of NOS, hemoglobin or methylene blue abolished both responses. These results strongly suggest that, in parallel with the release of $[Ca^{2+}]_i$, PAF stimulates NO production through specific PAF receptors located on the surfaces of EC.

PAF-INDUCED CALCIUM SIGNALING IN THE BRAIN

Recently, an impressive progress in brain research has been indicated an important role for PAF in the cellular function of the central nervous system. In rat brain slices, through a phospholipase C (PLC)-dependent mechanism, PAF was shown to rapidly increase [^{32}P] incorporation into phosphoinositides and phosphatidic acid (Catalan et al., 1992). PAF has been shown to induce an immediate and concentration-dependent elevation of $[Ca^{2+}]_i$ in fura-2-loaded NCB-20 cells (Yue et al., 1991a). This effect was inhibited in a concentration-dependent manner by PDBu, a PKC activator, whereas the biologically inactive PDD had no such an effect. PAF-induced release of $[Ca^{2+}]_i$ was inhibited by specific PAF antagonists, such BN 50739, WEB 2086, SRI 63-441 and BN 52021. Nifedipine and diltiazem had no effect on PAF-induced increase in $[Ca^{2+}]_i$; however depletion of extracellular Ca^{2+} $[Ca^{2+}]_e$ caused a marked reduction of PAF-induced increase in $[Ca^{2+}]_i$. The remainder release of $[Ca^{2+}]_i$ contributed from intracellular sources and was abolished by 10 mM 8-(N, N-diethylamine)octyl 3,4,5-trimethoxybenzoate HCl (TMB-8), a specific inhibitor of $[Ca^{2+}]_i$ release. NBC-20 cells exhibited homologous desensitization to sequential addition of PAF, and responded also to bradykinin or ATP. These data suggest that activation of neuronal PAF receptors results in an increase in $[Ca^{2+}]_i$ primarily through receptor-operated rather than voltage-dependent Ca^{2+} channels and to a lesser extent from $[Ca^{2+}]_i$ release.

Further analysis of the effect of PAF in myo-$[^3H]$inositol-labeled NCB-20 cells and neurohybrid NG108-15 (Yue et al., 1992a,b) on receptor-mediated cellular signal transduction mechanisms recently showed that PDBu, a PKC activator, and pertussis toxin (PTX) concentration-dependently inhibited PAF-induced phosphoinositide metabolism and IP_3 formation with a resulting $[Ca^{2+}]_i$ mobilization in correlation with an inhibition of $[^{32}P]$-labeling of the toxin substrate in the membranes. PAF (0.01-10 nM) stimulated guanosine 5-0-(3-$[^{35}S]$) triphosphate ($[^{35}S]$-GTP-γ-S) binding to $G\alpha$ i(1,2) proteins, which was inhibited by BN 50739, a hetrazepine-type PAF receptor antagonist. PAF-induced binding of $[^{35}S]$GTP-γ-S to $G\alpha$ s and $G\alpha$ o was also reduced by PTX. The effect of PDBu was inhibited by H-7, a PKC inhibitor. These results have revealed that neuronal cells possess PAF receptors linked to PLC through guanine nucleotide-binding proteins and receptor-operated Ca^{2+}-channels that are regulated by PKC. Both PTX-sensitive and -insensitive guanine-binding proteins appear to couple with the PAF receptors to induce activation of PLC and increase in $[Ca^{2+}]_i$.

Parallely with an increase in $[Ca^{2+}]_i$, PAF is able to enhance excitatory synaptic transmission in hippocampal neurones (Bito et al., 1992, Clark et al., 1992, Miller et al., 1991). The relationship between elevation of $[Ca^{2+}]_i$ and excitation exists not only in the brain, but also in a subset a peripheral, autonomic myenteric neurones (Willard, 1992). Excitation of the vagus nerve by PAF has been shown to correlate with breading pattern that might have an influence on PAF-induced bronchoconstriction (Clement et al., 1992). These findings point to the close relationship between Ca^{2+} signaling and nerve cell function, again suggesting that PAF-induced release of $[Ca^{2+}]_i$ is a unique cellular response.

Exposure of 20-day old fetal brain to PAF, but not lyso-PAF, has been found to stimulate AA release with concomitant increase in TXB_2 synthesis (Kunievsky & Yavin, 1992). Repeated PAF administration induced desensitization and the PAF response was inhibited by BN 52021, BN 50726, BN 50727, or indomethacin, and depended on $[Ca^{2+}]_i$ rather than $[Ca^{2+}]_e$, because chelation of $[Ca^{2+}]_i$ by Fluo-3/AM effectively reduced eicosanoid production.

PAF-INDUCED CALCIUM SIGNALING IN THE HEART

PAF affects cardiac function, released in early reperfusion after coronary occlusion, and its antagonists have been examined in cardiac anaphylaxis and myocardial ischemia (Koltai et al., 1989). Recently, the cellular electrophysiological effects of PAF in guinea-pig atria and papillary muscle were examined (Kecskemeti & Braquet, 1992). This study has to be considered as highly significant, because stimulation of cardiomyocytes could release of large quantity of $[Ca^{2+}]_i$ (Tseng & Hoffmann, 1989), and these changes may profoundly affect cardiac ion channels. PAF did not modify resting membrane potential or the maximum rate of depolarization, but it increased the amplitude of action potential. The repolarization phase was concentration-dependently shortened, whereas the action potential duration (APD) was somewhat increased. Interestingly, the PAF antagonist BN 52021 shortened APD duration, an effect that reminds of the effects induced by ATP-sensitive K^+ channel openers (Edwads & Weston, 1993). This provides indirect evidence that PAF may release $[Ca^{2+}]_i$ in the heart through which activation or blockade of some K^+ currents may occur which might be considerable future interest to shed more light on Ca^{2+} signaling in the heart.

DOWNREGULATION OF PAF RELEASE IN THE CELL

It has been clearly demonstrated that PAF induces desensitization of its own effect (Braquet et al., 1987, Prescott et al., 1990). As discussed above, PAF induces a transient increase in $[Ca^{2+}]_i$ release which in turn may activate PLA_2 to release AA, resulting in eicosanoid production. Deactivation and receptor downregulation are independent of PTX-sensitive G proteins and PKC, and appear to depend on the sequential release of various mediators (Schleiffenbaum & Fehr, 1990). For example, eicosanoids, particularly prostacyclin, have been shown to elevate intracellular cyclic AMP level which in turn can effectively decrease eicosanoid production in platelets (Gorman et al., 1977, Tateson et al., 1977), and vascular endothelial cells (Abigail et al., 1982). Therefore, PAF may downregulate its own release through activation of tthe phosphoinositide cycle and IP_3-induced release of $[Ca^{2+}]_i$.

CONCLUSION

The data presented here clearly indicate that PAF-induced intracellular signaling via the release of $[Ca^{2+}]_i$ is a highly responsive and extremely sensitive mechanism which may play a fundamental role both under normal and pathophysiological conditions. This underlies that the phospholipid mediator governs a great variety of cellular physiological processes. Downregulatory processes point to the significance of the modulatory role of PAF in its own release and in the generation of other cellular mediators. The equilibrium of this signaling machinery may maintain the healthy state of cell function. PAF may however trigger harmful amplification of mediator release through interacting with various cytokines (Braquet et al., 1989, 1990). An interaction with LPS and cytokines has also been shown to be an important pathway for up-regulation of mediator release in which PAF may also play a central role (Aepfelbacher et al., 1992, Nakamura et al., 1992). To recognize the borderline where normal cellular metabolism may turn to be pathologic and harmful to cell survival may clarify better the significance of PAF-induced signaling and will be the subject of future research in this imporant field of cell signaling.

References

Abigail, F., Brotherton, A., and Hoack, J.C. (1982). Role of Ca^{2+} and cyclic AMP in the regulation of the production of prostacyclin by the vascular endothelium. Proc. Natl. Acad. Sci. USA 79: 495-499.

Aepfelbacher, M., Ziegler-Heitbrock, H.W., Lux, I., and Weber, P.C. (1992). Bacterial lipopolysaccharide up-regulates platelet-activating factor stimulated Ca^{2+} mobiliz=ation and eicosanoid release in human Mono Mac 6 cells. J. Immunol. 148: 2186-2193.

Berridge, M.J. (1993a). Inositol triphosphate and calcium signalling. Nature 361: 315-325.

Berridge, M.J. (1993b). A tale of two messengers. Nature 365: 388-389.

Billah M.M. anf Lapetina, E.G. (1983). Platelet-activating factor stimulates metabolism of phosphoinositides in horse platelets: Possible relationship to Ca^{2+} mobilization during stimulation. Proc. Natl. Acad. Sci. USA 80: 965-968.

Billah, M.M. and Siegel, M.I. (1984). Calmodulin antagonists inhibit formation of platelet-activating factor in stimulated human neutrophils. Biochem. Biophys. Res. Commun. 118: 629-635.

Bito, H., Nakamura, M., Honda, Z., Izumi, T., Iwatsubo, T., Seyama, Y., Ogura, A., Kudo, V., and Shimizu, T. (1992). Platelet-activating factor (PAF) receptor in rat brain: PAF metabolizes intracellular Ca^{2+} in hippocampal neurons. Neuron 9: 285-294.

Braquet, P., Hosford, D., Koltz, P., Gilbaud, J., and Paubert-Braquet, M. (1990). Effect of platelet-activating factor on tumor necrosis factor-induced superoxide generation from human neutrophils. Possible involvement of G proteins. Lipids 26: 1071-1074.

Braquet, P., Paubert-Braquet, M., Bourgain, R., Bussolino, F., and Hosford, D. (1989a) PAF/cytokine autogenerated feedback networks in microvascular immune injury: consequences in shock, ischemia and graft rejection. J. Lipid Med. 1: 75-112.

Braquet, P., Paubert-Braquet, M., Koltai, M., Bourgain, R., Bussolino, F., and Hosford. D. (1989b). Is there a case for PAF antagonists in the treatment of ischemic states? Trends Pharmacol. Sci. 10: 23-30.

Braquet, P., Touqui, L., Shen, T.S., and Vargaftig, B.B. (1987). Perspectives in platelet-activating factor research. Pharmacol. Rev. 39: 97-145.

Catalan, R.E., Martinez, A.M., Aragones, M.D., Fernandez, I., Lombardia, M., and Miguel, B.G. (1992). PAF-induced activation of polyphosphoinositide-hydrolyzing phospholipase C in cerebral cortex. Biochem. Biophys. Res. Commun. 183: 300-305.

Chiba, Y., Mikoda, N., Kawasaki, H., and Ito, K. (1990). Endothelium-dependent relaxant action of platelet activating factor in the rat mesenteric artery. Naunyn-Schmiedeberg's Archs Pharmacol. 341: 68-73.

Clark, G.D., Happel, L.T., Zorumski, C.F., and Bazan, N.G. (1992). Enhancement of hippocampal excitatory synaptic transmission by platelet-activating factor. Neuron 9: 1211-1216.

Clement, M.G., Albertini, M., Dimori, M., and Aguggini, G. (1992). PAF and the role of vagus nerve in the breathing pattern of the pig. Prostagl. Leuk. Essent. Fatty Acids 45: 143-149,

Collins, S. (1993). Molecular structure of G-protein-coupled receptors and regulation of their expression. DN&P 6: 480-487.

Conrad, G.W. and Rink, T.J. (1986). Platelet-activating factor raises intracellular calcium ion concentration in macrophages. J. Cell Biol. 103: 439-450.

Edwards, G. and Weston, A.H. (1993). The pharmacology of ATP-sensitive K-channels. Annu. Rev. Pharmacol. Toxicol. 33: 597-637.

Forstermann, U., Pollock, J.S., Schmidt, H.H.H.W., Heller, M., and Murad, F. (1991). Calmodulin-dependent endothelium-derived relaxing factor/nitric oxide synthase activity is present in the particulate and cytosolic fractions of bovine aortic endothelial cells. Proc. Natl. Acad. Sci. USA 88: 1788-1792.

Gorman, R.R., Bunting, S., and Miller, O.V. (1977). Modulation of human platelet adenylate cyclase by prostacyclin (PGX). Prostaglandins 13: 377-388.

Hall, I.P. (1993). Inositol phosphates, cyclic AMP and signal transduction. DN&P 6: 5-11.

Hanahan, D.J., Demopoulos, C.A., Liehr, J., and Pinckard, R.N. (1980). Identification of platelet-activating factor isolated from rabbit. J. Biol. Chem. 255: 5514-5516.

Harnett, M.M. and Klaus, G.G.B. (1988). G protein coupling of antigen receptor-stimulated phosphatidylinositol hydrolysis in B cells. J. Immunol. 140: 3135-3139.

Homma, H. and Hanahan, D.J. (1988). Attenuation of platelet-activating factor (PAF)-induced stimulation of rabbit GTPase by phorbol ester, dibutyryl cAMP,and desensitization: concomitant effects on PAF receptor binding charecteristics. Arch. Biochem. Biophys. 262: 32-39.

Hwang, S.-B. and Lim, M.-H. (1986). Ionic and GTP regulation of binding of platelet-activating factor to receptors and platelet-activating factor-induced activation of GTPase in rabbit platelet membranes. J. Biol. Chem. 261: 532-537.

Kamata, K., Mori, T., Shigenobu, K., and Kasuya, Y. (1989). Endothelium-dependent vasodilator effects of platelet activating factor on rat resistance vessels. Br. J. Pharmacol. 98: 1360-1364.

Kecskemeti V. and Braquet, P. (1992). Cellular electrophysiological effects of platelet-activating factor (PAF) and its antagonist BN 52921 in cardiac preparations. Drugs Exptl. Clin. Res. 18: 23-27.

Koltai, M., Hosford, D., Guinot, P., Esanu, A., and Braquet, P. (1991a). Platelet-activating factor (PAF): a review of its effects, antagonists and possible future clinical applications. Drugs 42 (Part I): 9-29.

Koltai, M., Hosford, D., Guinot, P., Esanu, A., and Braquet, P. (1991b). Platelet-activating factor (PAF): a review of its effects, antagonists and possible future clinical applications. Drugs 42 (Part II): 174-204.

Koltai, M., Tosaki, A., Guillon, J.-M., Hosford, D., and Braquet, P. (1989). PAF antagonists as potential therapeutic agents in cardiac anaphylaxis and myocardial ischemia. Cardiovasc. Drugs Rev. 7: 177-198.

Kuijpers, T.W., Hakkert, B.C., Hoogerwerf, M., Leeuwenberg J.F.M., and Roos, D.(1991). Role of endothelial leukocyte adhesion molecule-1 and platelet-activating factor in neutrophil adherence to IL-1-prestimulated endothelial cells: Endothelial Leukocyte adhesion molecule-1-mediated CD18 activation. J. Immunol. 147: 1369-1376.

Kunievsky, B. and Yavin, E. (1992). Platelet-activating factor stimulates arachidonic acid release and enhances thromboxane B_2 production in intact fetal rat brain ex vivo. J. Pharmacol. Exp. Ther. 263: 562-568.

Lee, N.H. & Kerlavage, A.R. (1993). Molecular biology of G-protein-coupled receptors. D.N.&P. 6: 488-497.

Mazer, B.D., Domenico, J., Sawami, H., and Gelfand, E.W. (1991). Platelet-activating factor induces an increase in intracellular calcium and expression of regulatory genes in human lymphoblastoid cells. J. Immunol. 146: 1914-1920.

Mazer, B.D., Sawami, H., Tordai, A., Gelfand, E.W. (1992). Platelet-activating factor-mediatedtransmembrane signaling in human B lymphocytes is regulated through a Pertussis- and Cholera toxin-sensitive pathway. J. Clin. Invest. 90: 759-765.

Miller, L.G., Bazan, N.G., Roy, R.B., Clostre, F., Gaver, A., and Braquet, P. (1991). Platelet activating factor antagonists interact with GABA$_A$ receptors. Res. Commun. Chem. Pathol. Pharmacol. 74: 253-256.

Maudsley, D.J. and Morris, A.G. (1987). Rapid intracellular calcium changes in U937 monocyte cell line: transient inncrease in response to platelet-activatng faactor aand chemotactic peptide but not interferon or lipopolysaccharide. Immunology 61: 189-194.

Moritoki, H., Hisayama, T., Takeuchi, S., Miyano, H., and Kondoh, W. (1992). Involvement of nitric oxide pathway in the PAF-induced relaxation of rat thoracic aorta. Br. J. Pharmacol. 107: 196-201.

Nakamura, M., Honda, Z., Waga, T., Matsumoto, T., Noma, M., and Shimizu, T. (1992). Endotoxin transduces Ca^{2+} signalling via platelet-activating factor receptor. FEBS-Lett. 314: 125-129.

Nussler, A.K., Di Silvio, M., Billiar, T.R., Hoffman, R.A., Geller, D.A., Selby, R., Madariaga, J., and Simmons, R.L. (1992). Stimulation of nitric oxide synthase pathway in human hepatocytes by cytokines and endotoxin. J. Exp. Med. 176: 261-264.

Paubert-Braquet, M., Hosford, D., Koltz, P., Guilbaud, J., and Braquet, P. (1991). Tumor necrosis factor primes PAF-induced superoxide production by human neutrophils: possible involvement of G proteins. J. Lipid Mediators 2: S1-S14.

Pedemonte, C.H. (1993). Structure-function relationship of membrane ion pumps. D.N.&P. 6: 498-507.

Prescott, S.M. Zimmerman, G.A., and McIntyre, T.M. (1990). Platelet-activating factor. J. Biol. Chem. 265: 1781-1784.

Rees, D.D., Palmer, R.M.J., Schulz, R., Hodson, H.F., and Moncada, S. (1990). Characterization of three inhibitors of endothelial nitric oxide synthase in vitro and in vivo. Br. J. Pharmacol. 101: 746-752.

Rink, T.J. and Sage, S.O. (1990). Calcium signaling in human platelets. Annu. Rev. Physiol. 52: 431-446.

Schleiffenbaum, B. and Fehr, J. (1990). The tumor necrosis factor receptor and human neutrophil function. Deactivation and cross-deactivation of tumor necrosis factor-induced neutrophil responses by receptor down-regulation. J. Clin. Invest. 86:184-95.

Schulam, P.G., Putcha, G., Franklin-Johnson, J., and Schearer, W.T. (1990). Evidence for a platelet-activating factor receptor on human lymphoblastoid B cells activation of phosphoinositol cycle and induction of calcium mobilization. Biochem. Biophys. Res. Commun. 166: 1047-1052.

Shukla, S.D. and Hanahan, D.J. (1983). An early decrease in phosphatidylinositol 4,5-biphosphate upon stimulation of rabbit platelets with acetylglycerylether phosphorylcholine (platelet activating factor). Arch. Biochem. Biophys. 227: 626-629.

Tateson, J.E., Moncada, S., and Vane, J.R. (1977). Effects of prostacyclin (PGX) on cyclic AMP concentrations in human platelets. Prostaglandins 13: 389-397.

Tseng, G.-N. and Hoffmann, B.F. (1989). Two components of transient outward current in canine ventricular myocytes. Circ. Res. 64: 633-647.

Van Willingen, G. and Akkerman, J.-W.N. (1991) Protein kinase C and cyclic AMP regulate reversible exposure of binding sites for fibrinogen on the glycoprotein IIB-IIIA complex of human platelets. Biochem. J. 273: 115-120.

Ward, S.G. and Westnick, J. (1988). Antagonism of the platelet activating factor-induced rise of the intracellular calcium ion concentration of U337 cells. Br. J. Pharmacol. 93: 769-774.

Willard, A.L. (1992). Excitatory and neurotoxic actions of platelet-activating factor on rat myenteric neurons in cell culture. Ann. N. Y. Acad. Sci. 664: 284-292.

Yue, T.L., Gleason, M.M., Hallenbeck, J., and Feuerstein, G. (1991a). Characterization of platelet-activating factor-induced elevation of cytosolic free calcium level in neurohybrid NCB-20 cells. Neuroscience 41: 177-185.

Yue, T.L., Gu, J.L., and Feuerstein, G. (1992a). Protein kinase C activator phorbol 12, 13-dibutyrate inhibits platelet activating factor-stimulated Ca^{2+} mobilization and phosphoinositide turnover in neurohybrid NG108-15 cells. Neurochem. Res. 17: 997-1004.

Yue, T.L., Stadel, J.M., Sarau, H.M., Friedman, E., Gu, J.L., Powers, D.A., Gleason, M.M., Feuerstein, G., and Wang, H.Y. (1992b). Platelet-activating factor stimulates phosphoinositide turnover in neurohybrid NCB-20 cells: involvement of pertussis toxin-sensitive guanine nucleotide-binding proteins and inhibition of protein kinase C. Mol. Pharmacol. 41: 281-289.

NMDA RECEPTOR-STIMULATED RELEASE OF ARACHIDONIC

ACID: MECHANISMS FOR THE BAZAN EFFECT

Lloyd A. Horrocks and Akhlaq A. Farooqui

Department of Medical Biochemistry
The Ohio State University
Hamilton Hall, Room 465
1645 Neil Avenue
Columbus, Ohio   43210-1218   U.S.A.

SUMMARY

Excitatory amino acids, acting through the NMDA type of glutamate receptors, cause the release of arachidonic acid. Long-term potentiation in the hippocampus causes this release, part of the normal physiological process essential for memory. Overactivation of these receptors occurs in stroke, epilepsy, and other neurodegenerative diseases.

The activation of phospholipases and lipases, and increased membrane phospholipid hydrolysis with accumulation of free fatty acids (FFAs) and diacylglycerol, is harmful to neurons in three ways: (1) Loss of essential phospholipids from the membrane structure, with accumulation of FFAs, diacylglycerols, lysophospholipids, and platelet activating factor (PAF). This is the Bazan effect (Bazan, 1970). FFAs, diacylglycerols, and lysophospholipids have a detergent-like effect on neuronal membranes. They can uncouple oxidative phosphorylation and produce changes in membrane permeability by regulating ion channels (Keyser and Alger, 1990). PAF is proaggregatory and hemostatic; it may cause adhesion and activation of leukocytes and hence produce an inflammatory reaction at the blood-endothelial cell interface and open Ca^{2+} channels on presynaptic sites, causing increased release of glutamate (Clark et al., 1992). (2) The accumulation of FFAs can trigger an uncontrolled "arachidonic acid cascade". This includes the synthesis of prostaglandins, leukotrienes, and thromboxanes. An uncontrolled arachidonic acid cascade sets the stage for the increased production of free radicals and hence for lipid peroxidation and oxidative damage to membrane proteins. (3) Finally, calcium influx through increased phospholipid degradation may lead to sustained activation of protein kinase C and its translocation from cytosol to plasma membrane (Asaoka et al., 1992; Melloni et al., 1985; Nishizuka, 1992). The stimulation and translocation of protein kinase C may also be involved in neurodegeneration (Asaoka et al., 1992; Nishizuka, 1992).

Cell Signal Transduction, Second Messengers, and Protein Phosphorylation in Health and Disease
Edited by A.M. Municio and M.T. Miras-Portugal, Plenum Press, New York, 1994

INTRODUCTION

In neural membranes arachidonate is exclusively located in the *sn-2* position of membrane phospholipids. It is released by a number of enzymic pathways (Axelrod, 1990; Farooqui et al., 1992). A direct pathway involves the stimulation of phospholipase A_2. An indirect pathway requires the activation of phospholipase C, followed by diacylglycerol and monoacylglycerol lipases. Another pathway for the release of arachidonate utilizes a lysophospholipase preceded by a phospholipase A_1 (Chang et al., 1987; Farooqui et al., 1992; Glaser et al., 1993). Although the relative contributions of these various pathways to the release of arachidonate are still unknown, the importance of phospholipase A_2 and phospholipase C, plus the diacylglycerol lipase pathway, has been clearly demonstrated (Chang et al., 1987; Farooqui et al., 1992; Glaser et al., 1993).

ENZYMES FOR ARACHIDONIC ACID RELEASE

Numerous phospholipase A_2 activities have been described that exhibit biochemical heterogeneity with respect to substrate preference, optimum pH, Ca^{2+} requirement, and molecular weight (Chang et al., 1987; Farooqui et al., 1992; Glaser et al., 1993). Secretory forms of phospholipase A_2, such as those found in pancreatic and inflammatory fluids, have a low molecular weight (12-18 kDa), require millimolar Ca^{2+} concentrations, demonstrate no particular preference for arachidonate over other *sn-2* fatty acids, and contain critical disulfide bonds that are essential for enzymic activity (Dennis, 1987).

Less information exists about intracellular phospholipases A_2. These enzymes have higher molecular weight (70-110 kDa), are generally localized in cytosol, prefer arachidonate over other *sn-2* acyl groups, and are resistant to inactivation by reducing agents (Clark et al., 1991; Farooqui et al., 1992; Mayer and Marshall, 1993). Some intracellular phospholipases require Ca^{2+} (100-1000 nM) and are called Ca^{2+}-dependent phospholipases A_2; others do not require Ca^{2+} and are called Ca^{2+}-independent phospholipases A_2 (Gronich et al., 1990; Hazen et al., 1990; Farooqui et al., 1992; Glaser et al., 1993; Mayer and Marshall, 1993).

Using pyrenesulfonyl-labeled ethanolamine glycerophospholipids (Hirashima et al., 1990), we have purified two Ca^{2+}-independent phospholipases A_2 from bovine brain. One is active on 1,2-diacyl-*sn*-glycero-3-phosphoethanolamine and has a molecular weight of 110 kDa; the other is active on 1-alkenyl-2-acyl-*sn*-glycero-3-phosphoethanolamine (plasmenylethanolamine, ethanolamine plasmalogen) and has a molecular weight of 39 kDa (Hirashima et al., 1992). The roles of intracellular Ca^{2+}-dependent and Ca^{2+}-independent phospholipases A_2 are not fully understood. According to Keith Glaser, "Cytosolic phospholipase A_2 appears to be a more physiologically important enzyme based upon its ability to be regulated by normal cellular mechanisms" (Glaser et al., 1993). We propose that Ca^{2+}-dependent as well as Ca^{2+}-independent phospholipases may be involved in cellular injury. Ca^{2+}-independent phospholipase A_2 may occupy a proximal position in the pathway, initiating cell injury and amplifying the Ca^{2+}-dependent processes (including the activation of Ca^{2+}-dependent phospholipase A_2) that are important triggers of more distal events leading to cell death (Goligorsky et al., 1993).

We have also purified two diacylglycerol lipases from bovine brain to homogeneity (Farooqui et al., 1989b). Calcium ions have no effect on the activities of purified diacylglycerol lipases. However, the activities of membrane-bound enzymes are markedly stimulated by Ca^{2+} (Farooqui et al., 1985).

The mechanisms of regulation of receptor-mediated phospholipase A_2 and diacylglycerol lipase are not well-understood (Chang et al., 1987; Farooqui et al., 1992; Glaser et al., 1993). However, the following possibilities have been proposed: (1) Covalent modification (phosphorylation/dephosphorylation) of these enzymes and their regulatory proteins (Farooqui et al., 1992; Glaser et al., 1993; Mayer and Marshall, 1993); (2) Regulation of enzymic activities by G-proteins (Burch et al., 1986, 1988; Billah, 1987); and (3) Regulation by a transient increase in intracellular calcium that may stimulate the activities of these enzymes in the membrane-bound state (Farooqui et al., 1992; Glaser et al., 1993; Mayer and Marshall, 1993).

ALZHEIMER DISEASE

Recent findings related to the neurochemistry of Alzheimer disease (AD) are the discovery of markedly elevated activities of diacylglycerol and monoacylglycerol lipases in the nucleus basalis and hippocampus of autopsy brain samples (Farooqui et al., 1988a, 1990), and the reduced activity of protein kinase C in the brain homogenate of AD patients (Cole et al., 1988). Diacylglycerol activates protein kinase C. This enzyme also requires Ca^{2+} and a phospholipid for its activity (Nishizuka, 1986). However, without diacylglycerol, it is inactive at the resting concentration of Ca^{2+} in the cytosol. The above observations on increased activity of diacylglycerol lipases (Farooqui et al., 1988a, 1990) and decreased activity of protein kinase C (Cole et al., 1988) may be interrelated because the levels of diacylglycerol are regulated by diacylglycerol lipases and diacylglycerol kinases (Farooqui et al., 1989b).

A potential animal model for AD has been developed. Direct administration of ibotenate into the nucleus basalis magnocellularis of rat brain causes extensive neurodegeneration by destroying cholinergic cell bodies. We have injected ibotenate into the nucleus basalis magnocellularis region of rat brain to study whether an elevation of lipases is associated with the degeneration of cholinergic neurons in this potential animal model of AD. Two plasma membrane fractions were prepared from different regions of ibotenate-injected (right hemisphere) and control (left hemisphere) rat brain. One plasma membrane fraction was from synaptosomes (SPM) and the other from glial cell and neuronal cell bodies (PM). The activities of diacylglycerol and monoacylglycerol lipases in the PM and SPM fractions prepared from various regions of pooled rat brain (10 rats) were increased (2- to 4-fold) in the toxin-treated right hemisphere compared to the sodium chloride-injected left hemisphere (Fig. 1). PM and SPM fractions prepared from brains of animals killed 1 day and 3 days after ibotenate injection did not show any increase in diacylglycerol and monoacylglycerol activities. However, membrane preparations from rats killed 6 months after injection of ibotenate showed 3- to 4-fold increases in the specific activities of diacylglycerol and monoacylglycerol lipases. The prolonged elevation of diacylglycerol and monoacylglycerol lipases in rat

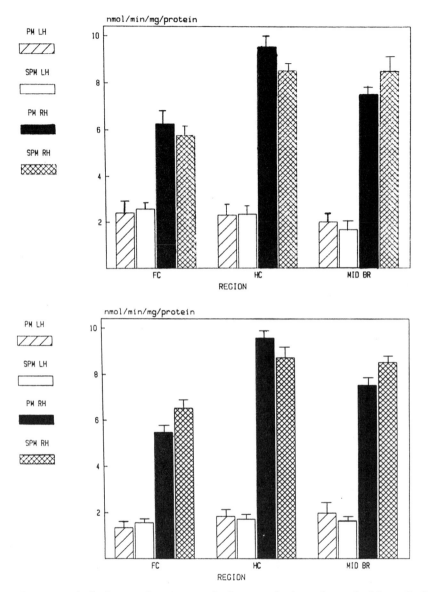

Fig. 1. Activities of monoacylglycerol (top) and diacylglycerol (bottom) lipases in PM and SPM fractions obtained from
different regions of rat brain injected with ibotenic acid.
Membranes were prepared 10 days after ibotenate injection.
LH - left hemisphere (NaCl-injected) and RH - right
hemisphere (ibotenate-injected).

brain found at 6 months after ibotenate injection may be similar to that found in AD brains.

Levels of glycerophospholipids, particularly ethanolamine plasmalogens, are markedly decreased in autopsy samples from AD patients compared to age-matched controls (Horrocks et al., 1978; Farooqui et al., 1988b, 1990; Nitsch et al., 1992; Söderberg et al., 1992). This decrease in glycerophospholipids in AD brain is accompanied by marked elevation of phospholipid degradation metabolites such as glycerophosphocholine, phosphocholine, and phosphoethanolamine (Barany et al., 1985; Miatto et al., 1986; Pettegrew et al., 1988a, 1988b; Pettegrew, 1989). Furthermore, marked increases have been reported in levels of prostaglandins and lipid peroxides in AD brain (Iwamoto et al., 1989; Jeandel et al., 1989; Subbarao et al., 1990; Volicer and Crino, 1990). The marked changes observed in phospholipids and their catabolic metabolites may be coupled to the elevated activities of lipolytic enzymes in AD brain (Farooqui et al., 1988a, 1990). A 63% lower phospholipase D activity in brain homogenate from AD patients, compared to controls, has been shown (Kanfer et al., 1986). To date no one has determined the activities of phospholipases A_2 and C in AD. Both enzymes are coupled to the excitatory amino acid receptors (Farooqui and Horrocks, 1991, 1992) and are involved in the turnover of neural membrane phospholipids (Dumuis et al., 1988, 1990; Lazarewicz et al., 1990) and with the generation of diacylglycerols, free fatty acids, and lipid peroxides. High levels of these lipid metabolites are cytotoxic and may contribute to neuronal injury in AD (Farooqui and Horrocks, 1991).

Although there is not an excessive release of glutamate in AD, the number of NMDA and other types of excitatory amino acid receptors is altered in the neocortex and hippocampus, compared to age-matched controls (Greenamyre and Young, 1989; Dewar et al., 1990). Because excitatory amino acid receptors are linked to lipases and phospholipases (Farooqui and Horrocks, 1991, 1992), abnormalities in phospholipid metabolism are observed in AD.

PHARMACOLOGY OF ARACHIDONIC ACID RELEASE

The interactions of glutamate and NMDA with excitatory amino acid receptors produce a marked increase in the release of arachidonate from the membrane phospholipids of the striatal, hippocampal, and hypothalamic neurons and cerebellar granule cells (Dumuis et al., 1988, 1990; Sanfeliu et al., 1990; Patel et al., 1990; Rage et al., 1991; Lazarewicz et al., 1988, 1990). This elevation can be blocked in a dose-related manner by the NMDA receptor antagonist 2-amino-5-phosphovalerate (Sanfeliu et al., 1990; Sucher et al., 1991). The enzymic mechanisms responsible for the liberation of arachidonate are not fully understood. However, based on the effects of mepacrine and nordihydroguaiaretic acid, phospholipase A_2 stimulation may be involved in this process. In the above studies, controls for the time-dependent increase in arachidonate were not included and glutamate and NMDA release of arachidonate occurred in the first minute (Wells, Anderson, Farooqui, and Horrocks, unpublished), long before any increase in the diacylglycerol lipase activity was observed (see below). This also suggests that the release of arachidonate is brought about by a receptor-mediated phospholipase A_2. Unlike the α-adrenergic receptor where phospholipase A_2 is coupled to the receptor complex through a G-protein (Axelrod, 1990), in cerebellar granule cells the coupling of the NMDA receptor to phospholipase A_2 is not mediated by a G-protein (Lazarewicz et al., 1990). Thus it seems likely that the elevated intracellular concentration of Ca^{2+}, triggered by the opening of NMDA receptor channels, may serve the second messenger function in activating phospholipase A_2 directly or through some other signal transduction mechanism.

Our studies (Farooqui et al., 1993) with neuron-enriched cultures from fetal mouse spinal cord have indicated that kainate has no effect, whereas glutamate and NMDA stimulate the activities of diacylglycerol and monoacylglycerol lipases in a dose- and time-dependent manner, and this increase can be blocked by dextrorphan or MK-801 (Farooqui et al., 1993).

Treatment of neuron-enriched cultures with glutamate caused concentration-dependent and time-dependent stimulations of diacylglycerol and monoacylglycerol lipase activities. Thus exposure to 50 µM glutamate for 15 min produced a 3-fold increase in the specific activities of diacylglycerol and monoacylglycerol lipases (Fig. 2). Higher glutamate concentrations resulted in lower specific activities of these enzymes. However, even at 100 µM glutamate, diacylglycerol and monoacylglycerol lipase activities were higher than the basal activities of these enzymes in neuron-enriched cultures. The time-dependence of the effect of glutamate on diacylglycerol and monoacylglycerol lipase activities is shown in Fig. 2. The activities of diacylglycerol and monoacylglycerol lipases were not affected during the initial 4 min by the glutamate treatment. A significant stimulation of enzymic activities was observed at 6 min, with maximal stimulation of lipase activities between 12 and 15 min. At longer times (25 min), enzymic activities decreased (30%).

Like L-glutamate, NMDA also stimulated the activities of diacylglycerol and monoacylglycerol lipases in a dose- and time-dependent manner. Thus, 15 min treatment with NMDA (50 µM) caused 5-fold and 3-fold increases in the specific activities of diacylglycerol and monoacylglycerol lipases, respectively (Fig. 3A). Maximal stimulation was obtained between 12 and 15 min (Fig. 3B). As for glutamate, enzymic activities decreased with longer exposure to NMDA.

Unlike glutamate and NMDA, the treatment of neuron-enriched cultures with kainate (100 µM) had no effect on diacylglycerol and monoacylglycerol lipase activities. Quisqualate (75 µM) produced a 2-fold to 4-fold stimulation of lipase activities in neuron-enriched cultures.

Dextrorphan (75 µM) and MK-801 (150 µM) had no effect on diacylglycerol and monoacylglycerol lipase activities of fetal mouse spinal cord cultures. These NMDA receptor antagonists prevented the stimulation of diacylglycerol and monoacylglycerol lipase activities by glutamate or NMDA (Figs. 4A, 4B).

At present it is not clear how the NMDA receptor is coupled to diacylglycerol and monoacylglycerol lipases. However, markedly increased polyphosphoinositide turnover is observed in striatal neurons after NMDA treatment. This suggests the involvement of phospholipase C. Diacylglycerol lipase probably hydrolyzes much of the diacylglycerol released by the action of phospholipase C. Thus, both phospholipase A_2 and the diacylglycerol lipase and monoacylglycerol lipase pathways participate in the release of arachidonate in response to glutamate and its analogs. However, the relative contributions of these pathways are still obscure. Phospholipase A_2 may be associated with the signal transduction process and

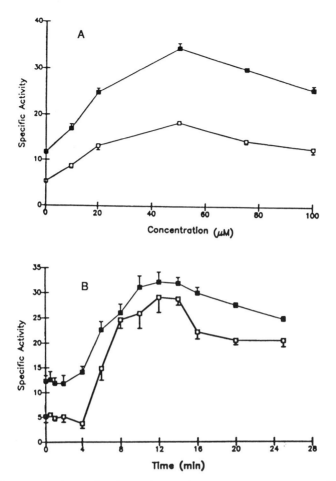

Fig. 2. Dose- (A) and time-dependence (B) curves of the
effect of glutamate on the activities of diacylglycerol (□)
and monoacylglycerol (■) lipases in neuron-enriched cultures
from fetal mouse spinal cord. Dose-dependence was studied at
15 min. Time-dependence was studied at 50 µM. Specific
activity is expressed as nmol/min/mg protein. Reprinted with
permission from Elsevier Science Publishers.

diacylglycerol lipase activity may be involved in neuronal
plasticity (Farooqui and Horrocks, 1993; Dennis et al., 1991).
        The treatment of cultured chick neurons with a mixture of
gangliosides produced various effects on the activities of
diacylglycerol and monoacylglycerol lipases. However, the
addition of the gangliosides directly to the incubation mixture
used to determine the lipase activities had no effect. The
time-dependence of the effect of gangliosides on monoacylglycerol lipase is shown in Fig. 5. In control cells cultivated
in CDM, the specific activity of this enzyme increased up to 24
h and remained constant thereafter. The treatment of the cells
with gangliosides produced a time-dependent stimulation in the
specific activity of monoacylglycerol lipase. Treatment with
10^{-8} M gangliosides resulted in an increase of the specific
activity of this enzyme only after 24 h, whereas with 10^{-5} M
gangliosides, a significant stimulation was observed after 15
minutes and increased up to 48 h.

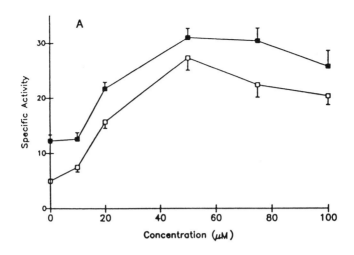

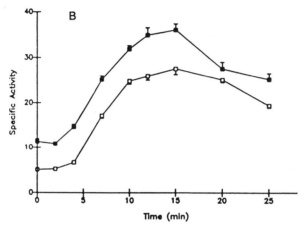

Fig. 3. Dose- (A) and time-dependence (B) curves of the effect of NMDA on the activities of diacylglycerol (□) and monoacylglycerol (■) lipases in neuron-enriched cultures from fetal mouse spinal cord. Dose-dependence was studied at 15 min. Time-dependence was studied at 50 µM. Specific activity is expressed as nmol/min/mg protein. Reprinted with permission from Elsevier Science Publishers.

The effect of gangliosides on diacylglycerol lipase activity in chick neurons is reported in Fig. 6. In contrast to monoacylglycerol lipase, the specific activity of the diacylglycerol lipase remained constant during 48 h in control cells cultivated in CDM. This enzyme also showed a significant increase in specific activity after ganglioside treatment of the neurons. In cells treated with a concentration of 10^{-8} M gangliosides, the time-dependent increase was observed over 24 h. After 48 h, the specific activity of the enzyme was similar to that of the control cells. Furthermore, the increase in ganglioside concentration from 10^{-8} M to 10^{-5} M produced a rapid stimulation of the specific activity of the enzyme for 15 min. Thereafter, the specific activity decreased over 24 h to a value significantly below that of the control cells; whereas after 48 h, the activity in both treated and control neurons

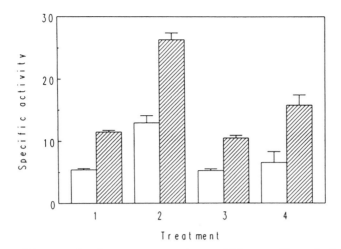

Fig. 4A. Effect of glutamate and MK-801 on diacylglycerol
(□) and monoacylglycerol (▨) lipases. Specific activity is
expressed as nmol/min/mg protein. Incubations were for 15
min. 1 - Control; 2 - Glutamate (100 µM for 15 min); 3 - MK-
801 (150 µM for 15 min); 4 - Glutamate (100 µM) + MK-801 (150
µM). Reprinted with permission from Elsevier Science
Publishers.

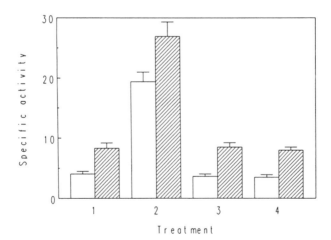

Fig. 4B. Effect of NMDA and dextrorphan on diacylglycerol
(□) and monoacylglycerol (▨) lipases. Specific activity is
expressed as nmol/min/mg protein. Incubations were for 15
min. 1 - Control; 2 - NMDA (50 µM); 3 - Dextrorphan (75 µM);
4 - NMDA (150 µM) + dextrorphan (75 µM). Reprinted with
permission from Elsevier Science Publishers.

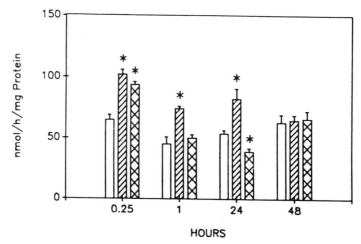

Fig. 5. Effect of gangliosides on the monoacylglycerol
lipase activity of chicken neurons in culture. The activity
was determined in control cells (CDM) (▢) and cells treated
with 10^{-8}M (▨) or 10^{-5}M (▨) gangliosides.

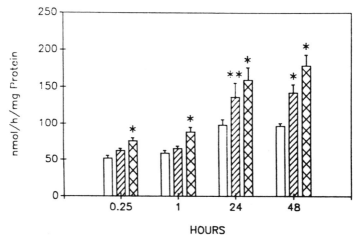

Fig. 6. Effect of gangliosides on the diacylglycerol lipase
activity of chicken neurons in culture. The activity was
determined in control cells (CDM) (▢) and cells treated with
10^{-8}M (▨) or 10^{-5}M (▨) gangliosides.

was similar. Our studies suggest that gangliosides regulate the activities of diacylglycerol and monoacylglycerol lipases in a time- and dose-dependent manner. The molecular mechanisms of this stimulation remain unknown. However, phosphorylation of many enzyme proteins is regulated by gangliosides.

CALCIUM IONS AND EXCITOTOXICITY

A recent advancement in understanding the role of Ca^{2+} in delayed neurotoxicity is the observation that inhibitors of nitric oxide synthase block the development of delayed NMDA-induced neuronal death in cortical neurons (Mattson, 1991; Dawson et al., 1991; Hoyt et al., 1992) and hippocampal slices (Izumi et al., 1992). Earlier studies have indicated that NMDA receptor activation promotes the Ca^{2+}-dependent release of nitric oxide in CNS neurons (Garthwaite et al., 1988). Nitric oxide is a short-lived, intra- and intercellular second messenger whose release is dependent upon increases in intracellular Ca^{2+}. For nitric oxide to play an important role in delayed neurodegeneration would require that this second messenger activate other longer-lived processes or be released in an ongoing fashion for a critical period of time. The NMDA-induced increase in phosphatidylinositol 4,5-bisphosphate hydrolysis observed in the neonatal cerebellum is also mediated by nitric oxide (Smith and Li, 1992). The stimulatory effect of NMDA on phosphatidylinositol hydrolysis is completely prevented by 100 µM L-N-monomethylarginine or nitro-arginine (Smith and Li, 1992), suggesting that NO production is linked to NMDA-induced hydrolysis of phosphatidylinositol. Nitric oxide also stimulates guanylate cyclases, leading to increases in intracellular cGMP. Thus, cGMP synthesis and phosphatidyl-inositol hydrolysis could provide a means for longer lasting changes in intracellular function initiated by nitric oxide (Moncada et al., 1991). Nitric oxide can also induce neuronal degeneration by virtue of its free radical property. It remains to be seen how nitric oxide affects membrane phospholipid degradation.

Thus, two major processes may be involved in neuronal injury caused by the overstimulation of excitatory amino acid receptors. One is the large Ca^{2+} influx (neuronal injury occurring when a certain threshold intracellular Ca^{2+} concentration is attained for a certain duration) and the other is the accumulation of free radicals and lipid peroxides as a result of neural membrane phospholipid degradation (Farooqui and Horrocks, 1991). Free radicals can disrupt membrane integrity by reacting with proteins and unsaturated lipids in the plasma membrane. These reactions lead to a chemical cross-linking of membrane proteins and lipids and a reduction in membrane unsaturated lipid content. This depletion of unsaturated lipids may be associated with alterations in membrane fluidity and permeability and changes in activities of membrane-bound enzymes and receptors (Farooqui et al., 1988b, 1989a, 1990). It has been suggested that calcium and free radicals may act in concert to induce neuronal injury (Pellegrini-Giampietro et al., 1990).

REFERENCES

Asaoka, Y., Nakamura, S., Yoshida, K., and Nishizuka, Y., 1992, Protein kinase C, calcium and phospholipid degradation, *Trends Biochem. Sci.* 17:414.

Axelrod, J., 1990, Receptor-mediated activation of phospholipase A_2 and arachidonic acid release in signal transduction, *Biochem. Soc. Trans.* 18:503.

Barany, M., Chang, Y.C., Arus, C., Rustan, T., and Frey, W.H., 1985, Increased glycerol-3-phosphorylcholine in post- mortem Alzheimer's brain [letter], *Lancet* 1:517.

Bazan, N.G., Jr., 1970, Effects of ischemia and electroconvulsive shock on free fatty acid pool in the brain, *Biochim. Biophys. Acta* 218:1.

Billah, M.M., 1987, Regulation of phospholipase A_2, in: "Annual Reports in Medicinal Chemistry, Vol. 22," 223-233, R.W. Egan, ed., Academic Press, Inc., New York.

Burch, R.M., Luini, A., and Axelrod, J., 1986, Phospholipase A_2 and phospholipase C are activated by distinct GTP-binding proteins in response to alpha 1-adrenergic stimulation in FRTL5 thyroid cells, *Proc. Natl. Acad. Sci. USA* 83:7201.

Burch, R.M., Ma, A.L., and Axelrod, J., 1988, Phorbol esters and diacylglycerols amplify bradykinin-stimulated prostaglandin synthesis in Swiss 3T3 fibroblasts, *J. Biol. Chem.* 263:4764.

Chang, J., Musser, J.H., and McGregor, H., 1987, Phospholipase A_2: Function and pharmacological regulation, *Biochem. Pharmacol.* 36:2429.

Clark, G.D., Happel, L.T., Zorumski, C.F., and Bazan, N.G., 1992, Enhancement of hippocampal excitatory synaptic transmission by platelet-activating factor, *Neuron* 9:1211.

Clark, J.D., Lin, L-L., Kriz, R.W., Ramesha, C.S., Sultzman, L.A., Lin, A.Y., Milona, N., and Knopf, J.L., 1991, A novel arachidonic acid-selective cytosolic PLA_2 contains a Ca^{2+}- dependent translocation domain with homology to PKC and GAP, *Cell* 65:1043.

Cole, G., Dobkins, K.R., Hansen, L.A., Terry, R.D., and Saitoh, T., 1988, Decreased levels of protein kinase C in Alzheimer brain, *Brain Res.* 452:165.

Dawson, V.L., Dawson, T.M., London, E.D., Bredt, D.S., and Snyder, S.H., 1991, Nitric oxide mediates glutamate neurotoxicity in primary cortical cultures, *Proc. Natl. Acad. Sci. USA* 88:6368.

Dennis, E., 1987, Phospholipase A_2 mechanism: inhibition and role in arachidonic acid release, *Drug Dev. Res.* 10:205.

Dennis, E.A., Rhee, S.G., Billah, M.M., and Hannun, Y.A., 1991, Role of phospholipases in generating lipid second messengers in signal transduction, *FASEB J.* 5:2068.

Dewar, D., Chalmers, D.T., Shand, A., Graham, D.I., and McCulloch, J., 1990, Selective reduction of quisqualate (AMPA) receptors in Alzheimer cerebellum, *Ann. Neurol.* 28:805.

Dumuis, A., Sebben, M., Haynes, L., Pin, J.-P., and Bockaert, J., 1988, NMDA receptors activate the arachidonic acid cascade system in striatal neurons, *Nature* 336:68.

Dumuis, A., Pin, P., Oomagari, K., Sebben, M., and Bockaert, J., 1990, Arachidonic acid release from striatal neurons by joint stimulation of ionotropic and metabotropic quisqualate receptors, *Nature* 347:182.

Farooqui, A.A. and Horrocks, L.A., 1991, Excitatory amino acid receptors, neural membrane phospholipid metabolism and neurological disorders, *Brain Res. Rev.* 16:171.

Farooqui, A.A. and Horrocks, L.A., 1992, Excitotoxicity and neural degeneration: Involvement of membrane phospholipids, *Biomed. Res.* 3:215.

Farooqui, A.A. and Horrocks, L.A., 1993, Excitotoxicity and neurological disorders: involvement of membrane phospholipids, *Internatl. Rev. Neurobiol.*, in press.

Farooqui, A.A., Pendley, C.E., II, Taylor, W.A., and Horrocks, L.A., 1985, Studies on diacylglycerol lipases and lysophospholipases of bovine brain, *in:* "Phospholipids in the Nervous System, Vol. II: Physiological Role," 179-192, L.A. Horrocks, J.N. Kanfer, G. Porcellati, eds., Raven Press, New York.

Farooqui, A.A., Liss, L., and Horrocks, L.A., 1988a, Stimulation of lipolytic enzymes in Alzheimer's disease, *Ann. Neurol.* 23:306.

Farooqui, A.A., Liss, L., and Horrocks, L.A., 1988b, Neurochemical aspects of Alzheimer's disease: Involvement of membrane phospholipids, *Metabolic Brain Dis.* 3:19.

Farooqui, A.A., Liss, L., and Horrocks, L.A., 1989a, Lipolytic enzyme activities in different brain regions in Alzheimer disease, *in:* "Phospholipid Research and the Nervous System. Biochemical and Molecular Pathology," 123-132, N.G. Bazan, L.A. Horrocks, G. Toffano, eds., Fidia Res. Series, Liviana Press, Italy.

Farooqui, A.A., Rammohan, K.W., and Horrocks, L.A., 1989b, Isolation, characterization and regulation of diacylglycerol lipases from bovine brain, *Ann. N. Y. Acad. Sci.* 559:25.

Farooqui, A.A., Liss, L., and Horrocks, L.A., 1990, Elevated activities of lipases and lysophospholipases in Alzheimer's disease, *Dementia* 1:208.

Farooqui, A.A., Hirashima, Y., and Horrocks, L.A., 1992, Brain phospholipases and their role in signal transduction, *in:* "Neurobiology of Essential Fatty Acids," 11-25, N.G. Bazan, G. Toffano, M. Murphy, eds., Plenum Press, New York.

Farooqui, A.A., Anderson, D.K., and Horrocks, L.A., 1993, Effect of glutamate and its analogs on diacylglycerol and monoacylglycerol lipase activities of neuron-enriched cultures, *Brain Res.* 604:180.

Garthwaite, J., Charles, S.L., and Chess-Williams, R., 1988, Endothelium-derived relaxing factor release on activation of NMDA receptors suggests role as intercellular messenger in the brain, *Nature* 336:385.

Glaser, K.B., Mobilio, D., Chang, J.Y., and Senko, N., 1993, Phospholipase A_2 enzymes: regulation and inhibition, *Trends Pharmacol. Sci.* 14:92.

Goligorsky, M.S., Morgan, M.A., Lyubsky, S., Gross, R.W., Adams, D.T., and Spitz, D.R., 1993, Establishment of a hydrogen peroxide resistant variant of renal tubular epithelial cells: role of calcium-independent phospholipase A_2 in cell damage, *Arch. Biochem. Biophys.* 301:119.

Greenamyre, J.T. and Young, A.B., 1989, Excitatory amino acids and Alzheimer's disease, *Neurobiol. Aging* 10:593.

Gronich, J.H., Bonventre, J.V., and Nemenoff, R.A., 1990, Purification of a high-molecular-mass form of phospholipase A_2 from rat kidney activated at

physiological calcium concentrations, *Biochem. J.*
271:37.

Hazen, S.L., Stuppy, R.J., and Gross, R.W., 1990, Purification and characterization of canine myocardial cytosolic phospholipase A_2, *J. Biol. Chem.* 265:10622.

Hirashima, Y., Mills, J.S., Yates, A.J., and Horrocks, L.A., 1990, Phospholipase A_2 activities with a plasmalogen substrate in brain and in neural tumor cells: A sensitive and specific assay using pyrenesulfonyl-labeled plasmenylethanolamine, *Biochim. Biophys. Acta* 1074:35.

Hirashima, Y., Farooqui, A.A., Mills, J.S., and Horrocks, L.A., 1992, Identification and purification of calcium-independent phospholipase A_2 from bovine brain cytosol, *J. Neurochem.* 59:708.

Horrocks, L.A., Spanner, S., Mozzi, R., Fu, S.C., D'Amato, R.A., and Krakowka, S., 1978, Plasmalogenase is elevated in early demyelinating lesions, *Adv. Exp. Med. Biol.* 100:423.

Hoyt, K.R., Tang, L-H., Aizenman, E., and Reynolds, I.J., 1992, Nitric oxide modulates NMDA-induced increases in intracellular Ca^{2+} in cultured rat forebrain neurons, *Brain Res.* 592:310.

Iwamoto, N., Kobayashi, K., and Kosaka, K., 1989, The formation of prostaglandins in the postmortem cerebral cortex of Alzheimer-type dementia patients, *J. Neurol.* 236:80.

Izumi, Y., Benz, A.M., Clifford, D.B., and Zorumski, C., 1992, Nitric oxide inhibitors attenuate N-methyl-D-aspartate excitotoxicity in rat hippocampal slices, *Neurosci. Lett.* 135:227.

Jeandel, C., Nicolas, M.B., Dubois, F., Nabet-Belleville, F., Penin, F., and Cuny, G., 1989, Lipid peroxidation and free radical scavengers in Alzheimer's disease, *Gerontology* 35:275.

Kanfer, J.N., Hattori, H., and Orihel, D., 1986, Reduced phospholipase D activity in brain tissue samples from Alzheimer's disease patients, *Ann. Neurol.* 20:265.

Keyser, D.O. and Alger, B.E., 1990, Arachidonic acid modulates hippocampal calcium current via protein kinase C and oxygen radicals, *Neuron* 5:545.

Lazarewicz, J.W., Wroblewski, J.T., Palmer, M.E., and Costa, E., 1988, Activation of N-methyl-D-aspartate-sensitive glutamate receptors stimulates arachidonic acid release in primary cultures of cerebellar granule cells, *Neuropharmacology* 27:765.

Lazarewicz, J.W., Wroblewski, J.T., and Costa, E., 1990, N-methyl-D-aspartate-sensitive glutamate receptors induce calcium-mediated arachidonic acid release in primary cultures of cerebellar granule cells, *J. Neurochem.* 55:1875.

Mattson, M.P., 1991, Evidence for the involvement of protein kinase C in neurodegenerative changes in cultured human cortical neurons, *Exp. Neurol.* 112:95.

Mayer, R.J. and Marshall, L.A., 1993, New insights on mammalian phospholipase A_2(s); comparison of arachidonoyl-selective and -nonselective enzymes, *FASEB J.* 7:339.

Melloni, E., Pontremoli, S., Michetti, M., Sacco, O., Sparatore, B., Salamino, F., and Horecker, B.L., 1985, Binding of protein kinase C to neutrophil membranes in the presence of Ca^2 and its activation by a Ca^{2+}

requiring proteinase, *Proc. Natl. Acad. Sci. USA* 82:6435.

Miatto, O., Gonzalez, R.G., Buonanno, F., and Growdon, J.H., 1986, In vitro ^{31}P NMR spectroscopy detects altered phospholipid metabolism in Alzheimer's disease, *Can. J. Neurol. Sci.* 13:535.

Moncada, S., Palmer, R.M.J., and Higgs, E.A., 1991, Nitric oxide: Physiology, pathophysiology and pharmacology, *Pharmacol. Rev.* 43:109.

Nishizuka, Y., 1986, Studies and perspectives of protein kinase C, *Science* 233:305.

Nishizuka, Y., 1992, Intracellular signaling by hydrolysis of phospholipids and activation of protein kinase C, *Science* 258:607.

Nitsch, R.M., Blusztajn, J.K., Pittas, A.G., Slack, B.E., Growdon, J.H., and Wurtman, R.J., 1992, Evidence for a membrane defect in Alzheimer disease brain, *Proc. Natl. Acad. Sci. USA* 89:1671.

Patel, A.J., Sanfeliu, C., and Hunt, A., 1990, Development and regulation of excitatory amino acid receptors involved in the release of arachidonic acid in cultured hippocampal neural cells, *Dev. Brain Res.* 57:55.

Pellegrini-Giampietro, D.E., Cherici, G., Alesiani, M., Carla, V., and Moroni, F., 1990, Excitatory amino acid release and free radical formation may cooperate in the genesis of ischemia-induced neuronal damage, *J. Neurosci.* 10:1035.

Pettegrew, J.W., 1989, Molecular insights into Alzheimer disease, *Ann. NY Acad. Sci.* 568:5.

Pettegrew, J.W., Moossy, J., Withers, G., McKeag, D., and Panchalingam, K., 1988a, ^{31}P Nuclear magnetic resonance study of the brain in Alzheimer's disease, *J. Neuropathol. Exp. Neurol.* 47:235.

Pettegrew, J.W., Panchalingam, K., Moosy, J., Martinez, J., Rao, G., and Boller, F., 1988b, Correlation of phosphorus-31 magnetic resonance spectroscopy and morphology finding in Alzheimer's disease, *Arch. Neurol.* 45:1093.

Rage, F., Pin, J.P., and Tapia-Arancibia, L., 1991, Phospholipase A$_2$ and somatostatin release are activated in response to N-methyl-D-aspartate receptor stimulation in hypothalamic neurons in primary culture, *J. Neuroendocrinol.* 3:515.

Sanfeliu, C., Hunt, A., and Patel, A.J., 1990, Exposure to N-methyl-D-aspartate increases release of arachidonic acid in primary cultures of rat hippocampal neurons and not in astrocytes, *Brain Res.* 526:241.

Smith, S. and Li, J., 1992, Novel action of nitric oxide as mediator of N-methyl-D-aspartate-induced phosphatidyl-inositol hydrolysis in neonatal rat cerebellum, *Mol. Pharmacol.* 43:1.

Söderberg, M., Edlund, C., Alafuzoff, I., Kristensson, K., and Dallner, G., 1992, Lipid composition in different regions of the brain in Alzheimer's disease/senile dementia of Alzheimer's type, *J. Neurochem.* 59:1646.

Subbarao, K.V., Richardson, J.S., and Ang, L.C., 1990, Autopsy samples of Alzheimer's cortex show increased peroxidation in vitro, *J. Neurochem.* 55:342.

Sucher, N.J., Lei, S.Z., and Lipton, S.A., 1991, Calcium channel antagonists attenuate NMDA receptor-mediated neurotoxicity of retinal ganglion cells in culture, *Brain Res.* 297:297.

Volicer, L. and Crino, P.B., 1990, Involvement of free
    radicals in dementia of the Alzheimer type: A
    hypothesis, *Neurobiol. Aging* 11:567.

THE ROLE OF ß-ADRENERGIC RECEPTOR KINASE IN THE MODULATION OF SIGNAL TRANSDUCTION

Irene Garcia-Higuera, Cristina Murga, Petronila Penela,
Ana Ruyiz-Gomez, and Federico Mayor, Jr.

Centro de Biología "Severo Ochoa" (CSIC-UAM)
Universidad Autónoma
28049 Madrid, Spain

INTRODUCTION

G protein -coupled receptors mediate the actions of a wide variety of extracellular signals, including numerous neurotransmitters and hormones,chemoattractants,cytokines and sensory stimuli such as photons and odorants,which regulate key physiological functions. A general property of these receptor systems is that their acute or sustained exposure to agonists often leads to a rapid loss of receptor responsiveness, a process termed desensitization,tachyphylaxis or tolerance.Such regulatory processes are essential for modulating the sensitivity of cells to messengers and allow the adaptation of cells to different signaling environments,thus contributing to signal integration and plasticity (Lefkowitz, 1993a; Vaello et al., 1994).

Together with the phototransduction cascade in retinal rod cells,the ß-adrenergic receptor (ßAR)-adenylyl cyclase system has been used as a prototypic model for the study of the molecular mechanisms of desensitization.These studies have revealed the occurrence of a complex regulatory network superimposed to the "classical" transduction

Cell Signal Transduction, Second Messengers, and Protein Phosphorylation in Health and Disease
Edited by A.M. Municio and M.T. Miras-Portugal, Plenum Press, New York, 1994

cascade (i.e.,ßAR,the transducing protein Gs and adenylyl cyclase),which is triggered every time that the receptor is activated.In this regard,rapid phosphorylation of the receptor by a specific,agonist-dependent serine/threonine kinase termed ß-adrenergic receptor kinase (ßARK) appears to play a pivotal role in ßAR regulation, and emerging evidence indicates that similar mechanisms may operate in other members of the G protein-coupled receptor superfamily (Benovic et al.,1988;Palczewski and Benovic,1991;Kobilka,1992).The aim of our laboratory is to investigate the role of ßARK and of other associated regulatory proteins in the modulation of the adrenergic transduction system,in order to contribute to a better understanding of how cells and tissues respond to cathecholamines in physiological or pathological circumstances.

Our recent work has confirmed the participation of ßARK in desensitization processes both in nervous cells in culture and in peripheral tissues physiologically exposed to high catecholamine concentrations.Furthermore,we have recently described that ßARK,in addition to its previously reported presence in the cytosol and plasma membrane,can be found associated with microsomal membranes,thus suggesting new functional roles for this kinase and the existence of distinct, highly dynamic pools of ßARK within the cell.

MOLECULAR MECHANISMS OF ß-ADRENERGIC RECEPTOR DESENSITIZATION

The catecholamines epinephrine and norepinephrine are important extracellular messengers which contribute to the regulation of essential physiological functions in the brain,heart,lung,liver and other tissues.Many of such actions are mediated through its interaction with ß-adrenergic receptors.The activated ßAR interacts with the signal transducing protein Gs,which incorporates GTP and dissociates in its subunits αs and βγ, which in turn modulate the activity of adenylyl cyclases and probably additional plasma membrane effectors.The increased levels of cAMP and the subsequent activation of cAMP-dependent protein kinase (PKA) triggers a specific cellular response by changing the activity of channels,receptors,enzymes and the transcription of certain genes.Work form different laboratories has put forward that acute or continuated stimulation of ßAR can promote two different processes of receptor regulation (Dohlman et al.,1991;Kobilka,1992). Rapid or short-term regulation is noted within seconds of ßAR activation and is based on receptor phophorylation and uncoupling induced by agonist-promoted changes in the activity and subcellular localization of several regulatory proteins (see below).Long-term regulatory mechanisms operate after chronic exposure to ß-agonists and involve changes in receptor synthesis and degradation,usually leading to receptor down-regulation (Collins et al.,1992)

Rapid desensitization of ßAR involves both transient sequestration of the receptors away from the plasma membrane and functional uncoupling of ßAR from Gs, which can be accomplished by phosphorylation of cytoplasmic domains of the ßAR by either PKA or the specific ß-adrenergic receptor kinase (ßARK).Regulation by PKA directly leads to ßAR uncoupling by modifying a domain implicated in the interaction with αs, and provides an example of classical feedback regulatory loops (Dohlman et al.,1991).Since phosphorylation by PKA does not require receptor occupancy by an agonist, this kind of regulation can also be promoted by other messengers acting on the same cell,in a cross-talk process known as heterologous desensitization.On the contrary,ßARK is a cytoplasmic serine/threonine kinase that specifically phosphorylates only the agonist-occupied form of ßAR by transiently translocating to the plasma membrane in response to the presence of ß-agonists in the medium,thus promoting receptor- specific or homologous desensitization (Palczewski and Benovic,1991). Phosphorylation by ßARK facilitates the interaction with the receptor of an additional soluble protein,termed ß-arrestin (Lohse et al.,1990),which produces the uncoupling of the receptor from Gs and the cyclase (see Figure 1).Uncoupled receptors are then transiently internalized,probably by an endosomal pathway (von Zastrow and Kobilka,1992).It has been recently proposed that the main role of this transient sequestration of the receptors away from the plasma membrane would be to allow its dephosphorylation and recycling (Yu et al.,1993).

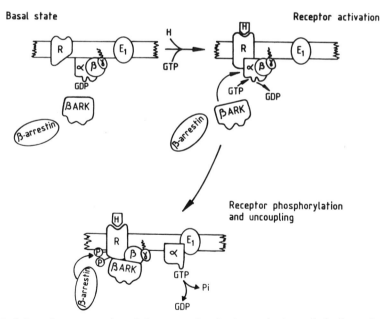

Figure 1.-Schematic representation of the proposed molecular mechanisms of ß-adrenergic receptor desensitization.R,ß-adrenergic receptor;H,hormone;αβγ,subunits of heterotrimeric G proteins;E1,plasma membrane effector;ßARK,ß-adrenergic receptor kinase.

A similar desensitization mechanism operates in the retina,where light-activated rhodopsin is phosphorylated by a cytosolic rhodopsin kinase,which in turns promotes the binding of a protein termed arrestin,thus precluding coupling to the heterotrimeric G protein transducin and terminating activation of the retinal cGMP phosphodiesterase (Hargrave and McDowell,1992).It is worth noting that these regulatory proteins are highly homologous members of the same gene families and display some activity in the other related transduction system.For instance,ßARK has been shown to phosphorylate rhodopsin in a light-dependent way (Benovic et al.,1986)and arrestin can partially substitute for ß-arrestin as uncoupling protein in a reconstituted ßAR /Gs experimental system (Lohse et al.,1990).

ßARK Translocation and the Role of βγ Subunits

A shared feature of rhodopsin kinase and ßARK is that they are cytoplasmic enzymes that phosphorylate membrane-bound receptors, so they must translocate to the membrane environment upon receptor stimulation.In fact, a rapid and transient increase in plasma membrane associated ßARK can be observed upon cell stimulation by ß-adrenergic and other agonists (Mayor,jr. et al.,1987 and references in Benovic and Gomez,1993). Very little is known about the mechanisms governing ßARK translocation.However,recent evidence indicate that ßARK interacts with heterotrimeric G proteins and isolated βγ subunits *in vitro*,and these subunits strongly stimulates ßARK activity towards activated receptors (Haga and Haga,1992;Pitcher et al.,1992;Koch et al.,1993).Such stimulation is not only a consequence of ßARK recruitment to the membrane,but implies direct activation of the kinase by βγ subunits,so ßARK can be considered as a effector coupled to G proteins (Clapham and Neer,1993) (Fig. 1).This proposed scheme,thas has yet to be demonstrated *in vivo*,would provide a mechanism for ßARK translocation and explain the close relationship between receptor activation and desensitization,since only activated receptors would release the free βγ subunits needed for kinase localization and activation.Furthermore,it opens the possibility that different βγ complexes may have differential affinities for ßARK and related kinases, thus modulating the substrate specificity of these enzymes (see below).However,other different mechanisms may be involved in modulating the translocation of other members of this kinase family.For instance,rhodopsin kinase is thus far unique within G protein coupled receptor kinases in being isoprenylated,and the farnesylated kinase is able to translocate to the membrane in a light -dependent manner (reviewed in Lefkowitz,1993a).

Multiple G Protein-Coupled Receptors can be Regulated by ßARK and/or Related G Protein Receptor Kinases

Several lines of evidence indicate that regulation by agonist-dependent kinases is not a unique characteristic of rhodopsin and ß-adrenergic receptors, but may be a common feature of all G protein-coupled receptors.This is indicated both by the fact that ßARK shows a broad substrate specificity and by the identification of additional members of a multigene family homologous to ßARK and rhodopsin kinase.To date,direct phosphorylation studies have shown that ßARK-1 (or ßARK-2) not only phosphorylate the ß2 adrenergic receptor to a high stoichiometry,but it can also phosphorylate the purified α2–adrenergic,the m2 muscarinic cholinergic and the substance P receptors in an agonist-dependent way (reviewed in Lefkowitz,1993a and Benovic and Gomez,1993),in addition to phosphorylate rhodopsin to a low level (Table I). Moreover, several extracellular mesengers (ß-agonists,prostaglandin E1,somatostatin and platelet activating factor) appear to promote ßARK translocation to the plasma membrane, thus suggesting the involvement of this kinase in the regulation of its receptors (Mayor jr.et al.,1987;Chuang et al.,1992).Overall,these data indicate that ßARK and/or related enzymes may be implicated in the regulation of a variety of receptors coupled to different G proteins and plasma membrane effectors, including the ubiquitous adenylyl cyclase and phospholipase C systems.

The complexity of these regulatory mechanisms is further increased by the existence of a family of G protein-coupled receptor kinases (GRKs).ßARK-1 is highly homologous to ßARK-2 and to a *Drosophila* kinase termed GPRK-1, whereas rhodopsin kinase and the recently described GRK-5 and GRK-6 are more divergent (Benovic and Gomez,1993).The cellular functions of these proteins,its substrate specificity and tissue distribution are important issues to be addressed.Furthermore,in addition to reported *in vitro* experiments with purified,reconstituted proteins (Lefkowitz,1993a),the participation of ßARK and related kinases in physiological processes of desensitization and receptor regulation needs to be more clearly established.

PHYSIOLOGICAL ROLES OF ßARK

Several studies in the ß2-adrenergic receptor system have suggested that, given its requirement for receptor occupancy, ßARK-mediated regulatory mechanisms would be specially important in the presence of high levels of catecholamines.Thus,it has been proposed that the role of this kinase would be specially relevant in the nervous tissue,where neurons and glial cells are exposed to high and rapidly changing

Table 1.-ßARK and related kinases may regulate different G protein-coupled receptors

Receptor	G protein	Effector/messenger	Experimental evidence
Rhodopsin	Gt	PDE/cGMP	*In vivo* regulation by RK**
Odorants	Golf	AC,PLC/cAMP,IP3	*In vivo* regulation by ßARK-2
ß2-adrenergic	Gs	AC/cAMP	*In vitro* phosphorylation by ßARK-1,ßARK-2*
α2-adrenergic	Gi	AC(-)/cAMP	*In vitro* phosphorylation by ßARK-1
m2-muscarinic	Gi/Gq	AC(-),PLC/cAMP,IP3	*In vitro* phosphorylation by ßARK-1,ßARK-2*
substance P	Gq/G11	PLC/IP3	*In vitro* phosphorylation by ßARK-1,ßARK-2
Prostaglandin E1	Gs	AC/cAMP	Agonists promote ßARK translocation
Somatostatin	Gi	AC(-)/cAMP	Agonists promote ßARK translocation
PAF	Gq	PLC/IP3	Agonists promote ßARK translocation

*Agonists also have been reported to promote ßARK translocation;**Also phosphorylated in vitro by ßARK1,ßARK2 and GRK6.
Abbreviations:AC,adenylyl cyclases;ßARK,ß-adrenergic receptor kinase;GRK,G protein -coupled receptor kinases;PDE, retinal cGMP phosphodiesterase;PLC,phospholipases C;RK,rhodopsin kinase

concentrations of messengers released at the synapses.This interpretation is consistent with the fact that expression of ßARK-1 mRNA and those of most other related kinases is higher in the brain and highly innervated tissues (Benovic et al.,1989,Benovic and Gomez,1993).We have recently reported that exposure of C6 glioma cells to ß-adrenergic agonists triggers a strikingly rapid, transient process of ßARK activity translocation from the cytoplasm to the plasma membrane, which takes place within the same time frame of ß-adrenergic receptor desensitization and sequestration (García-Higuera and Mayor,jr.,1992).These results suggest that ßARK-mediated receptor regulation is a suitable mechanism for rapid regulation of ß-adrenergic receptor function in the nervous tisssue. The fact that treatment of permeabilized olfactory cilia with antibodies to ßARK-2 blocks odorant-mediated desensitization also argue for a functional role of G protein-coupled receptor kinases in specialized sensory neurons (see Lefkowitz,1993a).

Interestingly, recent data from our laboratory indicate (in line with other reports (Chuang et al.,1992) showing high levels of ßARK expression in blood cells) that the role of ßARK as a rapid modulator of adrenergic receptor function is not restricted to the synapse,but may have functional relevance in peripheral tissues in response to high circulating or local levels of agonists.We have investigated whether rapid regulation of ß-adrenergic receptors occurs in the neonatal rat liver immediately after birth, a physiological situation characterized by a dramatic but transient increase in plasma catecholamines, which play a critical role in the adaptation of the neonate to extrauterine life (Lagercrantz and Slotkin,1986).Consistent with this acute stimulation,rapid desensitization and transient sequestration of ß-adrenergic receptors can be detected within the first minutes of life; rapid changes in ßARK specific activity in different subcellular fractions are also apparent immediately after birth, thus strongly suggesting a role for this kinase in desensitization mechanisms *in vivo* (García-Higuera and Mayor,jr.,1994).It is tempting to speculate that these mechanisms of receptor regulation may play a role in limiting and modulating the extent, kind and time frame of catecholamine actions during the perinatal period, and possibly in other physiological or pathological circumstances in which peripheral tissues are exposed to high circulating or local levels of messengers.

ßARK DISPLAYS A VARIETY OF INTRACELLULAR LOCATIONS

Until recently,it was presumed that ßARK was a soluble,cytosolic protein that transiently translocated to the cytoplasmic side of the plasma membrane in order to phosphorylate agonist- occupied receptors (Mayor,jr. et al.,1987;Chuang et al.,1992;García-Higuera and Mayor,jr.,1992;Pitcher et al.,1992). However,during our studies on ßAR desensitization in the neonatal rat liver,we unexpectedly found that a significant amount of total ßARK activity (35-50%) was apparent in rat liver microsomal membranes,which also displayed higher ßARK specific activity than any other subcellular fraction (García-Higuera and Mayor,jr.,1994).Immunoblot analysis with different antibodies raised in our laboratory against purified recombinant ßARK expressed in Sf9 cells, ßARK fusion proteins or synthetic peptides corresponding to different domains of the bovine ßARK-1 sequence (García-Higuera et al.,1994; Murga and Mayor,jr.,in preparation) confirmed the presence of ßARK protein in microsomal fractions of neonatal rat liver as well as in other tisues and cultured cells (García-Higuera et al.,1994). Cell-free association experiments indicate that ßARK is peripherally associated to the cytoplasmic side of microsomal membranes.The interaction of this kinase is reversible and appears to involve electrostatic interactions,in a way reminiscent of its transient association to the plasma membrane. The binding of ßARK approaches saturation and is inhibited by prior heat or protease treatment of the microsomal membranes,thus suggesting that ßARK associates with a specific protein component in the microsomes (García-Higuera et al.,1994).Gradient fractionation studies and both indirect immunofluorescence and immunogold electron microscopy localization experiments in living cultured cells using affinity- purified anti-ßARK antibodies confirm the association of ßARK with microsomal structures *in situ* ,with a preferential endoplasmic reticulum pattern.The same is observed in human kidney 293 cells stably overexpressing bovine ßARK-1(Ruiz-Gómez and Mayor,jr.,unpublished).

Our results are in line with recent reports showing that proteins involved in plasma membrane signal transduction, such as heterotrimeric G proteins, can also be found associated with membranes of different intracellular organelles,including the endoplasmic reticulum (Balch,1992;Barr et al.,1992).Since G protein βγ subunits have been recently reported to bind to recombinant ßARK in an *in vitro* assay (Pitcher et al.,1992;Koch et al.,1993),the presence in microsomal fractions of these potential candidates for ßARK anchoring make it tempting to suggest that these subunits are also involved in the interaction of ßARK with intracellular membranes,and that several interacting ßARK pools (i.e. microsome-bound,plasma membrane -bound and cyttosolic) may exist inside

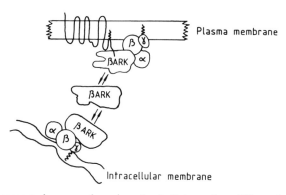

Figure 2.-ßARK appears to be a very dynamic molecule that can show different locations within the cell by means of specific interactions with membrane proteins.αβγ,subunits of heterotrimeric G proteins;ßARK,ß-adrenergic receptor kinase;an activated seven transmembrane-segments G protein-coupled receptor is shown in the plasma membrane.

the cell (Fig.2).However,the direct identification of the protein(s) implicated in ßARK association and of the mechanisms governing its subcellular location awaits further investigation.Other important point raised by our studies is the possible functional role of ßARK in such an intracellular location.It may be related to its translocation and receptor regulation function in the plasma membrane,the microsomal ßARK representing a reservoir or a necessary modification step on the way to (or from) the plasma membrane (see García-Higuera and Mayor,jr.,1994).Alternatively,or in addition,ßARK may serve other yet unknown cellular functions.For instance,heterotrimeric G proteins (and probably its related regulatory receptors) have been recently implicated in the regulation of protein export and vesicular trafficking within the cell (Bomstel and Mostov,1992;Schwaninger et al.,1992) and the possibility that ßARK may regulate some of these processes needs to be explored.

CONCLUSION

Adequate activation and deactivation mechanisms for plasma membrane signal transduction systems are essential for the normal functioning of intercellular communication.Given the numerous physiological roles of G protein-coupled receptors in neurotransmission,endocrine signalling and processing of sensory stimuli, it might be expected that alterations in such mechanisms underlie a variety of human pathologies,and some examples based on constitutively active receptors or mutated G protein α subunits have already been reported(Lefkowitz,1993b).In this context,the study of the regulatory network formed by ßARK,ß-arrestin and other associated regulatory proteins can shed new light in our understanding of physiological and pathological situations,and help to develop new drug targets.In particular,the study of the functional roles of ßARK in the cell,its relationships with G protein subunits and the factors modulating its activation and localization is an active and interesting field of research.

Acknowledgements

This work was supported by DGICYT grants PM890060 and PB920135,CAM grant C10591,Boehringer Ingelheim and Fundación Ramón Areces.C.M. and P.P are recipients of predoctoral fellowships from MEC and CAM,respectively.This work is dedicated to the memory of the late Dr.F.Sanabria.

REFERENCES

Balch,W.E.,1992,From G minor to G major,Curr.Biol.2:157

Barr,F.A.,Leyte,A.,and Huttner,W.B.,1992,Trimeric G proteins and vesicle formation,Trends Cell Biol. 2:91

Benovic,J.L.,Mayor,jr.,F.,Somers,R.,Caron,M.G.,and Lefkowitz,R.J.,1986,Light dependent phosphorylation of rhodopsin by the ß-adrenergic receptor kinase,Nature 322:869

Benovic,J.L.,Bouvier,M.,Caron,M.G.,and Lefkowitz,R.J.,1988,Regulation of adenylyl cyclase- coupled receptors,Ann.Rev.Cell Biol. 4:405

Benovic,J.L.,DeBlasi,A.,Stone,W.C.,Caron,M.G.,and Lefkowitz,R.J.,1989,ß-adrenergic receptor kinase: primary structure delineates a multigene family,Science 246:235

Benovic,J.L.,and Gomez,J.,1993,Molecular cloning and expression of GRK6,J.Biol.Chem. 268:19521

Bomsel,M.,and Mostov,K.,1992,Role of trimeric G proteins in membrane traffic,Mol.Biol.Cell 3:1317

Chuang,T.T.,Sallese,M.,Ambrosini,G.,Parruti,G.,and DeBlasi,A.,1992,High expression of ß-adrenergic receptor kinase in human peripheral blood leucocytes,J.Biol.Chem. 267:6886

Collins,S.,Caron,M.G.,and Lefkowitz,R.J.,1992,From ligand binding to gene expression:new insights into the regulation of G protein-coupled receptors,Trends Biochem.Sci. 17:37

Dohlman,H.G.,Thorner,J.,Caron,M.G.,and Lefkowitz,R.J.,1991,Model systems for the study of seven transmembrane-segment receptors,Annu.Rev.Biochem.60:653

García-Higuera,I.,and Mayor,jr.,F.,1992,Rapid agonist-induced beta-adrenergic receptor kinase translocation in C6 glioma cells,FEBS Lett. 302:61

García-Higuera,I.,and Mayor,jr.,F.,1994,Rapid desensitization of neonatal rat liver ß-adrenergic receptors. A role for ß-adrenergic receptor kinase,J.Clin.Invest.,in press

García-Higuera,I.,Penela,P.,Murga,C.,Egea,G.,Bonay,P.,Benovic,J.L.,and Mayor,jr.,F.,1994,Association of the regulatory ß-adrenergic receptor kinase with rat liver microsomal membranes,J.Biol.Chem.,in press

Haga,K.,and Haga,T.,1992,Activation by G proteins ßγ subunits of agonist-or light-dependent phosphorylation of muscarinic acetylcholine receptors and rhodopsin,J.Biol.Chem.267:2222

Hargrave,P.A.,and McDowell,H.,1992,Rhodopsin and phototransduction: a model system for G protein-linked receptors,FASEB J. 6:2323

Kobilka,B.K.,1992,Adrenergic receptors as models for G protein-coupled receptors,Annu.Rev.Neurosci. 15:87

Koch,W.J.,Inglese,J.,Stone,W.C.,and Lefkowitz,R.J.,1993,The binding site for the ßγ subunits of heterotrimeric G proteins on the ß-adrenergic receptor kinase,J.Biol.Chem.268:8256

Lagercrantz,H.,and Slotkin,T.A.,1986,The stress of being born,Sci.Am. 254:100

Lefkowitz,R.J.,1993a,G protein-coupled receptor kinases,Cell 74:409

Lefkowitz,R.J.,1993b,Turned on to ill effect,Nature 365:603

Lohse,M.J.,Benovic,J.L.,Codina,J.,Caron,M.G.,and Lefkowitz,R.J.,1990,ß-arrestin:a protein that regulates ß-adrenergic receptor function

Mayor,jr.,F.,Benovic,J.L.,Caron,M.G.,and Lefkowitz,R.J.,1987,Somatostatin induces translocation of the ß-adrenergic receptor kinase and desensitizes somatostatin receptors in S49 lymphoma cells, J.Biol.Chem.262:6468

Palczewski,K.,and Benovic,J.L.,1991,G protein-coupled receptor kinases,Trends Biochem.Sci.16:387

Pitcher,J.A.,Inglese,J.,Higgins,J.B.,Arriza,J.L.,Casey,P.J.,Kim,C.,Benovic,J.L.,Kwatra,M.M.,Caron,M. G.,and Lefkowitz,R.J.,1992,Role of ßγ subunits of G proteins in targeting the ß-adrenergic receptor kinase to membrane-bound receptors,Science 257:1264

Schwaninger,R.,Plutner,H.,Bokoch,G.M.,and Balch,W.E.,1992,Multiple GTP-binding proteins regulate vesicular transport from the ER to Golgi membranes,J.Cell Biol.119:1077

Vaello,M.L.,Ruiz-Gómez,A.,Lerma,J.,and Mayor,jr.,F.,1994,Modulation of inhibitory glycine receptors by phosphorylation by protein kinase C and cAMP-dependent protein kinase,J.Biol.Chem.,in press

vonZastrow,M.,and Kobilka,B.K.,1992,Ligand-regulated internalization and recycling of human ß2-adrenergic receptors between the plasma membrane and endosomes containing transferrin receptors, J.Biol.Chem.267:3530

Yu,S.S.,Lefkowitz,R.J.,and Hausdorff,W.P.,1993,ß-adrenergic receptor sequestration:a potential mechanism of receptor resensitization,J.Biol.Chem.268:337

SIGNAL TRANSDUCTION IN T CELLS THROUGH THE INTERLEUKIN 2 RECEPTOR

Eva Cano, Mª Angeles Muñoz-Fernández and Manuel Fresno

Centro de Biología Molecular "Severo Ochoa" (CSIC-UAM)
Universidad Autónoma de Madrid
Cantoblanco, 28049 Madrid, Spain

INTRODUCTION

Interleukin 2 (IL-2) plays a central role in the immune responses regulating the proliferation and differentiation of T lymphocytes[1,2]. The interaction of IL-2 with specific high affinity receptors (IL-2R) expressed on T cells upon activation induces T cell proliferation. However, the molecular bases of IL-2 induced signal transduction mechanism are poorly understood.

Thus, contradictory data on the ability of IL-2 to stimulate phosphoinositide hydrolysis[3-5] and Ca^{2+} fluxes[6] have been detected upon IL-2 binding to its receptor. Furthermore, the data on the participation of a GTP binding protein (G protein) in the action of IL-2 which could be responsible for the activation of the phosphoinositide metabolism[7], has been challenged[8] The translocation of the PK-C by IL-2, and activation of the phosphoinositide hydrolysis induced by IL-2 have reported[4,9]. In addition, phorbol esters which translocate PK-C in a similar manner[10], and mimic some of the IL-2 actions, such as the induction of c-myc gene[11] or the induction of T cell proliferation[12]. Therefore, it was suggested that the activation of PK-C by IL-2 as a crucial step in the signal transduction elicited by IL-2[11]. Recent studies have clearly shown that IL-2 activates several kinases. Thus, IL-2 induces the phosphorylation of several proteins in serine/threonine[13] as well as in tyrosine[14,15] including its own receptor[16,17]. However, the kinases involved are for from clear, although, pp56lck have been described associated to the IL-2R β chain[18,19].

We have investigated the IL-2 signal transduction pathway. Activation of PK-C is not required for IL-2 activity. On the contrary, cyclic nucleotide activated serin kinases and tyrosine kinases seems to be involved. We have found that IL-2 induces an activation of phospholipases A_2 (PLA2) and D (PLD) and a transient accumulation of phosphatidic acid (PA).

Cell Signal Transduction, Second Messengers, and Protein Phosphorylation in Health and Disease
Edited by A.M. Municio and M.T. Miras-Portugal, Plenum Press, New York, 1994

139

MATERIALS AND METHODS

Proliferation assay

The culture and measurement of DNA synthesis of CTLL-2 cells or PHA-activated human T cell blast kept 24 hr without IL-2 was carried out as described[22].

Lipid isolation and analysis

CTLL-2 cells (10^6/ml) were labelled with 5 μCi of [9,10 ^{3}H] palmitic o arachidonic acid (AA) per ml during 16h. The CTLL-2 were made resting by incubation for the last 12h in IL-2-free RPMI medium, but containing the corresponding radioactive compounds, prior to all lipid assays. Alternatively, the cells were incubated in IL-2-free medium for 12 h and then labelled with 100 μCi [^{3}H] choline for the last 4h. Aliquots of 5×10^6 labelled cells were incubated with recombinant IL-2 (10U/ml) for the indicated times. To measure PLD activity in side the cells, 0.4% ethanol was present in the cell cultures during the assay to allow accumulation of phosphatidyl ethanol (PEt). At the end of the incubations, lipids were extracted by treating the cells with 3.75 volumes of chloroform: methanol (1:2) and vortexed them for 90 seconds. Then, 1.25 volumes of chloroform plus 1.25 volumes of HCl 25mM were added. The samples were vortexed again during 60 and centrifuged to separate the phases. The organic phases were recovered and dried in a N_2 stream.

The levels of PA and phospholipids, were determined by thin layer chromatography (TLC) using H_2O: Acetic acid: Methanol: Acetone: Chloroform (8:12:13:15:40) as solvent with a prerunning overnight with Potassium Oxalate (1%) in Methanol: H_2O (40:60). The levels of diacylglycerol (DAG) and PEt were determined by TLC using Ethylacetate:Isooctane:Acetic acid:H_2O (100: (50:20:100) as a solvent. The levels of phosphatidylcholine (PC) and lyso-PC (LPC) were analyzed by TLC using Chloroform: Methanol: Acetic acid: H_2O (25:15:4:2).

Analysis of phosphorylated proteins

Resting CTLL-2 cells metabolically labelled with 100 μCi [^{32}P] orthophosphate for 12 hr and were treated with IL-2 in presence or not of the different kinase inhibitors or with kinase activators for 30 min. Then the cells were lysed and the phosphoproteins analyzed by 2-D gel electrophoresis as described[13].

Resting T cell blast were pelleted, washed three times in RPMI 1640 medium and then resuspended to a concentration of 2 x 10^7/ml in RPMI 1640 medium. Then, rIl-2 (20 U6ml) or the anti-CD3 UCHT-1 antibody (10 μg/ml) were added to cells. At various times after the addition of the stimuli, cells were lysed by addition of 1% SDS. The lysate was then immediately heated to 100°C for 5 min to inactivate kinases and phosphatases present. The proteins were precipitated and electrophoresed as described[21]. After electrophoresis, proteins were transferred by electroblotting onto a PVDF membrane and the blot was then probed with P-Tyr antibody (4G10), directly labeled with ^{125}I as described[21].

RESULTS

Protein kinases activated by IL-2

The mouse CTLL-2 T lymphocyte cloned cells can grown in culture medium only supplemented with transferrin, 2-mercaptoethanol, and recombinant IL-2 representing therefore a good model to study IL-2 signal transduction. In normal culture conditions 72 h after the last addition of IL-2 these cells were mostly in a "quiescent" state (90-98% in the G0/G1 phase of the cell cycle), and they did not incorporate [3H] thymidine into DNA. When they were restimulated with fresh rIL-2 they progressed through the cell cycle, and 16 h later they started to incorporate [3H] thymidine into DNA. This incorporation was dependent on the dose of IL-2 used. The same cells also responded to a different mitogenic stimulus provided by a phorbol ester, such a PDBu, a known activator of PK-C (12), although the cells never synthesized DNA in response to phorbol esters as well as to IL-2 (Figure 1A).

Treatment with high concentrations of phorbol esters of CTLL-2 for long incubation periods leads to the disappearance of the PK-C activity. The PK-C activity of cells treated with high dose of phorbol esters for 48 h was less than 3% of that of the control cells (22). As expected, these PK-C-depleted T cells no longer synthesized DNA in response to mitogenic doses of PDBu. However, they incorporated [3H] thymidine identically to control cells in response to IL-2 (Figure 1B).

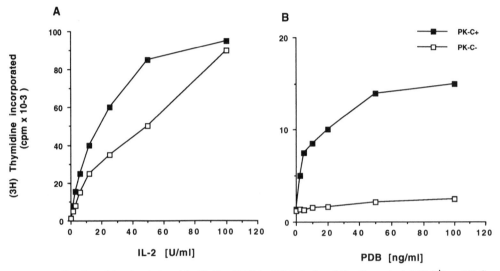

Figure 1. T cell proliferation induced by IL-2 or PDB in CTLL-2 cloned T cells control (PK-C+) or PK-C depleted (PK-C-).

The above results were corroborated by the use of specific kinase inhibitors. Thus, H-7 an inhibitor of PK-C did nor inhibit IL-2 induced T cell proliferation although it blocked phorbol ester induced proliferation. Surprisingly, H-8, H-9 and HA 1004 which are preferential inhibitors of cyclic nucleotide-dependent protein kinases inhibited T cell proliferation induced by IL-2 but not by phorbol ester (Figure 2).

The above results indicated that cyclic nucleotide dependent kinases but not PK-C were involved in the signal transduction pathway induced by IL-2. However, there could be some shared substrates between both types of kinases since PK-C activation also lead to T cell proliferation. To further investigate this point we analyzed the proteins phosphorylated in response to IL-2 and PDB. Treatment of CTLL-2 cells with IL-2 for 30 min induced several variations on the pattern of phosphorylated proteins (Cano et al., submitted for publication). Furthermore, PDB treatment induced a phosphorylation pattern only partially overlapping with the one observed with IL-2. This further corroborate that the kinase induced by IL-2 is not the same that the one activated by PDB. Among those proteins, there was one of 66 kDa whose phosphorylation status induced by different concentrations of IL-2 correlated well with the level of proliferation induced. Furthermore, pp66 was also phosphorylated by PDB. However, pp66 phosphorylation induced by IL-2 was blocked by H-8, H-9 or HA 1004 but not by PK-C inhibitor H-7. Moreover, treatment of CTLL-2 cells with dibutiryl-cAMP and in lesser extent with dibutiryl-cGMP induced pp66 phosphorylation (Figure 3). Furthermore, the pattern induced by cAMP in 30 min was remarkably similar to the one induced by IL-2, suggesting that IL-2 activated a kinase similar to PK-A.

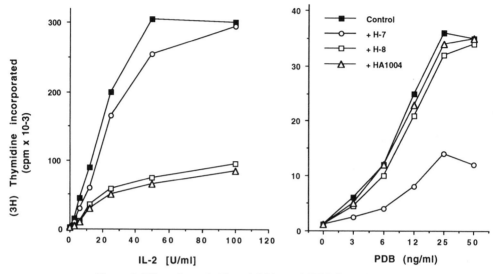

Figure 2. Effect of protein kinase inhibitors of CTLL-2 proliferation.

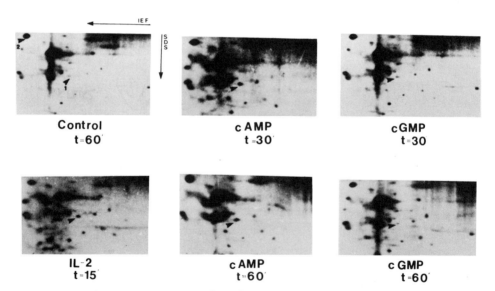

Figure 3. 2-D gel electrophoresis of CTLL-2 phosphorylated proteins treated as indicated.

Tyrosine phosphorylation was also required for IL-2 activity since genistein on inhibitor of Tyr-kinase strongly blocked IL-2 activity (Cano et al., submitted for publication) We also studied the proteins phosphorylated in tyrosine induced by IL-2 and compared them to the ones induced by activation with antigen. As shown in Figure 4, UCHT-1, an anti-CD3 antibody, stimulates the phosphorylation of several proteins being specially prominent those of 135, 110, 80 72, 70 and 55 kd. By contrast, the tyrosine phosphorylation induced by IL-2 was weaker and some of the proteins phosphorylated were of similar m.w. than the one phosphorylated by anti-CD3. However, some of them specially one around 68kDa was specific.

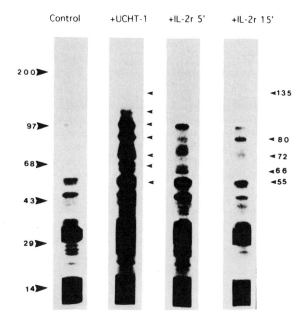

Figure 4. Analysis of proteins phosphorylated in tyrosine induced by IL-2 or anti-CD3 (UCHT-1) antibody in IL-2 dependent human T cells.

Effect of IL-2 on phospholipid metabolism

After activation of CTLL-2 cells with IL-2 were lysed and their phospholipid content analyzed by thin-layer chromatography. The results obtained from several experiments are summarized in Figure 5. An increase of (^{3}H) palmitic-labelled PA was observed as early as 1 min after IL-2 stimulation and, which was maximal at 1-5 min and then rapidly declined.

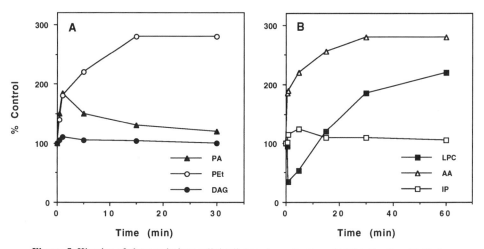

Figure 5. Kinetics of changes in intracellular lipids after activation of CTLL-2 cells with IL-2.

The accumulation of PA could derive from several metabolic pathways. One involves the PLC, which gives rise to DAG which is then phosphorylated by the DAG kinase. The other involves the direct action of PLD. However, the levels of DAG remained constant at any time point after addition of IL-2. Inositol phosphate levels also remained constant. To further corroborate the involvement of PLD, we tested more directly PLD activation by the transphosphatidylation reaction (measuring PEt formation) in (^{3}H) palmitic-labelled CTLL-2 cells cultured with 0.4% ethanol at different times after IL-2 activation. As shown in Figure 5, there was a increase in PEt levels, thus suggesting more directly that PLD is transiently activated. Furthermore, this PEt increase correlates with PA levels, therefore, strongly arguing in favour of PLD activation and against PLC activation.

Furthermore, IL-2 also induces a biphasic response in the levels of LPC. As early an 1 minute a sharp decrease in LPC levels was observed which recovered to control levels in 5-10 min. Thereafter, there was a constant increase on LPC with time up to the latest point measured, 3 hours. Moreover, then was also an increase in free arachidonic acid in the cytoplasm (Figure 5).

DISCUSSION

The interaction of IL-2 with its receptor plays a pivotal role in the activation and differentiation of T cells. However, despite recent advances on knowledge of the structure of the IL-2 and IL-2R[19] very little is known about the molecular mechanism of the signal transduction by IL-2. Although some initial reports suggested that IL-2 signal transduction pathway may operate through the activation of PI-PLC like many other growth factors, more recent evidence seems to discard elevations in the concentration of internal Ca^{2+} and inositolphosphates[3,5] as a second messengers of IL-2. More controversial is the activation of PK-C by IL-2. A translocation of PK-C by IL-2[9] and some reports of increases in DAG[5,22] have been taken as a support for PK-C involvement in IL-2 activity. However, we and other groups have failed to detect a DAG increase[3] and more importantly T cells which lacked PK-C either by depletion due to prolonged treatment with phorbol esters[20] or by mutation[23] were perfectly able to respond to IL-2, therefore, strongly suggesting that at least most isotypes of the PK-C, with the possible exception of PK-C ζ, are not induced by IL-2 and are not required for IL-2 activity.

It is becoming apparent that mitogenic stimuli activate cells through more than one transduction pathway. Recently, an accumulation of PA upon stimulation has been shown in a number of growth factor systems[24,25]. Furthermore, PA has been considered a second messenger since added exogenously induces the mitogenesis of several transformed cell lines, elicits an increase in the internal Ca^{2+} as well as in the internal pH and induces the expression of the mRNA for several protooncogenes and growth factors[26,27].

Our results shown indicate that IL-2 is able to induce PA accumulation, most likely through PLD activation, and strongly suggest that exogenous PA may act as a second messenger of IL-2 activity. This hypothesis is further supported by the fact that addition of PA and PLD mimics several IL-2 actions such us induction of c-myc, and upregulation of IL-2R[28]. However, the action of PA in cell proliferation is only a fraction of the one obtained with IL-2. Taken together these data indicate PA accumulation is not the only pathway elicited by IL-2 and suggest that this lymphokine may use other pathways in addition to PA to completely transduce its signal.

Theoretically, IL-2 may increase PA accumulation by different mechanisms. One possibility is that PA is formed by the coordinate activity of a PLC which renders DAG which is then phosphorylated by a DAG-Kinase. Recently, Merida et al[29] have reported an increase on (^{3}H) myristic labelled PA induced by IL-2. In addition, they also observed an increase in (^{3}H) myristic labelled but not (^{3}H) arachidonic labelled DAG. Similar increases on (^{3}H) myristic labelled DAG were recently reported on B lymphocytes upon IL-2 stimulation[30]. More interestingly, these two recent reports showed evidence for the involvement of PI-PLC-derived glycophosphatidylinositol (GPI) molecules in IL-2 activity[29,30]. However, several groups, including our own, have consistently failed to detect an increase in PLC activity, intracellular Ca^{2+} and DAG levels[3,5] even labelling the cells with (^{3}H) myristic (Cano et al., unpublished results). As suggested by Merida et al[29] the IL-2-activated PL could primarily hydrolyze GPI rendering inositol phosphoglycans and not inositolphosphates. However, in our hands, addition of exogenous PI-PLC, which hydrolyzes GPI-glycolipids on the surface of CTLL-2 cells, was unable to mimic IL-2 activity[28].

We think that IL-2 is activating PLD as other growth factors do on A341 and 3T3 cells[27]. The activity of PLD on phospholipids, mainly PC, renders directly PA. We favour this hypothesis which is supported by: 1) IL-2 activated cells contains higher levels of PLD activity detected by PEt formation. 2) The addition of PLD mimics the effect of PA on CTLL-2 cells[29].

Interestingly, we did not detect an increase in (^{32}P)-labelled PA, in agreement with previous data of Kozumbo et al[3]. Merida et al[29] have also failed to detected an increase in arachidonate-PA, induced by IL-2. These data are consistent with a model in which the coordinate activation of PLD and PLA$_2$ will release arachidonic acid (AA) and decrease LPC levels after in IL-2 activation. Our data indicates that a PC specific PLD is activated by IL-2.

If PA is a second messenger, how does it transduce the IL-2 signal? As mentioned before it is very unlikely that PA is dephosphorylated to produce DAG. Although, some groups have found increased DAG levels[5,22]. If DAG is produced, there would not be a concomitant activation of PK-C, which appears no to be the case[20,23]. Recently, it has been shown that PI-derived but not PC-derived DAG is able to activate PK-C[31], which could account for some of the above discrepancies. It is also possible that PK-C ζ which is not activated by DAG may be involved. Furthermore, only dimyristoyl-DAG levels but not other fatty acid containing DAG are increased by IL-2[29,30]. This represent a minor percent of the total DAG of the cells and may be a byproduct of a PA phosphohydrolase. Alternatively, the dimyristoil-DAG could activate an still unknown protein kinase. Our data, indicating that palmitic-labelled-DAG is not detected in IL-2-treated cells, despite an increase in similarly labelled PA, suggests that this pathway is not involved.

It has been proposed that PA-induced cell proliferation proceeds via a pertussis toxin sensitive G-protein which regulate cAMP levels[32,33]. Furthermore, PA stimulates cGMP formation in neuroblastoma cells[34]. Therefore it, is possible that IL-2 induced PA regulate cyclic nucleotide levels which then regulate T cell growth which will be consistent with our phosphorylation experiments. Moreover, IL-2 induce a transient increase in intracellular cAMP levels (Cano et al., submitted for publication). Furthermore, PA activate c-ras[26] which has been involved in IL-2 signal transduction[35].

In summary, we propose a model for IL-2 activity which is depicted in Figure 6. IL-2 activates a tyrosine kinase, probably pp56lck[19], which may be responsible for activation through tyr-phosphorylation of a PC specific PLD as it has been shown in other systems. Then, the coordinate action of PC-PLD together with PLA$_2$ may induce free arachidonic acid and a temporary decrease in LPC content giving raise to a concomitant increase in PA. Later the decrease in PC-PLD activity will lead to LPC accumulation which may be important also in maintaining proliferation[36]. This PA may in turn raise the cAMP level and activate the PK-A and ras protoncogenes which then transmit the signal to the nucleus. Similarly, activation of PK-C may in part mimic IL-2 activity since it is able to activate PL-PCD and also to phosphorylate similar substrates than IL-2 (PK-A) i.e pp66.

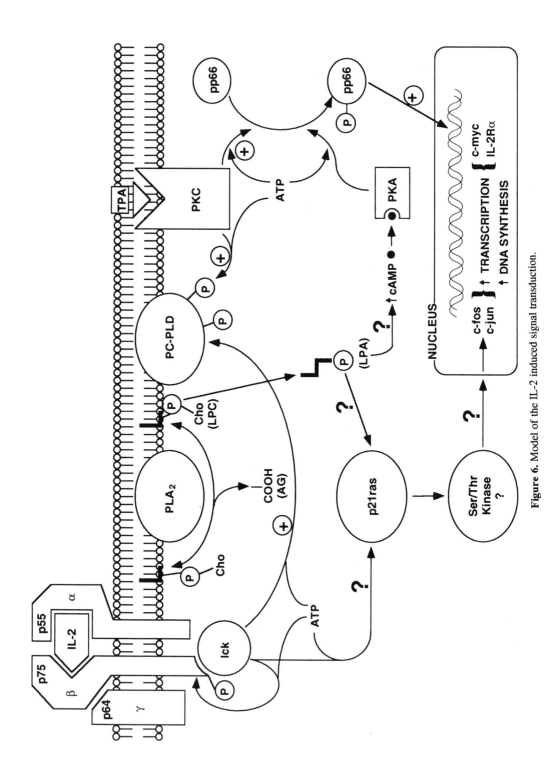

Figure 6. Model of the IL-2 induced signal transduction.

146

Acknowledgements

This work was supported by grants from DGICYT, Comunidad Autónoma de Madrid and Fundación Ramón Areces.

REFERENCES

1. K. A. Smith. "Interleukin 2". Annu Rev Immunol. 2:319. (1984)
2. D. A. Cantrell and K. A. Smith. "The interleukin-2 T-cell system: a new cell growth model". Science. 224:1312. (1984)
3. W. J. Kozumbo, D. T. Harris, S. Gromkowski, J. C. Cerottini and P. A. Cerutti. "Molecular mechanisms involved in T cell activation. II. The phosphatidylinositol signal-transducing mechanism mediates antigen-induced lymphokine production but not interleukin 2-induced proliferation in cloned cytotoxic T lymphocytes". J Immunol. 138:606. (1987)
4. E. Bonvini, F. W. Ruscetti, M. Ponzoni, T. Hoffman and W. L. Farrar. "Interleukin 2 rapidly stimulates synthesis and breakdown of polyphosphoinositides in interleukin 2-dependent, murine T-cell lines". J Biol Chem. 262:4160. (1987)
5. G. B. Mills, D. J. Stewart, A. Mellors and E. W. Gelfand. "Interleukin 2 does not induce phosphatidylinositol hydrolysis in activated T cells". J Immunol. 136:3019. (1986)
6. D. T. Harris, W. J. Kozumbo, P. Cerutti and J. C. Cerottini. "Molecular mechanisms involved in T cell activation. I. Evidence for independent signal-transducing pathways in lymphokine production vs proliferation in cloned cytotoxic T lymphocytes". J Immunol. 138:600. (1987)
7. S. W. Evans, S. K. Beckner and W. L. Farrar. "Stimulation of specific GTP binding and hydrolysis activities in lymphocyte membrane by interleukin-2. Nature. 325:166. (1987)
8. H. R. Bourne. "Interleukin-2 and GTP binding in lymphocytes". Nature. 326:833. (1987)
9. W. L. Farrar and W. B. Anderson. "Interleukin-2 stimulates association of protein kinase C with plasma membrane". Nature. 315:233. (1985)
10. Y. Nishizuka. "The role of protein kinase C in cell surface signal transduction and tumour promotion". Nature. 308:693. (1984)
11. W. R. Benjamin, P. S. Steeg and J. J. Farrar. "Production of immune interferon by an interleukin 2-independent murine T cell line". Proc Natl Acad Sci U S A. 79:5379. (1982)
12. D. K. Kim, G. Otten, R. L. Moldwin, D. E. Dunn, G. J. Nau and F. W. Fitch. "PMA alone induces proliferation of some murine T cell clones but not others". J Immunol. 137:2755. (1986)
13. T. Ishii, M. Kohno, M. Nakamura, Y. Hinuma and K. Sugamura. "Characterization of interleukin 2-stimulated phosphorylation of 67 and 63 kDa proteins in human T-cells". Biochem J. 242:211. (1987)v
14. I. Merida and G. N. Gaulton. "Protein tyrosine phosphorylation associated with activation of the interleukin 2 receptor". J Biol Chem. 265:5690. (1990)
15. E. M. Saltzman, K. White and J. E. Casnellie. "Stimulation of the antigen and interleukin-2 receptors on T lymphocytes activates distinct tyrosine protein kinases". J Biol Chem. 265:10138. (1990)
16. G. B. Mills, C. May, M. McGill, M. Fung, M. Baker, R. Sutherland and W. C. Greene. "Interleukin 2-induced tyrosine phosphorylation. Interleukin 2 receptor beta is tyrosine phosphorylated". J Biol Chem. 265:3561. (1990)
17. G. N. Gaulton and D. D. Eardley. "Interleukin 2-dependent phosphorylation of interleukin 2 receptors and other T cell membrane proteins". J Immunol. 136:2470. (1986)
18. M. Hatakeyama, T. Kono, N. Kobayashi, A. Kawahara, S. D. Levin, R. M. Perlmutter and T. Taniguchi. "Interaction of the IL-2 receptor with the src-family kinase p56lck: identification of novel intermolecular association". Science. 252:1523. (1991)
19. Y. Minami, T. Kono, T. Miyazaki and T. Taniguchi. Annu Rev Immunol. 11:245. (1993)

20. J. M. Redondo, R. A. Lopez, V. Vila, E. J. Cragoe and M. Fresno. "The role of protein kinase C in T lymphocyte proliferation. Existence of protein kinase C-dependent and -independent pathways". J Biol Chem. 263:17467. (1988)

21. S. C. Ley, A. A. Davies, B. Druker and M. J. Crumpton. "The T cell receptor/CD3 complex and CD2 stimulate the tyrosine phosphorylation of indistinguishable patterns of polypeptides in the human T leukemic cell line Jurkat". Eur J Immunol. 21:2203. (1991)

22. W. L. Farrar, J. L. Cleveland, S. K. Beckner, E. Bonvini and S. W. Evans. "Biochemical and molecular events associated with interleukin 2 regulation of lymphocyte proliferation". Immunol Rev. 92:49. (1986)

23. G. B. Mills, P. Girard, S. Grinstein and E. W. Gelfand. "Interleukin-2 induces proliferation of T lymphocyte mutants lacking protein kinase C". Cell. 55:91. (1988)

24. D. M. Salmon and T. W. Honeyman. "Proposed mechanism of cholinergic action in smooth muscle". Nature. 284:344. (1980)

25. J. J. Putney, S. J. Weiss, D. W. C. Van and R. A. Haddas. "Is phosphatidic acid a calcium ionophore under neurohumoral control?". Nature. 284:345. (1980)

26. C. L. Yu, M. H. Tsai and D. W. Stacey. "Cellular ras activity and phospholipid metabolism". Cell. 52:63. (1988)

27. W. H. Moolenaar, W. Kruijer, B. C. Tilly, I. Verlaan, A. J. Bierman and L. S. de. "Growth factor-like action of phosphatidic acid". Nature. 323:171. (1986)

28. E. Cano, M. A. Muñoz-Fernández and M. Fresno. "Regulation of interleukin-2 responses by phosphatidic acid". Eur J Immunol. 22:1883. (1992)

29. I. Merida, J. C. Pratt and G. N. Gaulton. "Regulation of interleukin 2-dependent growth responses by glycosylphosphatidylinositol molecules". Proc Natl Acad Sci U S A. 87:9421. (1990)

30. D. D. Eardley and M. E. Koshland. "Glycosylphosphatidylinositol: a candidate system for interleukin-2 signal transduction". Science. 251:78. (1991)

31. K. L. Leach, V. A. Ruff, T. M. Wright, M. S. Pessin and D. M. Raben. "Dissociation of protein kinase C activation and sn-1,2-diacylglycerol formation. Comparison of phosphatidylinositol- and phosphatidylcholine-derived diglycerides in alpha-thrombin-stimulated fibroblasts". J Biol Chem. 266:3215. (1991)

32. C. E. van Corven, A. Groenink, K. Jalink, T. Eichholtz and W. H. Moolenaar. "Lysophosphatidate-induced cell proliferation: identification and dissection of signaling pathways mediated by G proteins". Cell. 59:45. (1989)

33. T. Murayama and M. Ui. "Phosphatidic acid may stimulate membrane receptors mediating adenylate cyclase inhibition and phospholipid breakdown in 3T3 fibroblasts". J Biol Chem. 262:5522. (1987)

34. S. Ohsako and T. Deguchi. "Stimulation of phosphatidic acid of calcium influx and cyclic GMP synthesis in neuroblastoma cells". J Biol Chem. 256:10945. (1981)

35. J. Downward, J. Graves and D. Cantrell. "The regulation and function of p21ras in T cells". Immunol Today. 13:89. (1992)

36. Y. Asaoka, M. Oka, K. Yoshida, Y. Sasaki and Y. Nishizuka. "Role of lysophosphatidylcholine in T-lymphocyte activation: involvement of phospholipase A2 in signal transduction through protein kinase C". Proc Natl Acad Sci U S A. 89:6447. (1992)

A POSSIBLE MECHANISM FOR THE STIMULATION OF CELL DNA SYNTHESIS BY VIRAL INFECTION

Nieves Villanueva[1] and Jesús Avila[2]

[1]Servicio de Virología, CNMVIS, Instituto de Salud Carlos III
Majadahonda 28220, Madrid, Spain
[2]Cento de Biología Molecular
Universidad de Madrid, 28049 Madrid, Spain

It is known that viral infection can modify the DNA replication process in animal cells. For example, infection of quiescent cells by human cytomegalovirus (CMV) resumes cell DNA synthesis by an unkown mechanism.

There is a report indicating that taxol, a drug which stabilizes intact polymerized microtubules, inhibits that stimulation of cell DNA synthesis by cytomegalovirus infection[1]. A correlation between the proportion of depolymerized, or polymerized microtubule protein, with the initiation of DNA synthesis in resting cells has been suggested[2-3]. It has been shown that addition of colchicine (a drug that increases the level of unpolymerized tubulin, the main component of microtubules) results in the initiation of DNA synthesis in fibroblasts[2-3], whereas taxol (a drug that increases the level of assembled microtubules, as previously indicated) prevents initiation of DNA synthesis[4].

A possible mechanism for the stimulation of cell DNA synthesis by CMV infection could be related to the fact that viral infection results in microtubule depolymerization in the host cells[5].

It is known that major factors responsible for the stabilization of polymerized microtubules are the microtubule associated proteins (MAPs)[6]. A superfamily of these proteins contains neural (MAP2 and tau) and non neural (MAP4) proteins. All of them share common tubulin binding motifs. These motifs are composed by imperfect tandem repeats containing homologous 18 amino acid motifs that have glycine-rich stretches at the carboxy terminal end[7-11]. These distinct motifs stabilize to a different extent the assembled microtubules[12-14] and may complete among them selves for the binding of tubulin.

Thus, a possible way to explain the effect of CMV in the depolymerization of cell microtubules[6] could be the presence of a viral protein containing a sequence related to that of the tubulin binding motifs found in MAPs. Such protein might interfere with the normal binding of the stabilizing MAPs to microtubules. This interference may finally result in the disassembly of the cell microtubules and as a consequence, in the stimulation of cell DNA synthesis.

Since a main characteristic of the tubulin binding motifs of the superfamily of MAPs composed by MAP2/tau/MAP4 is a stretch of glycines, we have look for the presence of

Cell Signal Transduction, Second Messengers, and Protein Phosphorylation in Health and Disease
Edited by A.M. Municio and M.T. Miras-Portugal, Plenum Press, New York, 1994

glycine streches among the CMV proteins. The search was done in human CMV strain AD169, since it is known that this virus promotes cell DNA synthesis in resting cells. The sequence analyses indicated that a glycine rich stretch occurs in a number of viral proteins from which amino acid sequences have been deduced from the corresponding open reading frames in viral DNA. These proteins are UL44, 56, 102, 112 and TRS/IRS1 (The names are designated according to the localization of their frames)[15]. One of these proteins UL44, has a region with a sequence related to that of the tubulin binding motifs of MAP2/tau/MAP4 superfamily. This sequence is indicated in Table 1. This protein may compete with MAP4 (the more abundant MAP in non nueral cells) for tubulin binding. The possible consequence of this sequestering of tubulin by the viral protein could be the destabilization of microtubules and, afterwads, the initiation of cellular DNA synthesis. In this way CMV, and perhaps other viruses, could stimulate DNA synthesis in the host cell.

Table 1. Comparison of the tubulin binding motifs of a neural (Tau) and a non-neural (MAP₄) microtubule binding protein with the residues 382 to 398 of protein UL44 from cytomegalovirus.

UL44	T	S	K	G	G	S	G	G	G	G	G	G	G	G	G	G
Tau	T	S	K	C	G	S	L	G	N	I	K	H	V	P	G	G
MAP4	S	S	K	C	G	S	K	A	N	I	K	H	K	P	G	G

REFERENCES

1 – Ball, R,L, Carney, D. and Albrecht, T.Taxol inhibits stimulation of cell DNA synthesis of human cytomegaloviruses. Exp. Cell Res. (1990) 191, 37-44.

2 – Crossin, K. and Carney, D. Evidence that microtubule depolymerization early in the cell is sufficient to initiate DNA synthesis. Cell (1981) 23, 61-71.

3 – Otto, A., Zumbe, A., Gibson, L., Kubler, A.M. and Jiménez de Asua, L. Cytoskeleton-disrupting dugs enhace effect of growth factors and hormones on initiation of DNA synthesis. Proc. Natl. Acad. Sci. USA. (1979) 76, 6435-6438.

4 – Crossin, K. and Carney, D. Microtubule stabilization by taxol inhibits initiation of DNA synthesis by thrombin and by epidermal growth factor. Cell (1981) 27, 341-350.

5 – Pfeiffer, G., Willutzki, D., Weder, D., Becker, B. and Radsak, K. Changes in cell cytoskeleton upon CMV infection. Arch. Virol. (1983) 76, 153-159.

6 – Kirschner, M. and Mitchinson, T. Beyond self assembly: from microtubules to morphogenesis. Cell (1986) 45, 329-342.

7 – Aizawa, H., Emori, Y., Mori, A., Munofishi, H., Sakai, H. and Suzuki, K. Functional analyses of the domain structure of microtubule-associated protein 4 (MAP-U). J. Biol. Chem. (1991) 266, 9841-9846.

8 – Goedert, M., Crowther, R.A. and Garner, C.C. Molecular characterization of microtubule-associated proteins tau and MAP2. TINS (1991) 14, 193-199.

9 –   Lee, G., Cowan, N. and Kirschner, M. The primary structure and heterogeneity of tau protein from mouse brain. Science (1988) 239, 285-288.

10 – Lewis, S.A., Wang, D. and Cowan, N.J.Microtubule associated protein MAP2 shares a microtubule binding motif with tau protein. Science (1988) 242, 936-939.

11 – West, R., Tenbarge, K. and Olmsted, J.B. A model for microtubule-associated protein 4 structure. J. Biol. Chem. (1991) 266, 21886-21896.

12 – Joly, J. and Purich, D.L. Peptides corresponding to the second repeated sequence in MAP2 inhibit binding of microtubule-associated protein to microtubules. Biochemistry (1990) 29, 8916-8920.

13 – Joly, J.C., Flynn, G.C. and Purich, D.L. The microtubule-binding fragments of microtubule-associated protein 2: localization of the protease-accesible site and identication of an assembly-promoting peptide. J. Cell Biol. (1989) 109, 2289-2294.

14 – Maccioni, K., Vera, R., Domínguez, J. and Avila, J. A discrete repeated sequence defines a tubulin binding domain on microtubule associated protein tau. Arch. Biochem. Biophys (1989) 275, 568-579.

15 – Chee, M., Bankier, A., Beck, S., Bohni, R., Brown, C., Cerry, R., Hornsnell, T., Hutchison III, C., Konzarides, T., Martignetti, J., Preddie, E., Satchwell, C., Tomlison, P., Weston, K. and Banell, B. Analysis of the protein coding content of the sequence of human cytomegalovirus strain ADI69. Current Topics in Microbiology and Immunology (1990) 154, 126-169.

MODULATION BY PHOSPHORYLATION OF MICROTUBULE PROTEIN FUNCTION IN THE DEVELOPMENT OF NEURAL PROCESSES

J. Avila, M.D. Ledesma, L. Ulloa, J. García de Ancos, M. García Rocha, J. Domínguez, C. Sánchez, E. Montejo, I. Correas and Javier Díaz Nido

Centro de Biología Molecular "Severo Ochoa" (CSIC-UAM), Universidad Autónoma de Madrid, E-28049-Cantoblanco, Madrid, Spain

Summary

The possible role of microtubule associated proteins (MAPs) in the development of cytoplasmic extensions resulting in the appearance of the axon and dendrites is indicated. That role is based in the association of MAPs to microtubules. The association of MAPs to microtubules is mainly modulated by phosphorylation and that association regulates the degree of microtubule stability. The presence of stable microtubules correlates with that of cytoplasmic extensions.

Introduction

The differentiation of a neuroblast into a neuron can be followed, as indicated by Cajal, long time ago, by looking at cell morphology[1].

The spherical form of a neuroblast is changed into the very complex morphology of a neuron. This morphological change correlates with differences in the assembly of the microtubule protein. Brain microtubules are composed of tubulin and microtubule associated proteins (MAPs), being the first the main component of microtubules, whereas MAPs stabilize the assembled microtubule[2].

Axonal growth

During neuroblastoma differentiation, in which axon-like neurites develop, a net microtubule assembly has been found[3,4]. This assembly is not the consequence of an increase of microtubule protein but correlates with the phosphorylation of tubulin and mainly, of MAP$_1$B, the major MAP present in neuroblastoma cells [3,4]. By doing kinetic analyses it was found that modification of MAP$_1$B precedes microtubule assembly and, consequently, neurite extension; whereas tubulin phosphorylation is probably a consequence of microtubule assembly [3,4,5,6]. In this way it has been found that inhibition of microtubule assembly by colcemid addition prevents tubulin, but not MAP$_1$B, phosphorylation.

Cell Signal Transduction, Second Messengers, and Protein Phosphorylation in Health and Disease
Edited by A.M. Municio and M.T. Miras-Portugal, Plenum Press, New York, 1994

153

The modified residues of both tubulin and MAP$_1$B, have been localized. In the case of tubulin as a single residues at serine 444 present in the sequence EESESQG localized in class III b tubulin. For MAP$_1$B more than 10 different serine or theonine residues were modified and one of them was present in the sequence YSYETSD that has been, identified. The previous sequences suggest that a casein kinase II like protein (CKII) could be the enzyme that modifies both microtubule proteins. This possiblity has been confirmed by addition of an antisense oligonucleotide which specifically deplets casein kinase II[7]: Upon depletion, phosphorylation of MAP$_1$B and neurite extension were prevented.

MAP$_1$B is also phosphorylated by a proline directed protein kinase (PDPK) as suggested by the identification of other modified residue present in the sequence KSPSLSPSPPSP. By using antibodies which recognize phosphorylatable epitopes [8,9,10] it was observed that PDPK phospho MAP$_1$B isoform are mainly located at the distal part of a growing neurite, the growth cone in Ramón y Cajal´s nomenclature[1]. On the other hand CKII phospho MAP$_1$B isoforms are mainly present along the neurite. By taking together all the previous data it can be suggested that PDPK phosphorylation of MAP$_1$B may correlate with axonal sprouting and that of CKII with stabilization of axonal microtubules. A cartoon indicating the localization of PDPK-phospho MAP1B and CKII-phospho MAP1B isoforms is shown in Figure 1. It could be of interest to indicate that during development the proportion of PDPK-phospho MAP$_1$B isoforms decrease and at the time in which cell contacts occur (synaptogenesis) essentially there are not PDPK phospho MAP$_1$B isoforms at the axonal localization.

It has been suggested that other MAPs, known as tau factors, may have also a role in the stabilization of axonal microtubules. To test for that role tau cDNAS expressing different tau isoforms have been cloned and used for transfection studies in non neuronal cells. Some of the isoforms containing three repeated regions which can bind to microtubules and some others contained four of these motifs[11,12,13]. It has been observed that microtubules from transfected cells are more stable than those of untransfected controls, being more stable those microtubules of cells transfected with cDNAs expressing four tubulin binding motif proteins. Moreover, in some cases that stabilization results in cytoplasmic extensions resembling the neurites of neuroblastoma cells.

The characterization of the expressed tau protein in transfected cells have indicated the presence of phosphorylated isoforms and by testing which isoforms are bound to microtubules it has been found that unphosphorylated or hyperphosphorylated tau do not bind to microtubules whereas tau with a middle level of phosphorylation binds to them[4]. Additionally, the level of phosphorylation may, regulate the localization of tau isoforms. In this way it has been found that underphosphorylated four repeat tau isoforms are mainly localized in axons. In this way the phosphorylation of tau protein is under analysis. At the present it is not known the modification that facilitates the interaction of tau with microtubules but there is some knowledge about the hyperphosphorylation of tau.

Hyperphosphorylation of tau has been observed in some neurological disorders like Alzheimer´s[15-19] disease. In this disease there is a neuronal degeneration in which neurites are retracted. This retraction could be correlated with a decrease in microtubule stabilization and with a decrease in the interaction of tau with microtubules. It has been observed that hyperphosphorylated tau, as previously indicated, does not bind to microtubules. This observation may do compatible the modification of tau with neurite retraction.

At the present two phosphorylation sites at proline-rich sequences one located before and the other one after the tubulin binding motifs of tau molecule have been described in AD. These modifications are probably due to a PDPK, and they may result in conformational changes of the molecule. Those changes may affect to the interaction of tau with microtubules and, probably, will facilitate the aggregation of tau into the aberrant polymers, found in AD, and known like paired helical filaments[20].

In summary, Table 1 indicates the different axonal microtubule proteins their characterists of phosphorylation and the possible function of the isoforms.

Table 1- Phosphorylation of axonal microtubule proteins. The modification of tubulin, MAP₁B and tau, and the characterizations of those modifications are indicated. P and MT indicate phosphorylation and microtubules.

Protein	Mode phosphory	Kinase	P-sites	Example P-site	Localization	Possible function on MT	Is required P. for MT binding?
Tubulin		CKII	1	EEESESQG	axon	stabilize MT	NO
MAP₁B	I	PDPK	morethan1	KSPSLSPS-PPSP	growth cone	axon sprouting, is absent after synapto-genesis	
	II	CKII	morethan1	YSYETSD	axon	stabilize MT	YES
Tau	I	?			axon	stabilize MT	YES
	II	PDPK	morethan1	PGSPGTPG	cell body	destabilize MT	NO
				KSPVVSG-DTS	cell body	destabilize MT	NO

Dendrite formation

Since this work mainly deals with axonal microtubule proteins little will be mentionated about dendritic microtubule proteins. Nevertheless, it should be indicated that the appearence of dendrites correlates with the binding of a dendritic specific microtubule protein (MAP2) to microtubules[21]. However, the presence of MAP2 is not enough for dendrite sprouting and a possible modifications of MAP2 by phosphorylation has been suggested[23] to explein dendrite branching. In this way, it has been indicated that unphosphorylated MAP2[25] or hyperphosphorylated[26] MAP2 bind with a lower affinity to microtubules than MAP2 with a midle level of phosphorylation. Also, it has been described that MAP2 phosphorylation parallels dendrite arborization in cultured neurons[27].

A main feature of dendritic microtubules is their absence of polarity orientation[28]. Microtubules from non neural cells or those present in the axon of a neuron show an uniform polarity orientation in which the microtubule ends (those with a higher probability to incorporate tubulin subunits) are distal to the centrosome (in non neural cells) or to the cell body (in neurons). However, dendrite microtubules show no uniformity in their polarity orientation. It is not known the possible causes for that difference in dendritic microtubules although non uniform models can be proposed (see Figure 1). In model A the cause of the two orientation could be based in the fact that MAP2 can form antiparallel dimers[29] and if those dimers bind to two different microtubules, those microtubules will show antiparallel orientations Model B is based in a recent observation indicating that *in vitro* microtubule assembly in the presence of MAP2 could result in the appearance first of curved microtubules

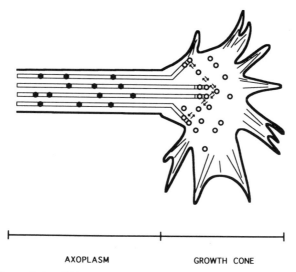

Figure 1.- Location of the different phospho MAP1B isoforms during axonal growth. PKPK-phospho MAP1B isoforms (O) are mainly present in the distal part of the axon (growth cone) and probably they have a role in axonal sprouting whereas CKII-phospho-MAP1B isoforms, (●) are mainly present in the axoplam and probably they have a role in the stabilization of axonal microtubules. MT indicates microtubules.

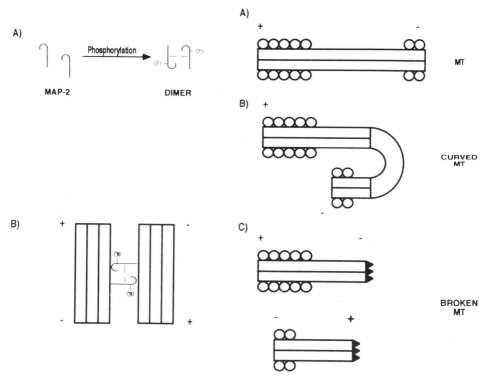

Figure 2.- Model for changes in microtubule polacity orientation in dendrites. In the left panel (model A) the cause of non uniform orientation is based in the formation of antiparallel MAP2 dimers[29]. In the right panel (Model B) the cause could be a microtubule cleavage by a MAP2- induced transition from a flexible to a rigid structure[31].

and afterwards of straigh microtubules[30]. It may due to a non uniform distribution of MAP2 along the microtubule at the time of its association with the microtubule. It may result in the appearance of curved regions. Once MAP2 is bound to the microtubule a reamangement of its distribution and the formation of rigid polymers[31] appear to take place. Thus, a possible mechanism to explain the presence of microtubules with different orientations could be the cleavage of curved microtubules as indicated Figure 2, if the formation of a rigid structure precedes the uniform distribution of MAP2 along the polymers. Alternative mechanisms could be based in a possible role of MAP2 molecules, directly or indirectly, as a microtubule organizing centers in the dendritic region of the neuron. At the present it is not known which one, if any, of the previous models is related to the changes in polarity orientation of dendritic microtubules.

REFERENCES

1   Ramón y Cajal, S. (1890) A quelle epoque apparaissent les expansions des cellule nerveuses de la moelle epinere du poulet. Anat. Anzerger 5, 609-613.

2   Kirschner, M.W. and Mitchison, T. (1986) Beyond self assembly: from microtubules to morphogenesis. Cell. 45, 329-342.

3   Gard, D.L. and Kirschner, M (1985) A polymer dependent increase in phosphorylation of b tubulin accompanies differentiation of a mouse neuroblastona cell line. J. Cell. Biol. 100, 764-774.

4   Díaz-Nido, J., Serrano, L., Méndez, E. and Avila, J. (1988) A casein kinase II related activity is involved in phosphorylation of microtubule associated protein MAP1B during neuroblastoma cell differentiation. J. Cell. Biol. 196, 2057-2065.

5   Díaz-Nido, J., Serrano, L., López-Otin, C., Vandekerchhove, J. and Avila, J. (1990) Phosphorylation of a neuronal-specific b tubulin isotope. J. Biol. Chem. 265, 13949-13954.

6   Serrano, L., Díaz-Nido, J., Wandosell, F. and Avila, J. (1987) Tubulin phosphorylation by casein kinase II is similar to that found *in vivo*. J. Cell. Biol.105, 1731-1739.

7   Ulloa, L., Díaz-Nido, J. and Avila, J. (1993) Depletion of casein kinase II by antisense oligonucleotide prevents neuritogenesis in neuroblastoma cells. EMBO J. 12, 1633-1640.

8   Ulloa, L., Avila, J. and Díaz-Nido, J. (1993) Heterogeneity in the phosphorylation of microtubule-associated protein MAP$_1$B during rat brain development. Journal of Neurochemistry. in press.

9   Mansfield, S.G., Díaz-Nido, J., Gordon-Weeks, P.R. and Avila, J. (1992) The distribution and phosphorylation of the microtubule-associated protein MAP$_1$B in growth cones. J. Neurocytol. 21, 1007-1022.

10 Díaz-Nido, J. and Avila, J. (1989) Characterization of proteins immunologically related to brain microtubule-associated protein MAP$_1$B in non-neural cells. J. Cell Sci. 92, 607-620.

11 Lee, G., Cowan, N. and Kirschner, M. (1988) The primary structure and heterogeneity of tau protein from mouse brain. Science 239, 285-288.

12 Goedert, M., Spillantini, M.G., Poiter, M.C., Ulrich, J. and Crowther, R.A. (1989a) Cloning and sequencing of the cDNA encoding an isoform of microtubule-associated protein tau containing foor tandem repeats: Differential expression of tau protein mRNAs in human brain. EMBO J. 8, 393-399.

13 Montejo de Garcini, E., Corrochano, L., Wischik, C.M., Díaz-Nido, J., Correas, I. and Avila, J. (1992) Differentiation of neruoblastoma cells correlates with an altered splicing pattern of tau RNA. FEBS Letters. 299, 10-14.

14 García de Ancos, J., Correas, I. and Avila, J. (1993) Differences in microtubule binding and self-association abilities of bovine brain tau isoforms. J. Biol. Chem. 268, 7976-7982.

15 Grundke-Iqbal, I., Iqbal, K., Quinlan, M., Tung, T.C. Zaidi, M.S., Wisniewski, H.M. and Binder, L.I. (1986a) Abnormal phosphorylation of the microtubule-associated protein tau in Alzheimer cytoskeletal pathology. Proc. Natl. Acad. Sci. USA. 83, 4913-4917.

16 Ihara, Y., Nukina, N., Miura, R. and Ogawara, M. (1986) Phosphorylated tau protein is integrated into paired helical filamants in Alzheimer's disease. J. Biochem. 99, 1807-1810.

17 Steiner, B., Mandelkow, E.M., Biernat, J., Gustke, N., Meyer, H.E., Schmidt, B., Mieskes, G., Söling, H.D., Drechsel, D., Kirschner, M.W., Godert, M. and Mandelkow, E. (1990) Phosphorylation of microtubule-associated protein tau: identification of the site for Ca^{2+} -calmodulin dependent kinase and relationship with tau phosphorylation in Alzheimer tangles. EMBO J. 9, 3539-3544.

18 Flament, S., Delacourte, A. and Mann, D.M.A. (1990) Phosphorylation of tau proteins: a major event during the process of neurofibrillary degeneration. A comparative study between Alzheimer's disease and Down's syndrome. Brain Res. 516, 15-19.

19 Ledesma, M.D., Correas, I., Avila, J. and Díaz-Nido, J. (1992) Implication of brain cdc2 and MAP2 kinases in the phosphorylation of tau protein in Alzheimer's disease. FEBS Letters. 308, 218-224.

20 González, P.J., Correas, I. and Avila, J. (1992) Solubilization and fractionation of paired helical filaments. Neuroscience 50, 491-499.

21 Matus, A. (1988) Microtubuel associated proteins. Ann Rev. Neurosci. 11, 29-44.

22 Higgins, D., Waxman, A. and Banker, G. (1987) The distribution of microtubule associated protein 2 changes when dendritic growth is induced in rat sympathetic neurons *in vitro*. Neuroscience 23, 121-130.

23 Friedrich, P. and Aszodi, A (1991) MAP2: a sensitive crosslinker and adjustable spacer in dendritic architecture. FEBS Letters 295, 5-9.

24 Díaz-Nido, J., Serrano, L., Hernández, M.A. and Avila J. (1990) Phosphorylation of microtubule protein in rat brain at different developmental stages. Comparison with that found in non neuronal cultures. J. Neurochem. 54, 211-222.

25 Brugg, B. and Matus, A. (1991) Phosphorylation determines the binding of microtubule associated protein (MAP2) to microtubules in living cells. J. Cell Biol. 114, 735-743.

26 Tsuyama, S., Terayama, Y. and Matsuyama, S. (1987) Numerous phosphates of microtubule associated protein in living rat brain. J. Biol. Chem. 262, 10886-10892.

27 Díez-Guerra, J. and Avila, J. (1993) MAP2 phosphorylation parallels dendrite arborization in hippocampal neurons in culture. NeuroReport 4, 419-422.

28 Baas, P.W., Deitch, J.S., Black, M.M. and Banker, G.A. (1988) Polarity orientation of microtubules in hippocampal neurons: Uniformity in the axon and non uniformity in the dendrite. Proc. Natl. Acad. Sci. USA 85, 8335-8339.

29 Wille, H., Mandelkow, E.M., Dingus, J., Vallee, R.B., Binder, L. I. and Mandelkow, E. (1992) Domain structure and antiparallel dimers of microtubule associated protein 2. J. Struct. Biol. 108, 46-63.

30 Kowalski, R.J. and Williams, R.C. (1993) Microtubule associated protein 2 alters the dynamic properties of microtubule assembly and disassembly. J. Biol. Chem. 268, 9847-9855.

31 Dye, R.B., Fink, S.P. and Williams, R.C. (1993) Taxol-induced flexibility of microtubules and its reversal by MAP2 and tau. J. Biol. Chem. 268, 6847-6850.

LOSS OF BASIC FIBROBLAST GROWTH FACTOR OF BRAIN EPENDYMA IN OLD SPONTANEOUSLY HYPERTENSIVE RATS

Pedro Cuevas[1], Diana Reimers[1], Fernando Carceller[1], Fu Xiaobing[1], and Guillermo Giménez-Gallego[2]

[1]Histology Department, Hospital Ramón y Cajal, 28034-Madrid
[2]Centro de Investigaciones Biológicas, CSIC, Velázquez 144, 28006-Madrid
Spain

SUMMARY

The immunochemical localization of basic fibroblast growth factor was studied in ventricular ependyma of aged-matched normotensive and spontaneously hypertensive rats at different ages using a polyclonal antibody against bFGF. The basic fibroblast growth factor-like immunoreactivity was observed in brain ependyma of young and old normotensive rats. However, a progressive loss of immunoreactivity was observed with age in spontaneously hypertensive rats. These results show a new neuroendocrine anomaly to be added to the many others previously observed in the hypothalamo-neurohypophyseal system of this rat strain, when they develop hypertension as they get old.

INTRODUCTION

Basic fibroblast growth factor (bFGF) is a very broad spectrum mitogen widely distributed in the organism,[1] isolated initially from the pituitary.[2] The protein is structurally related to nine different proteins that constitute the fibroblast growth factor family.[3] One of the member of this family, the acidic fibroblast growth factor (aFGF) is closely related to bFGF. The 55% the aligned sequence of both proteins is identical.[4] Thought initially isolated as mitogens, the fibroblast growth factors (FGFs) show also a wide spectrum of non-mitogenic activities.[3] So, it has been described that they modulate the endocrine effects of TRF on the anterior pituitary, and of FSH on the secretion of estrogen and progesterone by the granulosa ovarian cells, that they are implied in the regulation of food intake at the level of hypothalamus and in the secretion of tissue plasminogen activator by endothelial cells, and finally, that they are powerful vasodilators. It was reported that NO and ATP-sensitive K^+ channels are involved in FGFs vasodilatory activity when it was first described.[5] The actual relevance of FGFs in blood pressure homeostasis is unknown. FGFs have been localized in ependymal cells, furthermore, using cDNA probes for aFGF it has been shown that it is produced in this type of cells and that tanycytic processes of the third ventricle are in close contact with hypothalamic capillaries.[6, 7] Furthermore, it has been shown that ependymal cells release aFGF into the cerebrospinal fluid as a consequence of glucose increase after feeding. Released aFGF diffuses into the brain parenchyma an is taken by neurons.[7]

Spontaneously hypertensive rats (SHR) develop hypertension as they age. The hypertensive rats are characterized by deep neuroendocrine alterations. Thus, alterations in the levels of vasoactive intestinal polypeptide, vasopressin, adrenergic mines, angiotensin II have been observed.[8-12] These alterations appear accompanied of changes in the hypothalamo-neurohypophyseal system and higher levels of metabolic activity during the development of hypertension, compared with normal rats.[13, 14] It is obvious that many of this alterations could be originated by anomalies in the growth factor system responsible of neuronal homeostasis. Since FGFs are powerful neurotrophic factors,[15-19] we investigated whether alterations in the ependymal source of FGFs could be altered in the third ventricle. We found that the distribution of bFGF immunoreactivity in the ependymal cells of the third ventricle diminished with age in SHR. rats, oppositely to which happen in normal rats.

MATERIAL AND METHODS

Animals: Male Okamoto SHR, and their aged-matched controls Wistar-kyoto rats (WKR), (derived from the original Okamoto stock) were supplied by Centre d'Elevage Roger Janvier, Le Genest, France. Three groups of four rats 10, 14, and 18 month old were respectively studied per strain. Animals were maintained on a stock diet with water ad libitum.

Immunohistochemistry: Rats were anesthetized with an intraperitoneal injection of 1 ml anesthetic mixture (Ketamine hydrochloride 2.5 mg/ml, valium 2 mg/ml and 0.1 mg/ml atropine) and perfused retrogradely through the abdominal aorta with a washing solution (200 ml of saline containing 0.1% heparin) followed by fixation, first with 250 ml ice cold 4% paraformaldehyde in sodium acetate buffer (pH 6.5), and afterwards with 400 ml of an ice cold mixture of 4% paraformaldehyde and 0.2% glutaraldehyde in sodium tetraborate buffer (pH 9.59). After fixation, the brain was excised and processed for paraffin embedding.

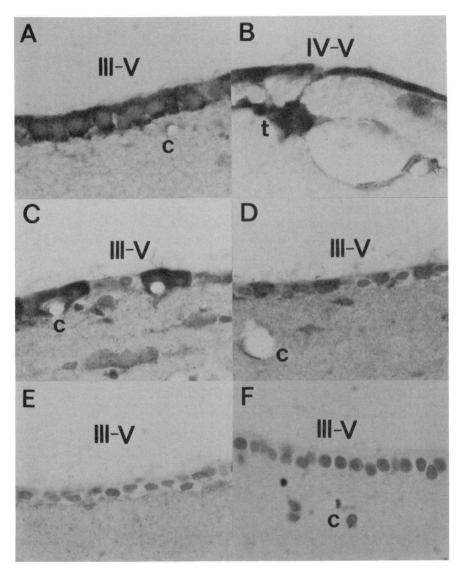

Figure 1. Representative images of the bFGF-like immunoreactivity in coronal hypothalamic sections of WKR and SHR rats. After immunostaining, sections were counterstained with hematoxylin and observed at x 400 original magnification. III-V, third ventricle, c, capillary, t, tanycyte. (A and B), 18-month old WKR. (C, D, E) 10-, 14-, and 18-month old SHR. (F), 10 month old WKR rat treated with immune serum pre-absorbed with bFGF.

For immunostaining, serial coronal sections of 10 μm from the selected brain areas, re-hydrated after deparaffinization, had their endogenous peroxidase activity quenched by a 30 min incubation, at room temperature, in 0.3% hydrogen peroxide, and their unspecific sites blocked by incubation in 1.5% goat serum (Vector Laboratories, Inc. Burlingame, CA) for 30 min. Treatment with polyclonal bFGF antibodies,[6] diluted 500 times in phosphate buffer saline containing 0.3% Triton and 5% bovine serum albumin, was carried out for 20 hours at 4°C. The sections were then treated with a 1:200 dilution of biotinylated goat anti-rabbit IgG (Vector Laboratories, Inc.) for 1 h, followed by a 30 min incubation with a biotin-avidin-peroxidase complex (Vector Laboratories, Inc.) and finally, for 5 min in 0.5 mg/ml of diaminobenzidine (Sigma Chemical Co). All steps were separated by buffer washes consisting in PBS-0.3% Triton. The sections were counterstained in Harris' hematoxylin, dehydrated, cleared and mounted.

RESULTS

We have shown[6] that abundant bFGF-immunoreactivity (bFGF-IR) is associated to ependymal cells of the rat lateral and third ventricles. Fig. 1 A is a representative immnunostaining, using antibodies against bFGF (anti-bFGF) of the region of the third ventricle of 18 month old normotensive rats. As shown in the figure, an intense immunore-activity, concentrated uniformly in the ependymal cells was observed. The staining was usually stronger in this region than in the rest of ventricular ependyma. No appreciable differences in staining intensity and pattern was observed between the 10- and 18-month old rats (not shown).

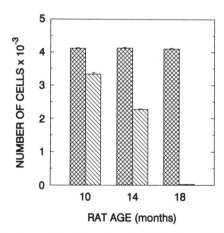

Figure 2. Change with age in the number of ependymal cells showing bFGF-like immunoreactivity into the third ventricle of SHR rats. Cross-hatched bars, total number of ependymal cells. Hatched bars, cells showing bFGF-IR. Bars represent the average + SD of the cells counted in four rats. The number of cells per animal is the total number counted in three different seriated sections (1 every 40 μm), at the level of the mammillary recess. Histological evaluation was performed using a x40 objective magnification.

The specificity of the staining was demonstrated by preincubating the antibody with an excess of either bFGF or aFGF [3.4 µg of bFGF per µl of serum, approximately a molar excess of growth factor to immunoglobulin of 3.5][20] at 4 °C for 24 hours. In the first case no immunostaining of the brain sections were observed (Fig. 1 F), while the activity of the antibody remained unaffected when it was treated with aFGF (not shown). Fig. 1 A also shows that anti-bFGF failed to stain the round-shaped nuclei of ependymal cells. At the same time, conspicuous bFGF-immunoreactive protrusions emerging from the basal portion of the ependyma typical of tanycyte processes, that some times appear in close association with capillaries were readily observable (Fig. 1 B). Immunostaining of the linear ependymal microvilli is also evident in the luminal portion of ependymal cells lining the third ventricle (not shown).

A remarkably different pattern of bFGF-immunoreactivity was observed in hypertensive rats. Sections of the third ventricle of 10-month old rats showed that ependymal immunostaining is often interrupted by patches of unstained cells, as illustrated in Fig. 1 C. The number of bFGF-immunonegative ependymal cells increased significantly in 14-month old rats (Fig. 1 D) to became dominant in 18-month old rats (Fig 1 E). Disappearance of bFGF-immunoreactivity in ependymal cells of the third ventricle of hypertensive rats appeared associated to a flattening of the ependyma and to a shape change of the cell nuclei (Fig. 1 C, D and E). A quantitative study of the decrease in the number of bFGF-immunopositive cells in hypertensive rats as they aged, is summarized in Fig. 2. ANOVA of the results showed a significant decrease $(F_{(2, 9)}=30342$ $p<0.00001$. Subsequent post-hoc Newman Keuls multiple comparison test revealed significant differences among the 10-, 14-, and 18-month old rats ($p<0.01$).

DISCUSSION

The ependyme forms a layer of simple cuboidal or columnar cells which lines the cavities of the brain and spinal cord. Many ependymocytes, known as tanycites[21] have processes which lead from the basal pole, extend between the subependymal cells, and come into contact with the blood brain capillaries, neurons and glial cells.[22] Tanycytes are associated with secretion into, and uptake and transport from, the cerebrospinal fluid (CSF), functions that have been explored mainly in 3rd ventricle where they are involved in transport of hormone-controlling factors in the median eminence and pituitary.[23, 24] Our findings that ependymal cells lining brain ventricles in hypertensive rats get impoverished in bFGF as the rat ages, show a new neuroendocrine anomaly of these rats that may have important implications in the functionality of the hypothalamo-neurohypophyseal system. Since it has been shown that a prime source of the endocrine changes leading to the development of hypertension as the rats age is the hypothalamus, it tempting to speculate that the decrease of a neurotrophic factor like bFGF could be linked to the appearance of these anomalies. This hypothesis obviously stresses the role of FGFs as regulators of the arterial blood pressure. It is evident that the results reported here do not provide by themselves any precise explanation of the development of high blood pressure in SHR rats.

Nevertheless, they may contribute to a deeper understanding of the etiology of hypertension and the physiology of FGFs.

Acknowledgments

We would like to express our gratitude to Chantal Bourdier and M. Carmen Partearroyo for her help in the preparation of this manuscript. Partially founded by the Dirección General de Investigación Científica y Técnica, and Fundación Gregorio Marañón-Boehringer Ingelheim España S. A. agreement.

REFERENCES

1. A. Baird and P. Böhlen, Fibroblast growth factors, *in* "Peptide growth factors and their receptors", M. B. Sporn and A. B. Roberts, eds, Springer, Berlin (1990).
2. F. Esch, A. Baird, N. Ling, N. Ueno, F, Hill, L. Denoroy, R. Klepper, D. Gospodarowicz, P. Bölen and R. Guillemin, Primary structure of bovine pituitary basic fibroblst growth factor (FGF) and comparison with the amino-terminal sequence of bovine brain acidic FGF, *Proc. Natl. Acad. Sci. USA* 82:6507 (1985).
3. G. Giménez-Gallego and P. Cuevas, Fibroblast growth factor, a protein with a broad spectrum of biological activities. *Neurological Res. in press.*
4. K. A. Thomas and G. Giménez-Gallego, Fibroblast growth factors: broad spectrum mitogens with potent angiogenic activity, *Trends in Biol. Sci.*11:1 (1986).
5. P. Cuevas, F. Carceller, S. Ortega, M. Zazo, I. Nieto and G. Giménez-Gallego, Hypotensive activity of fibroblast growth factor, *Science* 254:1208 (1991).
6. P. Cuevas, G. Giménez-Gallego, R. Martínez-Murillo and F. Carceller, Immunohisto-chemical localization of basic fibroblast growth factor in ependymal cells of the rat lateral and third ventricles, *Acta Anatom.* 141:307 (1991).
7. Y. Oomura K. Sasaki, K. Suzuki, T. Muto, A. Li, Z-I Ogita, K. Hanai, I. Tooyama, H. Kimura and N. Yanaihara, A new brain glucosensor and its physiological significance, *Am. J. Clin. Nutr.* 55:278S (1992).
8. R. Avidor, R. Eilam, R. Malach and I. Gozes, VIP-mRNA is increased in hypertensive rats, *Brain Res.* 503:304 (1989).
9. R. E. Lang, W. Rascher, T. Unger and D. Ganten, Reduced content of vasopresin in the brain of spontaneously hypertensive as compared to normotensive rats, *Neurosci. Lett.* 23:199 (1981).
10. M. Morris and M Keller, A specific deficiency in para ventricular vasopressin and oxytocin in the spontaneous hypertensive rat, *Brain Res.* 249:173 (1982).
11. A. Nagaoka and W Lovenberg, Regional changes in the activities of aminergic biosynthetic enzymes in the brain of hypertensive rats, *Eur. J. Pharmacol.* 43:297 (1977).
12. M. Morris, J. A. Wren, D. K. Sundberg, Central neural peptides and catecholamines in spontaneous and DOCA/salt hypertension, *Peptides*2:207 (1981).

13. D. O. Nelson and J. A. Boulant, Altered CNS neuroanatomical organization of spontaneous hypertensive rats, *Brain Res.* 226:119 (1981).

14. T. L. Krukoff and M. A. Weigel, Metabolic alteration in discrete regions of the rat brain during development of spontaneous hypertension, *Brain Res.* 499:1 (1989).

15. K. J. Anderson, D. Dam, S. Lee, C. W. Cotman, Basic fibroblast growth factor prevents death of lesioned cholinergic neurons in vivo, *Nature* 332:360 (1988).

16. P. Cuevas, F. Carceller, A. Esteban, A. Baird and R. Guillemin, Basic fibroblast growth factor (bFGF) enhances retinal ganglion cell survival and promotes growth of rat transected optic nerve, *Acta Nerol. Scand.* 79:263 (1988).

17. D. Otto, K. Unsicker and C. Grothe, Pharmacological effects of nerve growth factor and fibroblast growth factor applied to the transectioned sciatic nerve on neuron death in adult rat dorsal root ganglia, *Neurosci. Lett.* 83:156 (1987).

18. P. A. Walicke and A. Baird, Neurotrophic effects of basic and acidic fibroblast growth factors are not mediated through glial cells, *Dev. Brain Res.* 40:71 (1988).

19. D. Blottner, R. Westermann, C. Grothe, P. Böhlen and P. Unsicker, Basic fibroblast growth factor in the adrenal gland: Possible trophic role for preganglionic neurons in vivo, *Eur. J. Neurosci.* 1:471 (1989).

20. Jan Klein. "Immunology. The science of self-nonsense discrimination". John Wliley & Sons. New York, 1982.

21. O. E. Millhouse, A Golgi study of third ventricle tanycites in the adult rodent brain, *Z. Zellforsch. Mikrosk. Anat.* 121:1 (1971).

22. R. Bleier, The relations of ependyma to neurons and capillaries in the hypothalamus. A Golgi-Cox-study. *J. Comp. Neurol.* 142:439 (1971).

23. D. E. Scott, G. Krobisch-Dudley, K. M. Knigge, The ventricular system in neuroendocrine mechanisms-II. In vivo monoamine transport by ependyma of the median eminence, *Cell Tissue Res.* 154:1 (1974).

24. P. M. Ma, Tanycytes in the sunfish brain: NADH-diaphorase histochemistry and regional distribution, *J. Comp. Neurol.* 336:77 (1993).

DIADENOSINE POLYPHOSPHATES FROM NEURO-SECRETORY GRANULES: THE SEARCH FOR RECEPTORS, SIGNALS AND FUNCTION

M. Teresa Miras-Portugal, Jesus Pintor, Enrique Castro, Fernando Rodriguez-Pascual and Magdalena Torres

Departamento de Bioquimica
Facultad de Veterinaria
Universidad Complutense de Madrid
28040 Madrid- Spain

SUMMARY

The diadenosine polyphosphates -Ap_nA- are stored in secretory granules together with ATP and aminergic compounds: with serotonin in the dense granules of platelets, with acetylcholine in torpedo synaptic vesicles and with adrenaline and noradrenaline in chromaffin granules of adrenal medulla. The vesicular content in neural tissues is released to the extracellular media in a Ca^{2+} dependent way and in equimolecular ratios with respect to the other costored compounds. Cultured bovine neurochromaffin cells, torpedo synaptosomes and rat brain synaptic terminals, show the existence of specific binding sites for diadenosine polyphosphates. Two high affinity binding sites are found in these neural models with K_d values ranging from 0.1 to 0.7 nM for the first site and 5 to 6 nM for the second binding site. Displacement studies with P_2-purinoceptor ligands present a particular potency order, suggesting the presence of a new receptor subtype in rat brain synaptic terminals , designated as P_{2d}. The diadenosine polyphosphate receptors in neurochromaffin cells and vascular endothelial cells from adrenal medulla are coupled to Ca^{2+} release from internal stores and stimulation of Protein Kinase C. These intracellular signals cause the inhibition of catecholamine secretion from bovine chromaffin cells and induce a negative feedback for excitation in rat hippocampus. Both Ap_4A and Ap_5A evoke a fast rise in the $[Ca^{2+}]_i$ in endothelial cells from bovine adrenal medulla, the effect is concentration dependent, and the EC_{50} is in the range 1-10 μM. The extracellular destruction of adenine dinucleotides is carried out by an ecto-dinucleoside polyphosphate hydrolase with high affinity, the K_m values being close to 2 μM for the different diadenosine polyphosphates. The adenine mononucleotides produced are degraded to adenosine by the ecto-nucleotidases cascade. This nucleoside can be considered as the last extracellular product of purinergic transmission.

INTRODUCTION

The first dinucleoside polyphosphates reported in biological materials were the guanosine(5')tetraphospho-(5')guanosine (Gp_4G) and the guanosine(5')triphospho(5')-guanosine (Gp_3G), both discovered in large quantities in the encysted embryos of Artemia and related crustaceous, where they constitute a reservoir of purinergic

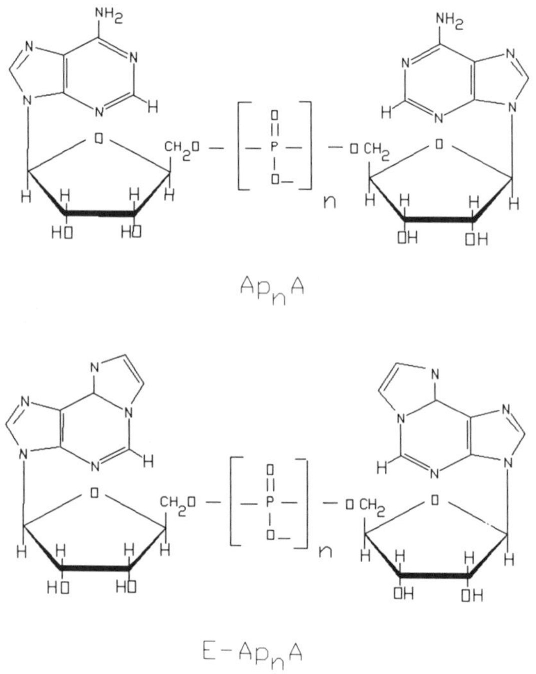

Figure 1.- **Structure of diadenosine polyphosphates, Ap_nA, and the etheno derivative, $\epsilon-Ap_nA$.** (n=3-6)

170

nucleotides (Finamore and Warner, 1963). Figure 1 shows the molecular structure of Ap_nA and their dietheno-derivatives.

The most ubiquitous and so far relevant of these compounds is the adenosine(5')tetraphospho(5')adenosine (diadenosine tetraphosphate, Ap_4A), which is formed in the reaction catalyzed by certain aminoacyl-tRNA synthetases (Zamecnik et al., 1966). The Ap_4A binds to the DNA polymerase-α and increases the initiation of DNA replication, playing a role of intracellular regulatory molecule (Rapaport and Zamecnick, 1976; Grummt et al., 1979). In addition to its nuclear action, Ap_4A is a potent inhibitor of the cytosolic adenosine kinase and an intracellular modulator of adenosine transporters present at the plasma membrane (Rotllán and Miras-Portugal, 1985; Casillas et al., 1993).

The discovery of the presence of diadenosine polyphosphates, Ap_3A and Ap_4A, in the dense granules of platelets, costored with serotonine, ADP and ATP, opened the possibility of an extracellular role of these compounds, (Flodgaard and Klenow, 1982, Lüthje and Ogilvie, 1983). Recently Ap_5A and Ap_6A have also been identified in platelets (Schlüter et al., 1994). The chromaffin granules from adrenal medulla were the first neuro-secretory granules where diadenosine polyphosphates were found. In these storage granules they are together with catecholamines and ATP (Rodriguez del Castillo et al., 1988). Thus the possibilities of purinergic signalling are largely increased and make even more complex and appealing this, still developing, field of research (Abbracchio, 1993 ; Burnstock, 1993).

Nevertheless , in order that diadenosine polyphosphates may be considered as a new type of purinergic transmitters, the classical criteria must be fulfilled: a) presence in neurosecretory vesicles, b) release in a Ca^{2+} dependent way, c) specific receptors coupled to a signal transduction mechanism, and d) the termination of the action by extracellular destruction and subsequent transport into cells.

PRESENCE OF DIADENOSINE POLYPHOSPHATES IN NEUROSECRETORY GRANULES

The presence of diadenosine polyphosphates in secretory granules has been studied in three different neural models: chromaffin cells, torpedo synaptosomes and rat brain synaptosomes.

Chromaffin cells are the homologous of sympathetic neurones and can be considered to be a pure catecholaminergic model, with little cellular heterogeneity and no glial contamination. Besides, the chromaffin granule was the first discovered secretion organelle and its composition one of the best characterized (Winkler and Carmichael, 1982). The catecholamines, ATP, the noradrenaline synthetising enzyme, namely the dopamine-ß-hydroxylase, enkephalins, as well as other active peptides and their proteic precursors, are costored in these granules (Simon and Aunis, 1989). ATP is the most abundant nucleotide, reaching intragranular concentrations of about 0.17 M. Other nucleotides such as ADP, GTP and GDP are also present.

Table I. Diadenosine Polyphosphates Content and Release from Neural Preparations

NEURAL PREPARATION	Ap_4A	Ap_5A	Ap_6A
Chromaffin granules (nmol/mg)	47 ± 7	47 ± 6	48 ± 9
Chromaffin cells (nmol/10^6 cells)	1.5 ± 0.3	1.5 ± 0.2	1.4 ± 0.2
Chromaffin cells secretion (pmol/10^6 cells)		30 ± 17 each	
(carbachol 100 µM)			
Torpedo synaptic vesicles (nmol/mg)	33 ± 7	23 ± 4	----
Rat brain synaptosomes (pmol/mg)	169 ± 25	159 ± 22	100 ± 25
Synaptosomal secretion (pmol/mg)	20 ± 3	17 ± 3	----
(veratridine 10 µM)			
In vivo push-pull (nM)			
Control	ND	ND	----
Amphetamine (5mg/Kg)	64.5 ± 5	57.5 ± 5	----
Amphetamine + haloperidol (3mg/Kg)	ND	ND	----

Ap_nA quantification was made by HPLC technique. The secretion experiments were done in the presence of 2.5 mM extracellular Ca^{2+} . Values are the means $\pm$ SD of 6 experiments.

The push-pull samples were collected every 10 min, the volume being 200 µl. The amphetamine values reported are at the maximal effect both in dose and time, that is 30 min after injection , and are the means $\pm$ SD of 4 experiments. ND means not detectable by HPLC technique.

Three adenine dinucleotides, namely diadenosine tetraphosphate (Ap_4A), diadenosine pentaphosphate (Ap_5A) and diadenosine hexaphosphate (Ap_6A) are costored in chromaffin granules. They were identified and quantified by H.P.L.C. technique, combined with enzymatic treatment with phosphodiesterase and analysis of the degradation products (Rodriguez del Castillo et al., 1988; Pintor et al., 1992c). The granular content is very similar for the three dinucleotides (table I), and their ratios to CA and adenine mononucleotides are around 90 and 20 respectively (Rodriguez del Castillo et al., 1988; Pintor et al., 1992c). Since the intragranular concentration of CA and ATP has been estimated to be 0.55 M and 0.17 M respectively, the intragranular concentration of each Ap_nA is about 5 mM.

The torpedo electric organ is a rich source of cholinergic vesicles, where ATP and acetylcholine are costored. Pure synaptic vesicles can be obtained and an HPLC chromatographic profile of their nucleotidic content is shown in figure 2. Ap_4A and Ap_5A are present with similar concentrations to that reported for chromaffin granules (Table I). Since Ap_6A is a very unstable compound, its presence was not possible to confirm in these preparations (Pintor et al., 1992b).

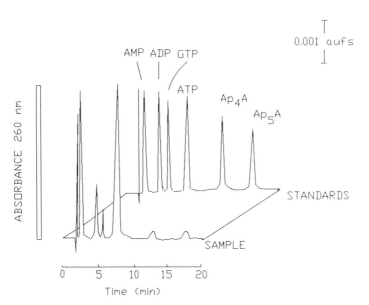

Figure 2.- H.P.L.C. elution profile of the adenine nucleotide and dinucleotide content in *Torpedo* synaptic vesicles.

Chromatographic profiles were obtained using a μBondapak C_{18} column and a buffer containing 10 mM KH_2PO_4, 2 mM tetrabutyl ammonium and 10 % acetonitrile, pH 7.5.

Ahead: Sample from *Torpedo* vesicles, where Ap_4A and Ap_5A were identified as constituents of the cholinergic vesicle nucleotide pool together with ATP, ADP, AMP, and GTP.

Behind: Standard mixture of 125 pmol AMP, ADP, GTP, ATP, Ap_4A and Ap_5A under the same elution conditions.

Apart from chromaffin granules and torpedo synaptic vesicles, there are no other pure neural aminergic or cholinergic systems available in order to study their costorage with adenine dinucleotides; from the results in these models and from their costorage with serotonine in platelets, a broad presence throughout the aminergic and cholinergic pathways in central nervous system can be infered. The presence of Ap_4A, Ap_5A and Ap_6A in rat brain synaptic terminals confirms this supposition. Due to the great heterogeneity of these synaptic terminals, their concentration values are more than two orders of magnitude lower than in pure vesicle preparations (Table I).

STORAGE OF DIADENOSINE POLYPHOSPHATES IN SECRETORY GRANULES

Chromaffin granules from bovine adrenal medulla can be employed to study the diadenosine polyphosphates transport into storage granules. Figure 3 shows an HPLC chromatogram, with the granular nucleotidic content compared with standards, and the radioactivity distribution in HPLC after the $[^3H]$-Ap_4A transport experiments.

Due to the structural similarities between the Ap_nA and ATP, it is necessary to mention that the ATP transport has been largely characterized in chromaffin granules. The vesicular ATP transporter is coupled to an electrochemical gradient (Weber and Winkler, 1981). The granule carrier is not specific for ATP, since ADP, UTP and GTP have similar affinities and are present in the granular content at a ratio similar to those in the cytosol (Weber and Winkler, 1981; Winkler and Carmichael, 1982). The transport of diadenosine tetraphosphate is inhibited by ATP with an $IC_{50} = 200$ μM indicating that the diadenosine polyphosphates are internalized by the vesicular nucleoside transporter.

The diadenosine tetraphosphate transport shows a saturation kinetic. The affinity has a K_m value of 70 μM, when measured at 37°C with an extra-vesicular pH of 7.2. The granular transport of Ap_nA by the nucleotide transporter, however, does not exclude the possibility of an intragranular synthesis, due the peculiar internal conditions of storage granules.

Diadenosine polyphosphates are also stored in torpedo cholinergic vesicles and in dense granules of platelets, but the absence of data on their vesicular transport does not allow comparison.

Ca^{2+} DEPENDENT RELEASE OF DIADENOSINE POLYPHOSPHATES FROM NEURAL TISSUES

Perfused bovine adrenal glands and isolated chromaffin cells release diadenosine polyphosphates upon stimulation of the nicotinic acetylcholine receptors, in the presence of extracellular Ca^{2+} (Pintor et al., 1991b; Pintor et al., 1992c). The quantification of released dinucleotides from the perfused gland is problematic due to the presence of

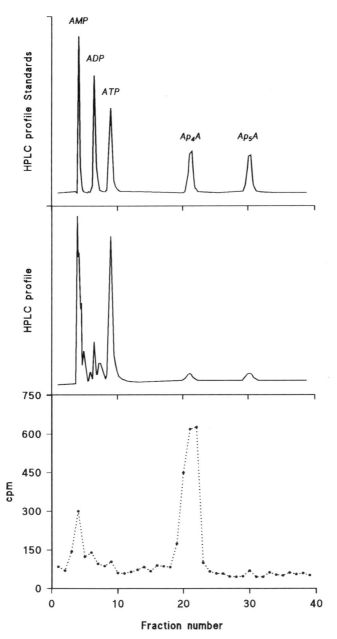

Figure 3.- Ap₄A transport into chromaffin granules. HPLC chromatogram of chromaffin granules nucleotidic content and Incorporation of [³H]-Ap₄A to the granular dinucleotides.

(A) chromatogram of 50 pmol nucleotide and dinucleotide standards. (B) HPLC chromatogram of the nucleotides and dinucleotides present in 0.2 mg protein of chromaffin granules. (C) The same quantity of chromaffin granules incubated in the presence of 1.5 μCi of [³H]-Ap₄A, (1 μM) at 37 °C. After 30 min incubation the granules were isolated by centrifugation in 1.6 M sucrose, as a pellet. Dotted lines represent the radioactivity of fractions collected fron the HPLC eluates.

ecto-dinucleotide hydrolases in neural and vascular cells, which can hydrolyze large amounts of released dinucleotides (Goldman et al., 1986; Ogilvie et al., 1989; Rodriguez-Pascual et al., 1992b).

Carbachol stimulation of isolated chromaffin cells induce the release from secretory granules. Catecholamines, ATP and diadenosine polyphosphates are released in the same equimolecular amounts as stored in the chromaffin granules. In table I the adenine nucleotides released per 1 million of cells are summarized; these values are highly variable depending on the cellular culture and can oscillate between 1-5 % of the total cellular content.

The Ca^{2+} dependent release from rat brain synaptic terminals is evoked by depolarization. Veratridine induces the release of 10-12 % of the total synaptosomal content of Ap_4A and Ap_5A, as summarized in table I.

The extracellular diadenosine polyphosphate levels reached after exocytosis are difficult to evaluate. In the model of blood platelets, assuming a complete release from the dense granules, an initial extracellular concentration of about 100 μM or even higher could exist in the surrounding area of the damaged vessel wall and the thrombus. The levels of diadenosine polyphosphates after distribution in blood can reach values of 1 μM (Ogilvie, 1992). In the case of chromaffin cells an extracellular concentration of 27 μM can be calculated (Pintor et al., 1991 b). Similar concentrations, or even higher, can be present at the synaptic cleft, where the extracellular space is very reduced.

The push-pull perfusion technique permits the approach to the central nervous system in "in vivo" situations. In basal conditions the Ap_nA compounds are under the detection level by HPLC technique in samples from the caudate putamen area that are very rich in aminergic terminals. Amphetamine administration increases the diadenosine polyphosphate levels in the perfusion samples. The concentration values for both compounds, Ap_4A and Ap_5A, at their maximal level are close to 60 nM each and summarized in table I (Pintor et al., 1993c). The real values at the nerve terminals can be expected to be two or three orders of magnitude higher. These values need to be taken into consideration to analyze the physiological relevance of diadenosine polyphosphate binding sites. The haloperidol inhibition of diadenosine polyphosphates release clearly demonstrates that these compounds are not directly released by amphetamine, but by the action of dopamine on its receptors, because haloperidol is a broad dopaminergic receptor antagonist (Table I).

DIADENOSINE POLYPHOSPHATE RECEPTORS AND TRANSDUCTION MECHANISMS

The presence of diadenosine polyphosphates costored with ATP in synaptic vesicles widely enlarges the extracellular signalling potencial of purinergic transmission.

In 1978 a basis for distinguishing two types of purinergic receptors was made (Burnstock, 1978). Since this pioneering article a great deal has been done to classify the nucleotidic receptors and their mechanisms of action.

High affinity binding sites for diadenosine polyphosphates are present in neural tissues and in all cases curvilinear Scatchard plot are observed. The kinetic parameters for the two binding sites in chromaffin cells, torpedo synaptosomes and rat brain synaptosomes are summarized in table II.

Table II: Binding Parameters of Diadenosine Polyphosphates

Binding parameters	chromaffin cells	torpedo synaptosomes	rat brain synaptosomes
K_{d1}(nM)	0.08 ± 0.01	0.4 ± 0.005	0.11 ± 0.02
B_{max1}	$5,420 \pm 450$ sites/cell	211 ± 22 fmol/mg prot.	3.9 ± 2.1 fmol/mg prot.
K_{d2} (nM)	5.6 ± 0.5	6.5 ± 0.7	18 ± 3.5
B_{max2}	$70,000$ sites/cell	734 ± 95 fmol/mg prot.	249 ± 30 fmol/mg prot.

Results of binding are the mean $\pm$ SD of 4 or more experiments in duplicate.

The presence of these two high affinity binding sites for Ap_4A needs to be analyzed in relation to the extracellular levels reached by the dinucleotides, discussed in the former section. The lowest affinity site could be saturated when exocytosis occurs and the highest affinity site could also be occupied when difusion occurs, or if the sites are far from the releasing areas. This aspect has special relevance at the vascular system where the effects of Ap_4A are known (Busse et al., 1989). In endothelial cells from bovine adrenal medulla the high affinity binding sites for Ap_4A have a K_d value of 0.045 ± 0.003 nM (Rodriguez-Pascual, unpublished results).

Displacement studies of [^{3}H]-Ap_4A with ATP synthetic analogs revealed a P_{2y}-like purinoceptor displacement order in chromaffin cells. In synaptic terminals from rat brain the displacement order does not correspond with the described P_2 purinoceptors, and the existence of a new purinergic receptor is suggested. Its high affinity for diadenosine polyphosphates was at the origin of its name P_{2d}-purinoceptor (Pintor et al., 1991a; Pintor et al., 1993a).

Nevertheless, a more complete characterization of the purinergic receptor family is necessary, and due to the lack of a fully developed pharmacology in this area, the cloning of receptors is a way to circumvent the problem. Recently, two nucleotidic receptors have been cloned, one with characteristics of a metabotropic P_{2u}-purinoceptor, and the other similar to the P_{2y}-purinoceptor (Lustig et al., 1993; Webb et al., 1993). Moreover, the purification of an ionotropic P_{2x}-purinoceptor indicates that complexity similar to that found in the case of acetylcholine or glutamate receptors can be expected.

Concerning the diadenosine polyphosphates, it is necessary to know the extent to which these compounds can interact with specific subpopulations of the P_2 purinoceptors, or if they are nucleotides with longer half-lives than ATP, a matter to be further discussed in the last section.

The presence of receptors for Ap_4A has been reported in membrane preparations of mouse brain and heart, with affinities in the μM range (Hilderman et al., 1991). A protease dependent step appears to be necessary to reveal the presence of an active receptor (Walter et al., 1993).

The transduction mechanisms coupled to diadenosine polyphosphate receptors have so far been studied in a reduced number of models. In neurochromaffin cells Ap_4A and Ap_5A evoked a concentration-dependent increase in cytosolic calcium ($[Ca^{2+}]_i$) in resting chromaffin cells, even in the absence of extracellular calcium. The EC_{50} values for this action were 28 ± 6 μM and 50 ± 7 μM respectively, (Castro et al., 1992). These concentrations are in the expected extracellular range after exocytosis. The activation of Protein Kinase C has also been reported to occur (Sen et al., 1993).

Studies carried out in single cell by fura-2 microfluorometry reveal a more complex picture. Different purinoceptor subtypes coupled to Ca^{2+} influx and intracellular Ca^{2+} release can be found in bovine adrenal chromaffin cells (Castro et al., 1994). These data appear more relevant considering that the different receptor distributions are related to the adrenergic and noradrenergic cellular populations present in chromaffin cells (Castro et al., 1993).

The vasoactive effects of ATP and the purinergic receptors on blood vessels have long been studied (review Olsson and Pearson, 1990). As this system is very rich in purinergic receptors, the actions of Ap_4A and Ap_5A on endothelial cells from bovine adrenal medulla were investigated. Cytosolic Ca^{2+} levels in single endothelial cells loaded with fura-2 probe were measured. Both Ap_4A and Ap_5A evoked a fast rise in the $[Ca^{2+}]_i$ of these cells. The response to these agonists was in the former of a fast and transient peak (about 500 nM) followed by a plateau which dropped upon withdrawal of the drug (Figure 4). When tested in the absence of extracellular Ca^{2+}, the peak response was maintained, while the plateau phase was abolished. This suggests that the response initiated by Ap_4A was mediated by Ca^{2+} release from intracellular stores by activation of phospholipase C and generation of inositol (1,4,5)-trisphosphate. The effect of Ap_4A and Ap_5A was concentration-dependent, with an EC_{50} in the range 1-10 μM. ATP, its analogues ADPßS and UTP mimicked the response to Ap_4A; but AMP and α,β-MeATP were inactive, indicating the involvement of a P_{2u}-like purinoceptor in the action of diadenosine polyphosphates on endothelial cells from adrenal medulla.

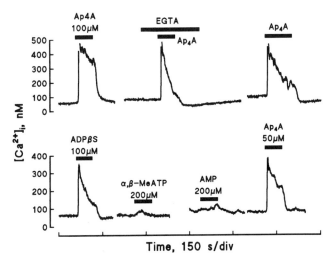

Fig 4.- Effect of Ap_4A and other purinergic agents in $[Ca^{2+}]_i$ levels of endothelial cells from bovine adrenal medulla. Fluorescence (fura-2) from individual cells was measured by microphotometry while continuously perfused. Each panel corresponds to a single cell representative of many experiments. Cells were challenged with the agonist during the time indicated by the bars. Washing between stimulations was 5-6 min. EGTA: normal extracellular medium ($[Ca^{2+}]_o = 2mM$) was replaced by $Ca^{2+}/EGTA$ buffered medium adjusted to $[Ca^{2+}]_o = 100nM$.

DIADENOSINE POLYPHOSPHATE EFFECTS

Actions of these compounds have been described in various systems. In blood plasma Ap_4A is known to be a competitive inhibitor for ADP in the platelet aggregation response (Lüthje and Ogilvie, 1983). Analogues of this compound are under study

179

because of their potential antithrombotic effect (Zamecnik et al., 1992).

On isolated non damaged mesenteric arteries, Ap_3A and Ap_4A exhibit considerable vasodilator activities at physiological concentrations (Pohl et al., 1991). Ap_5A and Ap_6A are vasopressive agents (Schlüter et al., 1994). The Ap_nA actions on vas deferens and urinary bladder also suggest a role in purinergic neuromuscular transmission (review Hoyle, 1990). The inhibitory effects on catecholamine secretion from isolated chromaffin cells enhance the modulatory role for these compounds in neural systems (Castro et al., 1990)

The presence of Ap_4A and Ap_5A in brain perfusates and their synaptosomal release by depolarization gives physiological relevance to the effects described at the central nervous system (Pintor et al., 1992a; 1993c). A weak depressant activity on neuronal firing has been noticed on rat cortical neurones (Stone and Perkins, 1981). Experiments in nodose ganglia neurones using voltage clamp techniques showed the presence of a cationic channel activated by these compounds with low ionic selectivity (Krishtal et al., 1988). In hippocampus the diadenosine polyphosphates are very effective exerting negative feedback onto excitatory synaptic transmission, this action being mediated through protein kinase C (Klishin et al., 1994).

EXTRACELLULAR DESTRUCTION OF DIADENOSINE POLYPHOSPHATES

Enzymatic activities degrading diadenosine polyphosphates have been identified in blood plasma and as ectoenzymes on the blood and vascular wall cells (Ogilvie, 1992). Endothelial cells from blood vessels have a high capacity to hydrolyze the Ap_3A and Ap_4A released from platelets . The affinity kinetic parameters obtained in cultured endothelial cells from diverse origins exhibit a K_m value about 10 μM for Ap_4A (Goldman et al., 1986; Ogilvie et al., 1989). The adenine nucleotides produced by enzymatic hydrolysis, at the surface of the cell, are subsequently degraded to adenosine by a cascade of ectonucleotidase enzymes.

In neural cells and synaptic terminals, ecto-ATPase, ecto-ADPase and ecto-5'-nucleotidase activities have been described (Richardson et al., 1987). Nevertheless, the activity of an ecto-diadenosine polyphosphate hydrolase (Ap_nA-hydrolase) has so far been reported only in isolated chromaffin cells (Miras-Portugal et al., 1990). This enzyme hydrolyses all the diadenosine polyphosphates in an asymmetric manner with high affinity , the K_m values being between 1-3 μM, in studies carried out by HPLC analysis (Rodriguez-Pascual et al., 1992). Recently, a continuous fluorimetric method has been developed to follow the hydrolysis of dinucleotides. This technique is based on the hydrolysis of the dietheno-derivatives of diadenosine polyphosphates (figure 1); these compounds are substrate analogues with similar affinity. Due to the interactions of the two etheno-adenosine rings, a notorious increase of the fluorescence emission is observed after hydrolysis, allowing a continuous enzymatic assay (Rotllán et al., 1991).

In chromaffin cells the hydrolysis of extracellular dinucleotides goes further to adenosine, because of the ectonucleotidases cascade (Torres et al., 1990). In table III the K_m and V_{max} values for these enzymes are summarized. It is important to emphasize the high affinity for Ap_4A hydrolysis, because this value is in the expected extracellular level range after exocytosis. Furthermore, the ecto-ATPase and ecto-ADPase activities present in these neural cells are about three orders of magnitude higher than the ecto-

Table III: Kinetic Parameters of Ectonucleotidase Activities in Cultured Chromaffin Cells

Ecto-enzyme	K_m (μM)	V_{max} (nmol/min. 10^6 cells)
Ecto-Ap_4A hydrolase	2.9 ± 0.5	0.012 ± 0.001
Ecto-ATPase	250 ± 18	18.2 ± 0.6
Ecto-ADPase	375 ± 40	12.6 ± 1.2
Ecto-5'-Nucleotidase	55 ± 5	2.6 ± 0.2

Activities were measured by HPLC. Ecto-Ap_4A hydrolase was studied using a radiometric technique.

Ap_nA-hydrolase. These results can explain the presence of Ap_4A and Ap_5A , but the absence of ATP and ADP, in brain perfusates after amphetamine stimulation (Pintor et al., 1993c). Thus, although ATP is released in much higher amounts than the Ap_nA compounds, the lower rate of hydrolysis allows them a longer extracellular half-life and new roles derived from this fact.

Figure 5 summarizes the more relevant aspects of diadenosine polyphosphates biology, that allow their presentation as new neurotransmitters or neuromodulators.

ACKNOWLEDGEMENTS

This work was supported by the Spanish DGICYT grant n°PB 92/0230. F. R-P is a fellow of the Rectorado of UCM.

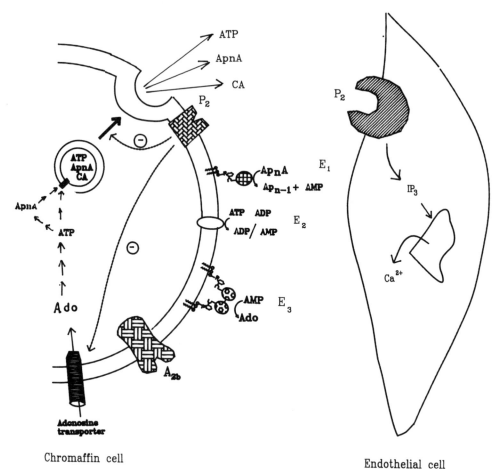

Chromaffin cell   Endothelial cell

Fig 5.- Diadenosine polyphosphates in neuro-chromaffin cells.

Diadenosine polyphosphates are transported into chromaffin granules by the vesicular nucleotide transporter. Their exocytotic release is induced by nicotinic agonists in a calcium dependent manner. The Ap_nA extracellular actions are mediated by P_2 purinoceptors. Neuro-chromaffin cells exhibit P_2 receptors coupled to metabotropic as well as ionotropic mechanisms. In these cells the extracellular Ap_nA inhibits both the adenosine transport and the catecholamine secretion. On the neighboring eendothelial cells, diadenosine polyphosphates are linked to P_{2u} metabotropic receptors. An ecto-dinucleoside polyphosphate hydrolase destroys the Ap_nA compounds, and the produced mononucleotides are subsequently degraded to adenosine by the ecto-nucleotidases cascade. Adenosine is transported into the cell to be transformed to adenine nucleotides.

REFERENCES

Abbracchio, M.P., Cattabeni, F, Fredholm, B.B., and Williams, M. (1993). Purinoceptor nomenclature: A status report. Drug. Development Research., 28, 207-213.

Burnstock, G. (1978). A basis for distinguishing two types of purinergic receptor. In Straub, R.W., Bolis, L. (eds). Cell membrane receptors for drugs and hormones: A multidisciplinary approach. New York, Raven Press. 107-118.

Burnstock, G. (1993). Physiological and pathological roles of purines: an update. Drug Development Research., 28, 195-206.

Busse, R., Ogilvie,A. and Pohl, U. (1988). Vasomotor activity of diadenosine triphosphate and diadenosine tetraphosphate in isolated arteries. Am. J. Physiol., 254, H828-H823.

Casillas, T., Delicado, E.G. and Miras-Portugal, M.T. (1993). Adenosine 5'-triphosphate modulation of nitrobenzylthioinosine binding sites in plasma membranes of bovine chromaffin cells. Neurosci. Lett., 159, 1-4.

Castro, E., Torres, M., Miras-Portugal, M.T. and Gonzalez, M.P. (1990). Effect of diadenosine polyphosphates on catecholamine secretion from isolated chromaffin cells. Br. J. Pharmacol., 100, 360-364.

Castro, E., Pintor, J. and Miras-Portugal, M.T. (1992). Ca^{2+}-stores mobilization by diadenosine tetraphosphate, Ap_4A, through a putative P_{2y} purinoceptor in adrenal chromaffin cells. Br. J. Pharmacol., 106, 833-837.

Castro, E. , Tomé, A.J.R., Miras-Portugal, M.T. and Rosario, L.M. (1993). Receptores purinérgicos en células cromafines: mecanismos de acción y distribución celular. XIV Reunión Nacional Grupo Español de la Celula Cromafin. pp-41.

Castro, E., Tomé, A.R., Miras-Portugal, M.T. and Rosario, L.M. (1994). Single cell fura-2-microfluorometry reveals different purinoceptor subtypes coupled to Ca^{2+} influx and intracellular Ca^{2+} release in bovine adrenal chromaffin and endothelial cells. Pflügers Arch. Eur. J. Physiol., (in press).

Finamore, F.J. and Warner, A.H. (1963). The occurrence of P_1,P_4-diguanosine 5'-tetraphosphate in brine shrimp eggs. J. Biol. Chem., 238, 344-348.

Flodgaard, H. and Klenow, H. (1982). Abundant amounts of diadenosine 5',5'''-P_1,P_4-tetraphosphate are present and releasable, but metabolically inactive in human platelets. Biochem.J., 208, 737-742.

Goldman, S.J., Gordon, E.L., and Slakey, L.L. (1986). Hydrolysis of diadenosine 5'-5''-p'-p''-triphosphate (Ap_3A) by porcine aortic endothelial cells. Circ. Res., 59, 362-366.

Grummt, F., Waltz, G., Jantzen, H.-M., Hamprecht, K., Huebscher, U., and Kuenzle, C.C. (1979). Diadenosine 5',5'''-P_1,P_4-tetraphosphate, a ligand of the 57-kilodalton subunit of DNA polymerase alfa. Proc. Natl. Acad. Sci. USA., 76, 6081-6085.

Hilderman, R.H., Martin, M., Zimmerman, J.K. and Pivorum, E.B. (1991). Identification of a unique membrane receptor for adenosine 5',5'''-P_1,P_4-tetraphosphate. J. Biol. Chem., 266, 6915-6918.

Hoyle,C.H.V. (1990). Pharmacological activity of adenine dinucleotides in the periphery: possible receptor classes and transmitter function. Gen.Pharmac., 21, 827-831.

Klishin, A., Lozovaya, N., Pintor, J., Miras-Portugal, M. and Krishtal, O.A. (1994). Possible functional role of diadenosine polyphosphates: Negative feedback for excitation in hippocampus. Neuroscience, 58, 235-236.

Krishtal, O.A., Marchenko, S.M., Obukhov, A.G. and Volkova, T.M. (1988). Receptors for ATP in rat sensory neurones: the structure-function relationship for ligands. Br. J. Pharmacol., 95, 1057-1062.

Lustig, K.D., Shiau, A.K., Brake, A.J. and Julius, D. (1993). Expression cloning of an ATP receptor from mouse neuroblastoma cells. Proc. Natl. Acad. Sci. USA., 90, 5113-5117.

Lüthje, J. and Ogilvie, A. (1983). Presence of diadenosine $5',5'''$-p_1,p_4-triphosphate (Ap_3A) in human platelets. Biochem. Biophys. Res. Commun., 115, 253-260.

Miras-Portugal, M.T., Pintor, J., Rotllán, P. and Torres, M. (1990). Characterization of ectonucleotidases in chromaffin cells. An. N.Y. Acad. Sci., 603, 523-526.

Ogilvie, A. (1992). Extracellular functions for Ap_nA. in "Ap_4A and other dinucleoside polyphosphates" Ed. McLennan, A.G. CRC press 230-273.

Ogilvie, A., Lüthje, J., Pohl, U. and Busse, R. (1989). Identification and partial characterization of an adenosine (5') tetraphospho(5')adenosine hydrolase on intact bovine aortic endothelial cells. Biochem. J., 259. 97-103.

Olsson, R.A. and Pearson, J.D.(1990). Cardiovascular purinoceptors. Physiol. Reviews., 70, 761-845.

Pintor, J., Torres, M., Castro, E. and Miras-Portugal, M.T. (1991a). Characterization of diadenosine tetraphosphate (Ap_4A) binding sites in cultured chromaffin cells: evidence for a P_{2y} site. Br. J. Pharmacol., 103, 1980-1984.

Pintor, J., Torres, M. and Miras-Portugal, M.T. (1991b). Carbachol induced release of diadenosine polyphosphates -Ap_4A and Ap_5A- from perfused bovine adrenal medulla and isolated chromaffin cells. Life Sci., 48. 2317-2324.

Pintor, J., Diaz-Rey, M.A., Torres, M. and Miras-Portugal, M.T. (1992a). Presence of diadenosine polyphosphates -Ap_4A and Ap_5A- in rat brain synaptic terminals. Ca^{2+} dependent release evoked by 4-aminopyridine and veratridine. Neurosci. lett., 136, 141-144.

Pintor, J., Kowalewski, H.J., Torres, M., Miras-Portugal, M.T. and Zimmermann, H. (1992b). Synaptic vesicle storage of diadenosine polyphosphates in the torpedo electric organ. Neurosci. Res. Commun., 10, 9-14.

Pintor, J., Rotllán, P., Torres, M. and Miras-Portugal, M.T. (1992c). Characterization and quantification of diadenosine hexaphosphate in chromaffin cells: Granular storage and secretagogue-induced release. Anal. Biochem., 200, 296-300

Pintor, J., Diaz-Rey, M.A. and Miras-Portugal, M.T. (1993a). Ap_4A and ADP-ß-S binding to P_2 purinoceptors present on rat brain synaptic terminals. Br. J. Pharmacol. 108, 1094-1099.

Pintor, J. and Miras-Portugal, M.T. (1993b). Diadenosine polyphosphates (Ap_xA) as new neurotansmitters. Drug Development Research., 28, 259-262.

Pintor, J., Porras, A., Mora, F. and Miras-Portugal, M.T. (1993c). Amphetamine-induced release of diadenosine polyphosphates -Ap_4A and Ap_5A- from caudate putamen of conscious rat. Neurosci. Lett. 150, 13-16.

Pohl, U., Ogilvie, A., Lamontagne, D., and Busse, R. (1991). Potent effects of Ap_3A and Ap_4A on coronary resistence and autacoid release of intact rabbit hearts. Am. J. Physiol., 260, H1692-H1697.

Rapaport, E. and Zamecnick P.C. (1976). Presence of diadenosine $5',5'''$-P_1,P_4-tetraphosphate (Ap_4A) in mammalian cells in levels varying widely with proliferative activity of the tissue:a possible positive "pleiotypic activator". Proc. Natl. Acad. Sci. USA., 73, 3984-3988.

Richardson, P.J., Brown, S.J., Bailyes, E.M. and Luzio, J.P. (1987). Ectoenzymes control adenosine modulation of inmunoisolated cholinergic synapses. Nature., 327, 232-234.

Rodriguez del Castillo, A., Torres, M., Delicado, E.G. and Miras-Portugal, M.T. (1988). Subcellular distribution studies of diadenosine polyphosphates -Ap_4A and Ap_5A- in bovine adrenal medulla: presence in chromaffin granules. J. Neurochem., 51, 1696-1703.

Rodriguez-Pascual, F., Torres, M. and Miras-Portugal, M.T. (1992a). Studies on the turnover of ecto-nucleotidases and ecto-dinucleoside polyphosphate hydrolase in cultured chromaffin cells. Neurosci. Res. Comm., 11, 101-107.

Rodriguez-Pascual, F., Torres, M., Rotllán, P. and Miras-Portugal, M.T. (1992b). Extracellular hydrolysis of diadenosine polyphosphates, Ap_nA, by bovine chromaffin cells in culture. Arch. Biochem. Biophys., 297, 176-183.

Rotllán, P. and Miras-Portugal, M.T. (1985). Adenosine kinase from adrenal medulla. Eur. J. Biochem., 151, 365-371.

Rotllán, P., Ramos, A., Pintor, J., Torres, M., and Miras-Portugal, M.T.(1991). Di(N1,N6-ethenoadenosine)$5',5'''$-P_1,P_4-tetraphosphate, a fluorescent enzymatically active derivative of Ap_4A. FEBS Lett., 280, 371-374.

Schlüter, H., Offers, E., Brüggemann, G., van der Glet, M., Topel, M., Nordhoff, E., Karas, M., Spieker, C., Witzel, H. and Zidek, W. (1994). Diadenosine phosphates and the physiological control of blood pressure. Nature., 367, 186-188.

Schmidt, R., Zimmermann, H. and Whittaker, V.P. (1980). Metal ion content of cholinergic vesicles isolated from the electric organ of torpedo: effect of stimulation induced transmitter release. Neuroscience., 5, 625-638.

Sen, R.P., Delicado, E.G., Castro, E. and Miras-Portugal, M.T. (1993). Effect of P_{2y} agonists on adenosine transport in cultured chromaffin cells. J. Neurochem., 60, 613-619.

Simon, J.P. and Aunis, D. (1989). Biochemistry of the chromogranin A protein family. Biochem. J., 262, 1-13.

Stone, T.W. and Perkins, M.N. (1981). Adenine dinucleotide effects on rat cortical neurones. Brain Research., 229, 241-245.

Torres, M., Pintor, J. and Miras-Portugal, M.T. (1990). Presence of ectonucleotidases in cultured chromaffin cells: hydrolysis of extracellular adenine nucleotides. Arch. Biochem. Biophys., 279, 37-44.

Walter, J., Lewis, T.E., Pivorum, E.P. and Hilderman, H.R. (1993). Activation of the mouse heart adenosine $5',5'''$-P_1,P_4-tetraphosphate receptor. Biochemistry., 32, 1264-1269.

Webb, T.E., Simon, J., Krishek, B.J., Bateson, A.N., Smart, T.G., King, B.F., Burnstock, G. and Barnard, E.A. (1993).Cloning and functional expression of a brain G-protein-coupled ATP receptor. FEBS Lett., 324, 219-225.

Weber, A. and Winkler, H. (1981). Specificity and mechanism of nucleotide uptake by adrenal chromaffin granules. Neuroscience., 6, 2269-2276.

Winkler, H. and Carmichael, S.W. (1982). The chromaffin granule. in "The secretory granule",(eds.Poisner and Trifaro). Elsevier Biomedical Press. 3-79.

Zamecnik, P.C., Stephenson, M.L., Janeway, C.M. and Randerath, K. (1966). Enzymatic synthesis of diadenosine tetraphosphate and diadenosine triphosphate with a purified lysyl-tRNA synthetase. Biochem. Biophys. Res. Comm., 24, 91-97.

Zamecnik, P.C., Kim, B., Gao, M.J., Taylor, G., and Blackburn, G.M. (1992). Analogues of diadenosine 5',5'''-P_1,P_4-tetraphosphate (Ap$_4$A) as potential anti-platelet-aggregation agents. Proc. Natl. Acad. Sci. USA., 89, 2370-2373.

MECHANISMS OF GABAergic REGULATION OF NEUROSECRETION IN CHROMAFFIN CELLS

Mónica Parramón, Mª Jesús Oset-Gasque, Mª Pilar González, and Mª Teresa Herrero

Departamento de Bioquímica y Biología Molecular II (Centro mixto C.S.I.C.-U.C.M.)
Facultad de Farmacia. Universidad Complutense
28040- Madrid
Spain

INTRODUCTION

γ-Aminobutyric acid (GABA), the major inhibitory neurotransmitter in the Central Nervous System (CNS), is also present in several peripheral tissues, where it has a functional role in the regulation of muscle contraction or hormonal secretion (for a review, see Tanaka & Taniyama, 1992).

Chromaffin cells from adrenal medulla are considered to be typical post-synaptic catecholaminergic neurons, because they have several embryological, structural, and functional properties in common with catecholaminergic neurons from brain. Like neurons, these cells originate from neural crest, a tissue of ectodermic origin, and possess secretory vesicles, i.e., the chromaffin granules. These granules synthesize, store, and release catecholamines (CAs) and a number of neuropeptides to the blood by calcium-dependent exocytotic mechanisms.

CA secretion is a process physiologically stimulated by acetylcholine (ACh), released from splanchnic nerves which supply the adrenal medulla. When ACh binds to nicotinic receptors located on the plasma membrane of chromaffin cells, it induces an entry of Na^+ through receptor-associated Na^+ channels which depolarizes the cells and allows the activation of voltage-dependent Ca^{2+} channels. The subsequent Ca^{2+} entry triggers the extrusion of the CA granule content to the extracellular space (Douglas, 1968).

However, the recent identification of several neurotransmitters in chromaffin cells, such as opioid peptides, substance P, somatostatin, vasoactive intestinal peptide, histamine, angiotensin, neuropeptide Y, atrial natriuretic peptide, and GABA (for a review, see Burgoyne, 1991), suggests a more complex picture of CA secretion by chromaffin cells, since, together with nicotinic ACh (nACh) receptors, these neurotransmitters could act as physiological modulators of CA secretion.

Cell Signal Transduction, Second Messengers, and Protein Phosphorylation in Health and Disease
Edited by A.M. Municio and M.T. Miras-Portugal, Plenum Press, New York, 1994

187

In relation to GABA regulation of chromaffin cell secretion, previous work by our group has demonstrated the presence of all enzymes implicated in GABA metabolism in the bovine adrenal medulla (Fernández-Ramil et al., 1982; 1983) as well as the existence of an active uptake (Oset-Gasque & Aunis, 1989), and exocytotic release of this neurotransmitter (Oset-Gasque et al., 1990). Moreover, the presence of the two subtypes of GABA receptors, one sensitive to muscimol (GABA$_A$) and the other insensitive to muscimol but sensitive to baclofen (GABA$_B$) has also been described (Castro et al, 1988), both having a role in the regulation of CA secretion by chromaffin cells (Figure 1).

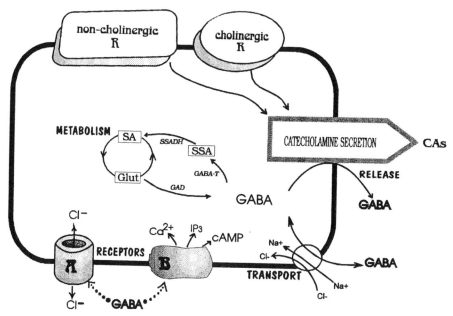

Figure 1. In adrenal chromaffin cells, all enzymes implicated in GABA metabolism (GAD, GABA-T and SSADH) are present. GABA is released by two mechanism, one calcium-dependent (exocytotic release) and other calcium-independent and sodium-dependent (inversion of GABA carrier). Both types of GABA receptors (GABA$_A$ and GABA$_B$) exist, which modulate CA secretion. In adddition to that, other receptors of different neurotransmitter systems are present in the plasma membrane (see text). GABA-T, 4-aminobutyrate-2-oxoglutarate aminotransferase; SSADH, semialdehyde succinic acid dehydrogenase; GAD, glutamate decarboxylase; SSA, succinic semialdehyde; SA, succinic acid; Glut, glutamate.

In this review we present our recent results concerning GABAergic modulation of CA secretion in chromaffin cells explainig the mechanisms by which both, ionotropic GABA$_A$ and metabotropic GABA$_B$ receptors exert such a regulatory action.

COEXISTENCE OF GABA$_A$ AND GABA$_B$ RECEPTORS IN ADRENAL MEDULLA MEMBRANES

The first evidence showing the possibility that adrenal medulla contains receptors for GABA was obtained by Kataoka et al. (1984), who found binding of [3H]muscimol to chromaffin cells in culture, and by Martínez et al. (1987), who observed that binding of [^{3}H]GABA was specifically displaced by muscimol and (-)-baclofen. These results suggested that the adrenal medulla contained both GABA$_A$ and GABA$_B$ receptors. Castro et al. (1988) showed that specific binding of [^{3}H]GABA to adrenal medulla membranes was biphasic, as reflected in the curved Scatchard plot, with a low and a high affinity component. These binding sites were specifically blocked by treatment with muscimol (low affinity) and baclofen (high affinity), respectively, indicating the existence of GABA$_A$ low affinity binding sites and GABA$_B$ high affinity binding sites. Computer-assisted analysis of data from blockade and nonblockade experiments gave kinetic constants of Kd$_1$=139$\pm$22 nM and Bmax=3.2$\pm$0.4 pmol/mg protein for GABA$_A$ sites of low affinity and Kd$_2$=41$\pm$6 nM and Bmax=0.35$\pm$0.26 pmol/mg protein for the GABA$_B$ sites of high affinity.

The existence of both sites was confirmed by experiments on calcium-sensitivity of binding. Specific [^{3}H]GABA binding obtained after muscimol blockade (GABA$_B$ sites), was enhanced when Ca^{2+} was added to the incubation medium, the effect reaching its maximum level at a Ca^{2+} concentration of 1 mM; however, [^{3}H]GABA binding obtained after blockade with (-)-baclofen (GABA$_A$ sites) was not significantly modified by Ca^{2+} ions until 1 mM (Castro et al., 1988).

FUNCTION AND MECHANISM OF ACTION OF GABA$_A$ RECEPTORS IN BOVINE CHROMAFFIN CELLS

Sangiah et al. (1974) reported for the first time that GABA caused release of CAs from isolated perfused bovine adrenals. Kitayama et al. (1984) found that GABA produced CA release from isolated perfused dog adrenals, this process being Ca^{2+}-dependent and independent of external Na$^+$ and Cl$^-$.

According to our results on bovine chromaffin cells (Castro et al., 1989), GABA stimulates basal CA release from these cells in a concentration-dependent manner. This release represents about 70 % of that obtained by the same doses of nicotine under similar experimental conditions. The fact that GABA stimulates basal CA release and that this stimulation can be mimicked by muscimol, a GABA$_A$ agonist, and partially blocked by bicuculline, a GABA$_A$ antagonist, indicates that this action is mediated in part by GABA$_A$ receptors. The EC$_{50}$ for the effect of GABA on basal CA secretion was 7.2$\pm$1.1 μM, which is well in line with GABA concentrations necessary to stimulate the GABA$_A$ receptor in brain preparations in vitro (Allan & Harris, 1986).

The CA release elicited by GABA was Ca^{2+}-dependent and was inhibited in the absence of Ca^{2+} in the extracellular medium and by blockers of the L-type of voltage-dependent calcium channels, such as nifedipine and Cd^{2+}, but not by blockers of the T-type voltage-dependent calcium channels, such as Ni^{2+} (Castro et al., 1989). This pattern of inhibition is in agreement with the known pathways for Ca^{2+} entry into chromaffin cells (Ladona et al., 1987). In addition, GABA and muscimol increased the intracelluar concentration of Ca^{2+} ([Ca^{2+}]i), as measured by fluorescent probe Fura-2AM. This increase was specifically blocked by the GABA$_A$ antagonist bicuculline and the calcium channel blocker nifedipine.

These results suggest that GABA evokes CA secretion in chromaffin cells by producing a depolarization due, maybe, to an increase in the membrane conductance to Cl$^-$, as occurs when the neurotransmitter binds to its specific site on the GABA receptor/Cl$^-$ channel complex in brain. Our results about Cl$^-$ fluxes measured by two independent methods, ^{36}Cl uptake and SPQ fluorescence probes showed that muscimol induces an increase in the efflux of Cl⁻ from cells in culture. This effect was blocked by picrotoxin, a specific blocker of the Cl⁻ channel associated with the $GABA_A$ receptor (González et al., 1992).

In addition to the GABA effect on the basal CA secretion, GABA, through its $GABA_A$ receptors, increases the CA release evoked by submaximal doses of nicotine but has an inhibitory effect on the nicotine-evoked CA secretion when nicotine is given in maximal doses (Castro et al., 1989). This is also true in the case of KCl-evoked CA secretion by chromaffin cells. Thus, at low concentration of KCl (between 5 and 15 mM) the effect of muscimol is excitatory but at higher concentrations the effect is inhibitory (González et al., 1992).

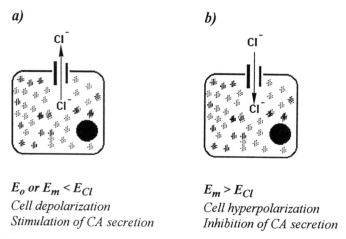

a)   *b)*

E_o or $E_m < E_{Cl}$ $E_m > E_{Cl}$
Cell depolarization   *Cell hyperpolarization*
Stimulation of CA secretion   *Inhibition of CA secretion*

Figure 2. (a) When GABA is bound to its $GABA_A$ receptor in chromaffin cells at resting membrane potential (E_o) or at a E_m below E_{Cl} (E_{Cl}= -51±2.1 mV) a Cl⁻-efflux is produced and, as consequence, a membrane depolarization and a stimulation of CA release occur. (b) By contrast, if chromaffin cells are depolarized, presenting a E_m above E_{Cl} (experimentaly achieved by stimulation with high KCl or nicotine concentrations), GABA causes a Cl⁻-infflux and, as a result, a membrane hyperpolarization and an inhibition of CA secretion.

These apparent contradictory effects could be explained by our results on the muscimol effect on membrane potential, measured by the fluorescent bisoxonol probe (Di-BAC 4(3)), and estimations of resting potential and equilibrium potential for Cl⁻ (González et al., 1992). At membrane potentials lower than the equilibrium potential for Cl⁻ (in our study -51±2.1 mV), muscimol induced a depolarization of chromaffin cells and, consequently, a stimulatory effect on CA secretion. Maximal stimulation was obtained when muscimol effect was assayed at resting conditions (in our study the resting membrane potential (E_o) was about -65 mV). However, when cells were stimulated by muscimol at membrane potentials greater than the equilibrium potential for Cl⁻ (obtained by high KCl or nicotine concentrations), the effect of the $GABA_A$ agonist was inhibitory (González et al., 1992) (Figure 2).

FUNCTION AND MECHANISM OF ACTION OF GABA$_B$ RECEPTORS IN BOVINE CHROMAFFIN CELLS

GABA$_B$ agonists modulate basal and nACh and GABA$_A$-receptor-stimulated CA secretion

The function of GABA$_B$ receptors in adrenal chromaffin cells has not been studied before. Our results show for the first time that the GABA$_B$ agonist baclofen and the most potent agonist available, 3-APPA increase basal CA secretion in a dose-dependent manner (EC$_{50}$=225±58 μM for 3-APPA and 151±35 μM for baclofen, respectively), this effect representing between 40-50 % of that obtained with nicotine at the same doses.

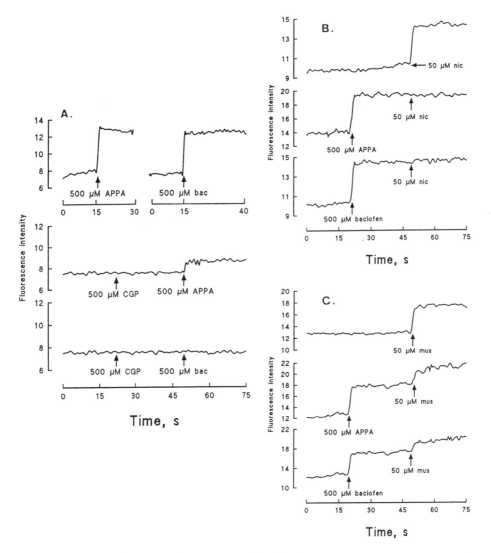

Figure 3. A.) Both GABA$_B$ agonists, 3-APPA and baclofen, increase membrane potential. This effect is blocked by pretreatment with the specific antagonist CGP 35348. The depolarization caused by GABA$_B$ agonists decreased or even blocked the nicotine (B) and GABA$_A$ agonist-induced depolarizations (C)

GABA$_B$ effect on CA secretion seems to be a true exocytotic and specific release. This is supported by the findings that:

(1) It was accompanied by a dose-dependent increase in [Ca^{2+}]i, with a EC$_{50}$=25$\pm$6 μM and 60$\pm$10 μM for 3-APPA and baclofen, repectively. Maximal increases in [Ca^{2+}]i are shown in Figure 5. The effect was almost completely blocked when CA secretion by these agonists was measured in a low Ca^{2+}-containing medium (200 nM).

(2) GABA$_B$ agonists depolarize the cells by increasing membrane potential in a specific form, since the effect was blocked by the specific antagonist CGP 35348 (Figure 3A).

(3) GABA$_B$ agonists were able to release other secretory granular components such as DBH to a similar extent as CAs, and

(4) Their effects are blocked by the specific antagonist CGP 35348, which at dose 0.5 mM inhibited 70 % to 100 % the agonist-evoked responses.

In addition to the effect of GABA$_B$ agonists on basal CA secretion, stimulation of chromaffin cells by these compounds also has an effect on the CA secretion evoked by nACh and GABA$_A$ agonists. When cells are preincubated with GABA$_B$ agonists before nicotine is added, these agonists strongly inhibit nicotine-evoked CA secretion in a dose-dependent manner. The IC$_{50}$ are of the same order of magnitude than those for activation of basal CA secretion (354$\pm$98 μM for baclofen and 240$\pm$59 μM for 3-APPA). This inhibitory effect was accompanied by a reduction of the increase of [Ca^{2+}]i (Oset-Gasque et al., 1993) and membrane potential induced by nicotine (Figure 3B), results which support the hypothesis of a true inhibitory modulation of nACh receptors by GABA$_B$ ones. Similar results were obtained for CA secretion evoked by GABA$_A$ agonists (IC$_{50}$=45$\pm$9 μM for baclofen and 21$\pm$7 μM for 3-APPA) as welll as for increases in [Ca^{2+}]i (Oset-Gasque et al., 1993) and in membrane potential (Figure 3C).

GABA$_B$ agonists increase cAMP levels

The molecular mechanism by which GABA$_B$ agonists modulate CA secretion in chromaffin cells seem to be mediated by the activation of the adenylate cyclase (AC). This conclusion is supported by the following experimental data:

(1) GABA$_B$ agonists increase intracellular levels of cyclic AMP (cAMP) in a dose-dependent fashion, effects which are well correlated with increases obtained in CA secretion (Figure 4).

(2) Several activators of AC including forskolin as well as cAMP analogues (pClp-cAMP) increase basal CA secretion in a dose-dependent manner, a direct correlation existing between both parameters, intracellular cAMP levels and CA secretion (Oset-Gasque et al., 1991)

(3) Pretreatment of cells with Pertussis toxin, which produces an inhibition of negative control of AC by ADP-ribosylation of Gi proteins , potentiates by 2 times cAMP levels and 4.5 times CA secretion induced by GABA$_B$ agonists (Oset-Gasque et al., 1993).

(4) Pretreatment of cells with Choleric toxin, which permanently activate AC, does not produce additional increases in either the cAMP levels or in the CA secretion induced by baclofen, indicating the activation of a common intracellular target, the Gs protein.

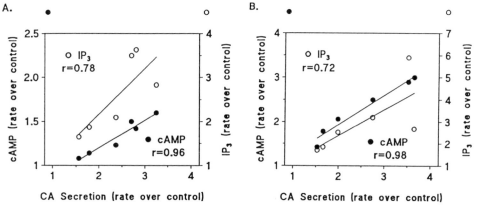

Figure 4. Correlation between the rates of CA secretion and cAMP or IP$_3$ levels induced by 3-APPA (A) or baclofen (B) in adrenal chromaffin cells. Data are means of 4 experiments performed in duplicate. A) (●) cAMP, y= 0.66x±0.60, r=0.98, P<0.001; (o) IP$_3$, y= 1.18x±0.004, r=0.72, ns. B) (●) cAMP, y= 0.31x±0.57, r=0.96, P<0.01; (o) IP$_3$, y= 1.04x±0.08, r=0.78, ns.

The inhibitory actions of GABA$_B$ agonists on CA secretion evoked by nicotinic and GABA$_A$ receptors could also be explained by this mechanism. This conclusion seems to be supported by both results which show that forskolin and pClp-cAMP have a dose-dependent inhibitory effect on CA secretion evoked by nicotine and muscimol (Oset-Gasque et al., 1991) and by others showing that cAMP or agonists that activate AC may produce an inactivation of nACh and GABA$_A$ receptor by phosphorylation (Alburquerque et al., 1986; Huganir et al., 1986; Harrison & Lambert, 1989; Leidenheimer et al., 1991). However, although GABA$_B$ agonists seem to mediate their effects on CA secretion by a mechanism implicating cAMP formation, we do not exclude the possibility of the existence of additional mechanisms for the action of these receptors (see below).

GABA$_B$ agonists mobilize calcium from intracellular stores and increase IP$_3$ levels in chromaffin cells

In order to determine the origin of [Ca^{2+}]i increased by GABA$_B$ agonists we measured the mobilization of calcium in response to GABA$_B$ agonist stimulation under different conditions, which are the absence of extracellular calcium or the blockade of different types of voltage-dependent calcium channels with specific antagonists. Results from Figure 5 show that the absence of calcium in the extracellular medium or its chelation by addition of EGTA reduces [Ca^{2+}]i increase by GABA$_B$ agonists in about a 70 %. The same result was obtained by specific blockade of L-type voltage-dependent calcium channels with nifedipine, while ω-conotoxin, a specific blocker of N-type calcium channels did not produce any inhibition of calcium entry. These results seem to indicate that the 70 % of [Ca^{2+}]i increase produced by GABA$_B$ agonists come from extracellular pools, but suggest that an additional mobilization of intracellular calcium pools could exist.

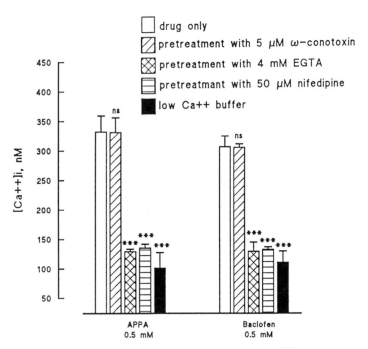

Figure 5. Effect of GABA$_B$ receptor agonists, 3-APPA and baclofen over [Ca^{2+}]i under different conditions as specified above. Data are means ±S.E.mean of 10 experiments. ns, non significant; ***, P<0.001.

So as to test the last possibility, we performed two types of experiments. First, we examined the intracellular calcium mobilization by GABA$_B$ agonists in an extracellular calcium-free medium previous depletion of caffeine and IP$_3$-sensitive stores. In these conditions, GABA$_B$ agonists were unable to produce any increase in [Ca^{2+}]i. Moreover, after caffeine pretreatment alone, 3-APPA was able to induce a little increase of [Ca^{2+}]i (Parramón et al., paper in preparation). Secondly, we examined whether GABA$_B$ agonists stimulate phospholopase C to produce IP$_3$. GABA$_B$ agonists cause a significant dose-dependent increase in IP$_3$ levels, reaching maximal effects at 100 μM concentration (Parramón et al., paper in preparation).

Taken together, results from intracellular calcium mobilization and IP$_3$ levels demonstrate that in addition to a very high extracellular calcium entry (70 %) mainly through L-type voltage-dependent calcium channel, there is an additional intracellular calcium mobilization, which could be implicated in the regulation of the GABA$_B$ evoked CA secretion (Figure 6). This contribution is minoritary and non-specific since in the absence of extracellular calcium although a little CA secretion exists, it is not dose-dependent. Moreover, the correlation between CA secretion in the presence of calcium and IP$_3$ levels is not statistically significant (Figure 4).

CONCLUDING REMARKS

The above results demonstrate that GABA, when binds to $GABA_A$ and $GABA_B$ receptors on the plasma membrane of chromaffin cells, can modulate both basal and evoked CA secretion, enhancing basal CA secretion and inhibiting secretagogue-evoked release.

The effect of $GABA_A$ modulation of CA secretion by chromaffin cells depends on the membrane potential, being depolarizing, excitatory, or hyperpolarizing, inhibitory, depending on the degree of change in the membrane potential produced by the stimulatory agent, below or above the equilibrium potential for Cl^- (Figure 2). These opposite actions could be mediated by an increase in the efflux or influx of Cl^- through Cl^- channel, respectively. The intermediate position of the E_{Cl} plays a crucial role in the dual effect of $GABA_A$ receptors on the control of the CA release on basal conditions. The opening of Cl^- channels depolarize the cells towards E_{Cl}, but if the opening coincides with a stimulus depolarizing the membrane above the E_{Cl}, this action will be dampened by the increased permeability to Cl^-, and a relative inhibition will be obtained.

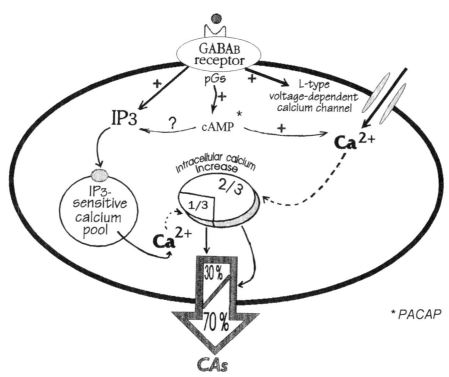

Figure 6. $GABA_B$ receptor activation produces an opening of L-type voltage-dependent calcium channels that is about two thirds the total $[Ca^{2+}]i$ induced by these agonists. This $[Ca^{2+}]i$ is the responsible of at least 70 % of $GABA_B$-evoked CA secretion. $GABA_B$ agonists produce also a calcium release from intracellular stores. Moreover, $GABA_B$ agonists produce an increase of cAMP levels. Since cAMP has similar effects over $[Ca^{2+}]i$ and subsequent CA secretion, we could hypothesize that cAMP is the main second messenger involved in $GABA_B$ receptor function in chromaffin cells. Similar effects on Ca^{2+}, IP_3 and CA secretion has been observed with PACAP (Pituitary Adenylate-Cyclase-Activating Polypeptide) in porcine adrenal medullary chromaffin cells (Isobe et al., 1993).

GABA$_B$ receptors mediate CA secretion by a completely different mechanism (Figure 6). GABA$_B$ receptor stimulation produces an increase of calcium entry from extracellular space mainly through the L-type of voltage-dependent calcium channels (70 %) but also mobilize calcium from intracellular stores (30 %). Moreover, GABA$_B$ stimulation presumibly activates AC and PLC producing cAMP and IP$_3$, respectively. Since (1) in the case of cAMP a direct correlation with CA secretion exists but the correlation between IP$_3$ and CA secretion is not significant, and (2) in the absence of extracellular calcium, although a little secretion of CA exists, it is not dose-dependent, we can hipothesize that both, calcium entry from extracellular medium and cAMP messenger system are two main signalling cascades which trigger GABA$_B$-stimulation of CA secretion. IP$_3$ production could be secundary to the calcium-dependent activation of PLC (Parramón et al., paper in preparation).

Moreover, cAMP can phosphorylate nACh and GABA$_A$ receptors, producing an inhibitory control of their actions on CA secretion. GABA$_A$ and GABA$_B$ receptors could interact each other in order to regulate CA secretion. In fact, our results seem to support a specific modulation of GABA$_A$ receptors by GABA$_B$ ones, since the IC$_{50}$ for the inhibition of GABA$_A$ receptor action by GABA$_B$ agonists was very much smaller than that for inhibition of nACh receptor action (Oset-Gasque et al., 1993).

In conclusion, GABA through both its GABA$_A$ and GABA$_B$ receptors, seems to play an important role in the modulation of chromaffin cell function anf therefore, GABA could be added to the long list of neurotransmitters and hormones implicated in the modulation of this complex function. The physiological significance of the results presented here is not known yet. It could be possible that ionotropic GABA$_A$ receptors, as well as nACh receptors, play a major role in the initiation of CA release *in vivo* while metabotropic receptors, including those for GABA$_B$, have a more important role in the modulation of the secretion evoked by ionotropic receptors.

ACKNOWLEDGMENTS

This work was supported by Research Grants CAM C168/91 from Comunidad Autónoma de Madrid and FISS 92/0241 from Fondo de Investigaciones Sanitarias de la Seguridad Social of Madrid (Ministry of Health, Spain).

REFERENCES

Alburquerque,E.X.,Deshipande,S.S., Aravaca,Y., Alkondon,M. and Daly, J.M., 1986, A possible involvement of cyclic AMP in the expression of desensitization of the nicotinic acetylcholine receptor. *FEBS Lett.*199: 113-120.

Allan,A.M. and Harris,R.A. , 1986, γ-Aminobutyric acid agonist and antagonists alter chloride flux across brain membranes. *Mol. Pharmacol* 29: 497-505.

Burgoyne, R.D., 1991, Control of exocytosis in adrenal chromaffin cells. *Biochim. Biophys. Acta* 1071: 174-202.

Castro,E., Oset-Gasque,M.J., Cañadas,S., Giménez,A. and González,M.P., 1988, GABA$_A$ and GABA$_B$ sites in bovine adrenal medulla membranes. *J. Neurosci. Res.*20: 241-245.

Castro,E., Oset-Gasque,M.J. and González,M.P.,1989, GABA$_A$ and GABA$_B$ receptors are functionally active in regulation of catecholamine secretion by bovine chromaffin cells. *J. Neurosci. Res.* 23: 290-296.

Douglas,W.W. and Rubin,R.P. , 1968, The role of calcium in the secretory response of the adrenal medulla to acetylcholine . *J Physiol* 331:577-597.

Fernández- Ramil,J.M., Sánchez-Prieto,J. and González,M.P., 1982, Presence of glutamate decarboxylase in bovine adrenal medullary cells. *Rev. Esp.Fisiol.* 38:91-96.

Fernández- Ramil,J.M., Sánchez-Prieto,J., Cañadas,S. and González,M.P., 1983, GABA-T in bovine adrenal medulla cells. Kinetic properties and comparison with GABA-T from other tissues. *Rev. Esp.Fisiol.* 39: 299-304.

González, M.P., Oset-Gasque, M.J., Castro, E., Bugeda, J., Arce, C. and Parramón,M., 1992, Mechanisms through which GABA$_A$ receptor modulate catecholamine secretion from bovine chromaffin cells. *Neuroscience,* 47: 487-494.

Harrison, N.L. and Lambert, N.A., 1989, Modification of GABA$_A$ receptor function by an analog of cyclic AMP. *Neurosci. Lett.,* 105: 137-142.

Huganir , R.L., Delcour , A.H., Greengard , P. and Hess, G.P. , 1986, Phosphorylation of the nicotinic acetylcholine receptor regulates its rate of desensitization. *Nature* 321: 774-776.

Isobe, K., Nakai, T. and Takuwa, Y., 1993, Ca^{2+}-dependent stimulatory effect of pituitary adenylate cyclase-activating polypeptide on catecholamine secretion from cultured porcine adrenal medullary chromaffin cells. *Endocrinology* 132:1757-1765.

Kataoka, Y. , Gutman Y. , Guidotti, A., Panula, P., Wrobleski, J. Cosenza-Murphy, D. , Wu, J. and Costa, E., 1984, Intrinsic GABAergic system of adrenal chromaffin cells. *Proc. Natl. Acad. Sci. U.S.A.* 81 : 3218-3222.

Kitayama, S., Morita, K., Dohi, T. and Tsujimoto, A., 1984 , The nature of the stimulatory action of γ-aminobutyric acid in the isolated perfused dog adrenals. *Naunyn Schmiedebergs Arch. Pharmacol.* 326: 106-110.

Ladona, M.G. , Aunis,D. , Gandia, L. and García, A., 1987, Dihydropyridine modulation of the chromaffin cell secretory response. *J. Neurochem.* 48: 483-490.

Leidenheimer, N.J., Browning, M.D. and Harris, R.A., 1991, GABA$_A$ receptor phosphorylation : multiple sites , actions and artifacts. *Trends Pharmacol. Sci.* 12: 84-87.

Martínez, P.M., Giménez, A., Castro, E., Oset-Gasque, M.J., Cañadas, S. and González, M.P., 1987, GABA binding to adrenal medulla membranes is sensitive to baclofen. *Comp. Biochem. Physiol.* 88C: 155-157.

Oset-Gasque,M.J. and Aunis, D., 1989, Gamma-aminobutyric acid uptake in bovine chromaffin cells in culture. *Biochem. Pharmacol.* 38: 2227-2232.

Oset Gasque, M.J., Castro, E. , and González,M.P., 1990, Mechanisms of [^{3}H] γ–aminobutyric acid release by chromaffin cells in primary culture. *J. Neurosci. Res.* 26: 181-187.

Oset-Gasque, M.J., Parramón, M. and González, M.P., 1991, XII Reunión del Grupo Español de la Célula Cromafin. Salamanca, Spain.

Oset-Gasque, M.J., Parramón, M . and González, M.P. , 1993, GABA$_B$ receptors modulate catecholamine secretion in chromaffin cells by a mechanism involving cyclic AMP formation. *Br. J. Pharmacol.* 110: 1586-1592.

Sangiah, S., Borowitz, J.L. and Yim, G.K. W, 1974, Action of GABA, picrotoxin and bicuculline on adrenal medulla. *Eur. J. Pharmac.* 27: 130-135.

Tanaka, C. and Taniyama, K., 1992, The role of GABA in the Peripheral Nervous System. In "*GABA outside the Central Nervous System*". Erdö, S.L., ed. Springer Verlag pp 3-17.

THE 5-HT$_{1D}$ RECEPTOR SUBTYPE AND MIGRAINE HEADACHE

Michael A. Moskowitz and Christian Waeber

Stroke Research Laboratory
Neurosurgery and Neurology
Massachusetts General Hospital
Boston, MA 02114

Serotonin (5-hydroxytryptamine, 5-HT) is involved in a wide variety of physiological functions and pathological states, including migraine. Fourteen subtypes of 5-HT receptors have been identified as of this writing (see Humphrey et al., 1993); while several of them might be the target of prophylactic anti-migraine drugs, the 5-HT$_{1D}$ subtype appears most relevant among the known receptor subtypes, and this is based on the clinical efficacy of the receptor agonists sumatriptan and ergot alkaloids in aborting migraine headache (Deliganis and Peroutka, 1991). Accordingly, this chapter will briefly review current knowledge of the 5-HT$_{1D}$ receptor subtype with particular emphasis on its ligand binding properties, second messengers, molecular biology and mechanism of action.

5-HT$_{1D}$ RECEPTORS

This subtype has been first characterized by Heuring and Peroutka (1987) using radioligand binding studies on bovine brain homogenates. Subsequent findings have revealed that the binding sites labelled under the conditions of this study might in fact be accounted for by up to 5 different subtypes (5-HT$_{1D\alpha}$, 5-HT$_{1D\beta}$, 5-HT$_{1E}$, 5-HT$_{1F}$ and 5-HT$_5$; see Beer et al, 1993 and Plassat et al, 1992). This section will thus focus on the characteristics of the "5-HT$_{1D}$-like" receptors, in order to define the target(s) of sumatriptan and ergot alkaloids.

The pharmacological profile of 5-HT$_{1D}$ receptors was established in binding studies using [^{3}H]5-HT as a radioligand, in the presence of blockers of the 5-HT$_{1A}$ and 5-HT$_{1C}$ subtypes (Heuring and Peroutka, 1987). 5-HT$_{1D}$ binding sites possess nanomolar affinities for 5-carboxamidotryptamine (5-CT), 5-HT, 5-methoxytryptamine and metergoline. Their affinity for ergotamine, dihydroergotamine, sumatriptan, methiothepin and RU24969 ranges from 15 to 50 nM. However, their affinity for ß-blockers is low. Sites with this profile have been detected in human, monkey, cat, dog, guinea-pig, rabbit, pig, calf and pigeon, whereas the corresponding sites in mouse, rat and opossum also show a nanomolar affinity for ß-blockers and RU24969 (the latter sites are known as 5-HT$_{1B}$ receptors) (see Zifa and Fillion, 1992).

Cell Signal Transduction, Second Messengers, and Protein Phosphorylation in Health and Disease
Edited by A.M. Municio and M.T. Miras-Portugal, Plenum Press, New York, 1994

The homogeneity of these labelled sites however can be questioned because the 5-HT_{1D} binding sites were defined by exclusion criteria and the ligands exhibit poor selectivity. Indeed, the presence of a low affinity component was detected by analysis of 5-CT and sumatriptan competition curves, indicating the existence of yet another receptor subtype, termed $5-HT_{1E}$. Recently however, O-carboxymethylglycyl-^{125}I-tyrosinamide ([^{125}I]GTI) has been reported to label a homogeneous population of $5-HT_{1D}$ sites in various species (Bruinvels et al., 1992). Moreover the availability of [^{3}H]5-CT and [^{3}H]sumatriptan should help to define the binding characteristics of the "5-HT_{1D}-like" receptors.

With suitable pharmacological tools lacking, molecular biological techniques have helped to establish with greater precision the identity of the $5-HT_{1D}$ receptor family and their localization within brain. For example, the cloning of the gene encoding each subtype and its selective transfection in cell lines free of other 5-HT receptor subtypes has allowed the characterization of the drug binding profile and coupling properties of pure receptor populations. Thus in humans, the receptors previously characterized as 5-HT_{1D} (i.e. with high affinity for 5-CT and sumatriptan) correspond to two pharmacologically indistinguishable proteins encoded by the $5-HT_{1D\alpha}$ and $5-HT_{1D\beta}$ genes (see Hartig et al, 1992). Rats and mice also possess the homologues of these receptors; while their $5-HT_{1D\alpha}$ genes encodes a receptor with pharmacological characteristics similar to those of the human $5-HT_{1D}$ receptors, a single mutation in their $5-HT_{1D\beta}$ gene seems to be responsible for the ß-blocker sensitivity of the encoded receptor, which has been named $5-HT_{1B}$. It should be mentioned that in rat as in human brain, most of the 5-$HT_{1B/1D}$ receptor sites are accounted for by the product of the $5-HT_{1D\beta}$ gene, the level of expression of the $5-HT_{1D\alpha}$ gene being very low throughout the brain (Bruinvels et al., submitted); this might explain the scarcity of drug binding or functional data displaying a $5-HT_{1D}$ profile (i.e. with low affinity for ß-blockers) in the rat.

As mentioned above, a subpopulation of $5-HT_{1D}$ binding sites with a low affinity for 5-CT has been demonstrated in different brain regions of all species investigated, including rats and pigeons. Two corresponding human genes have been cloned: one showing a low affinity for sumatriptan (termed $5-HT_{1E}$, previously $5-HT_{1E\alpha}$), the other showing comparable affinities for 5-HT and sumatriptan (termed $5-HT_{1F}$, previously 5-$HT_{1E\beta}$). The latter property makes the $5-HT_{1F}$ subtype an additional candidate to mediate some of the therapeutic actions of sumatriptan. All members of the 5-HT_{1D}-like family (5-$HT_{1B/1D//1E/1F}$) show about 50-60% sequence homology and are all negatively coupled to adenylate cyclase in transfected cells (see below).

Interestingly, two 5-HT receptor genes with a very low sequence homology to other $5-HT_1$ receptors have been cloned in rat and mouse (termed $5-HT_{5A}$ and $5-HT_{5B}$); the drug binding profiles of their encoded proteins in transfected cells is very similar to that of $5-HT_{1D}$ receptors (Plassat et al, 1992). Although their coupling mechanism is unknown and they have not been detected in other species, one has to consider that they may account, at least partly and in rodents, for some $5-HT_{1D}$ binding activity and functional correlates.

The 5-HT_{1D}-like receptors belong to the superfamily of receptors coupled to GTP-binding proteins (G proteins), as demonstrated for all the other 5-HT receptors except 5-HT_3 (see Humphrey et al, 1993). When expressed in transfected cells, $5-HT_{1D\alpha}$, 5-$HT_{1D\beta}$, $5-HT_{1E}$ and $5-HT_{1F}$ receptors are negatively coupled to adenylate cyclase, although other second messenger systems can also be modulated by these receptors. Thus, the human $5-HT_{1Dc}$ receptor and its dog homolog (previously termed RDC4) can both inhibit or activate adenylate cyclase in transfected cells (Van Sande et al., 1993).

Human $5-HT_{1D\alpha}$ and $5-HT_{1D\beta}$ receptors expressed in murine fibroblasts are coupled to both increases in intracellular calcium concentration and cyclase inhibiting activity (Zgombick et al., 1993); interestingly, this effect is larger for the $5-HT_{1D\beta}$

subtype, which has been shown to be the subtype present on human and bovine cerebral arteries (Hamel et al, 1993). The modulation of intracellular calcium concentration might underlie the constricting action of sumatriptan in the cerebral vasculature. *In vivo*, both the rat $5-HT_{1B}$ and the human or guinea-pig $5-HT_{1D}$ receptors inhibit the forskolin-stimulated adenylate cyclase activity in homogenates prepared from substantia nigra (see Zifa and Fillion, 1992). In the latter species, both species homology arguments and the low level of expression of the $5-HT_{1D\alpha}$ gene detected in the brain suggest that adenylate cyclase inhibition in the substantia nigra is accounted for by $5-HT_{1D\beta}$ receptors. The functional coupling and the role of the $5-HT_{1D\alpha}$, $5-HT_{1E}$ and $5-HT_{1F}$ receptors remains to be determined, as well as the *in vivo* significance of the additional second messenger systems described in transfected cells.

Since the characterization of $5-HT_{1D}$ binding sites, several functional correlates have been described which possess a $5-HT_{1D}$ pharmacological profile. As already mentioned, the lack of suitable compounds which discriminate between the $5-HT_{1D\alpha}$ and $5-HT_{1D\beta}$ subtypes does not permit assigning these correlates to one or the other subtype. One might use the pharmacological differences displayed by rat $5-HT_{1D}$ and $5-HT_{1B}$ (i.e. $5-HT_{1D\beta}$) receptors to study their differential functions and extrapolate the results to humans.

The presynaptic autoreceptor might provide a good example, This receptor controls 5-HT release and displays a $5-HT_{1D}$ pharmacological profile in human, pig, rabbit and guinea pig brain cortex (see Zifa and Fillion, 1992). In the rat cortex, both $5-HT_{1B}$ and $5-HT_{1D}$ receptors seem to function as presynaptic autoreceptors (Limberger et al, 1991). Several autoradiographic studies indicate that $5-HT_{1B/1D}$ receptors might be located on the terminals of raphé neurons but also on the terminals of non serotonergic neurons. Thus, the destruction of striatal neurons or the ablation of one eye results in the loss of $5-HT_{1D}$ binding sites in the ipsilateral projection region (substantia nigra and superior colliculus, respectively) (see Zifa and Fillion, 1992). Additionally, *in situ* hybridization studies have detected messenger RNA's for $5-HT_{1D\alpha}$ and $5-HT_{1D\beta}$ receptors in regions where no $5-HT_{1B/1D}$ binding sites can be detected, further indicating a localization of the receptor protein on the terminals of projection neurons (Bruinvels et al., submitted). Indeed, functional studies have shown that $5-HT_{1B}$ receptors inhibit acetylcholine release in rat hippocampus, while $5-HT_{1D}$ receptors have a similar role in the guinea pig hippocampus. $5-HT_{1B}$ (in rat vena cava) and $5-HT_{1D}$ (in human saphenous vein) inhibit noradrenaline release from sympathetic nerve endings (see Zifa and Fillion, 1991). The findings that $5-HT_{1B/1D}$ receptors control presynaptic transmitter release are interesting in view of the hypothesis that $5-HT_{1D}$-like receptors located on trigeminovascular terminals block neuropeptide release within the dura mater, explaining, at least partly, the abortive effect of sumatriptan and ergot alkaloids in migraine attacks (see MECHANISM OF ACTION).

Whereas most neuronal $5-HT_{1D}$-like receptors seem to be presynaptically located and modulate the release of neurotransmitters, postjunctional $5-HT_1$ receptors with a pharmacological profile resembling that of $5-HT_{1D}$ receptors have been described in several types of blood vessels. For example, these receptors have been shown to mediate endothelium-dependent relaxation to 5-HT in pig coronary artery (Schoeffter and Hoyer, 1990) and guinea-pig jugular vein (Gupta, 1992). In cerebral and pial arteries, $5-HT_{1D\beta}$ receptors are located on the vascular smooth muscle cells and mediate contraction, an effect which might contribute to the therapeutic effect of sumatriptan (Hamel et al, 1993). In contrast to the central nervous system, wherein the pharmacological profile of the functional effects of $5-HT_{1D}$ receptors correlates well with the pharmacological data obtained in transfected cells, the pharmacological characteristics of most of the vascular effects of $5-HT_{1D}$-like receptors only displays an imperfect fit with the available data. This might indicate that other receptor subtypes still have to be cloned or that differences in the cellular environment of these receptors might alter their pharmacological properties. In particular, the identity of the antimigraine site of action of sumatriptan and of the 5-HT$_{1D}$-like receptors inhibiting neuropeptide release from trigeminovascular nerve endings and blocking neurogenic plasma protein extravasation still remains to be established.

MECHANISM OF ACTION:

Sumatriptan, developed as a selective constrictor of intracranial blood vessels, blocks neurally-mediated sensory neurotransmitter release and nociceptive neurotransmission within the meninges (see Moskowitz, 1992). As noted above, the drug binds with high affinity to 5-HT$_{1B/1D}$ receptor subtype, but is inactive at receptor sites for dopamine, norepinephrine, glutamate, GABA and other 5-HT receptor types. Postjunctional receptors cause constriction within intracranial and in a few extracranial vessels. Sumatriptan crosses the blood brain barrier with great difficulty, hence the brain *per se* is not the most relevant site for drug activity. Extracranial cephalic vessels, another potential drug target, do not possess 5-HT$_1$-like receptors and therefore, are not featured as prominently in the etiology of pain as the dura, pia mater and arachnoid, at least in sumatriptan-responding patients.

Substance P, neurokinin A, and calcitonin gene-related peptide are contained within the axons of small caliber afferents projecting from the trigeminal ganglion to innervate meningeal blood vessels. Trigeminovascular fibers transmit nociceptive information and exhibit electrophysiological properties consistent with those of A-delta and C-fibers. Triggering mechanisms for migraine pain remain elusive but chemical events in blood, vessel wall and especially brain (see below) remain potential sources for the generation of nociceptive or sensitizing molecules (e.g., potassium, hydrogen ion, bradykinin, histamine, prostaglandins, leukotrienes, or serotonin itself). Following depolarization (or calcium-dependent local generator mechanisms), trigeminovascular axons release neuropeptides into the vessel wall whereupon they cause a sterile inflammatory response. Plasma protein extravasation, mast cell secretion and degranulation, platelet aggregation and adherence, and endothelial activation, as evidenced by numerous endothelial vesicles, microvilli and vacuoles within postcapillary venules of dural vessels. Extravasation, and the above noted tissue changes as well, continue to evolve over minutes to hours, despite brief (e.g., $\leq$ 5 minutes) periods of nerve stimulation. Inflammation "shifts" the stimulus-response of nociceptive fibers and by so doing, sensitizes them to respond to otherwise innocuous stimuli (e.g., vessel pulsations, changes in venous pressure). The formulation, although unproven in migraine, is consistent with both the clinical picture, and with the finding that plasma levels of calcitonin gene-related peptide become elevated within jugular venous blood in patients during an attack.

Agents that either destroy sensory fibers (neonatal capsaicin treatment), block action potentials (xylocaine), inhibit neuropeptide release (5-HT$_1$-like agonists), or degrade released neuropeptides (endopeptidase 24.11) inhibit the meningeal neuroinflammatory response. Non-steroidal antiinflammatory drugs such as aspirin, indomethacin and ketorolac, particularly useful in migraine, block neurogenic inflammation as well. The dosages required to block both the inflammatory response in animals and vascular head pain in man are similar. One may infer that sumatriptan and the ergot alkaloids do not block the etiology of migraine *per se*, but inhibit selectively the response of a final common pathway. If true, it follows that a favorable drug response does not by itself establish the diagnosis of migraine.

The importance of prejunctional mechanisms to the site of drug action was recently documented using the tools of molecular biology. Pharmacological experiments were also consistent with this finding. For example, sumatriptan and the ergot alkaloids blocked the increases in neuropeptide levels within sagittal sinus blood (presumably reflecting neuropeptide release) during electrical stimulation but did not block the extravasation induced by substance P. Using reverse transcriptase-polymerase chain reaction and specific probes to 5-HT_{1Dc} and $5\text{-HT}_{1D\beta}$ gene sequences, the 5-HT_{1Dc} but not $5\text{-HT}_{1D\beta}$ gene was selectively amplified from human trigeminal ganglion. However, the $5\text{-HT}_{1D\alpha}$ (or 5-HT_{1D}) mRNA was not expressed within vascular smooth muscle of pial arteries. Hence, the possibility exists that drugs can be developed which activate prejunctional mechanisms without constricting pial or coronary arteries. Recent laboratory data support the contention that sumatriptan and the ergot alkaloids may alleviate pain in the presence of meningeal irritation caused by a variety of other etiologies such as viral, bacterial meningitis, head injury or subarachnoid hemorrhage. Nozaki et al explored this question using immunohistochemical techniques by examining the expression of c-fos (or related antigens) in brain stem following the instillation of autologous blood or another noxious chemical, carrageenin, into the subarachnoid space of experimental animals. C-fos immunoreactivity was induced within cells in lamina I, II_o of trigeminal nucleus caudalis (TNC; a region analogous to the dorsal horn of spinal cord, and involved in the processing of nociceptive information). The number of expressing cells related to the amount of injected blood. The response was capsaicin-sensitive and also attenuated by surgical transection of meningeal afferents. Sumatriptan, dihydroergotamine and the analgesic morphine also attenuated cell labeling, however, sumatriptan did not decrease the response induced when the noxious chemical formalin was applied to the nasal mucosa. Again, 5-HT_1-like mechanisms have not been identified within extracranial cephalic vessels.

Recurrent spreading depression also increases c-fos labeling within the ipsilateral superficial lamina of TNC. [Spreading depression is in fact, not a depression but a contiguously spreading excitatory wave of depolarization moving at a rate of 2-6 mm/minute, and followed by inhibition (i.e., depression).] Spreading depression was elicited in the c-fos experiments by microinjections of KCl into posterior parietal cortex. SD-induced activation of TNC suggests that endogenous neurophysiological events in neocortex are capable of activating trigeminovascular fibers innervating meninges and penetrating cortical blood vessels. We suggest the following formulation: neurophysiologically-driven ionic and metabolic mechanisms in brain can promote the release of nociceptive substances from neocortex into the interstitial space. Within the perivascular space, released substances activate and sensitize trigeminovascular fibers which surround pial vessels supplying and draining neocortex within Virchow-Robin spaces. As a result, impulses are conveyed to TNC and then to more rostral centers. Pain may ensue.

Importantly, the c-fos response induced by SD was blocked by sumatriptan as well as by meningeal deafferentation; the occurrence of spreading depression itself however was not blocked by these treatments. The site of sumatriptan's effect is again most likely the primary afferent fiber inasmuch as spreading depression is not known to alter the properties of the blood brain barrier within brain stem. The human neurophysiological counterpart to spreading depression in experimental animals is unknown at the present time, although, some investigators have postulated that spreading depression itself underlies the focal neurological findings during the aura stage of migraine headaches.

Based on apparent differences that exist between the 5-HT_1-like trigeminovascular receptor and the smooth muscle receptor, we are encouraged that future drug discovery programs can target drugs for development which bind selectively and with high affinity to pre- versus postjunctional recognition sites.

REFERENCES

Beer, M.S., Middlemiss, D.N. and McAllister, G., 1993, 5-HT1-like receptor: six down and still counting, *Trends Pharmacol. Sci.*, 14:228.

Bruinvels, A.T., Lery, H., Nozulak, J., Palacios, J.M. and Hoyer, D., 1992, 5-HT$_{1D}$ binding sites in various species: similar pharmacological profile in calf, guinea-pig, dog, monkey and human brain membranes, *Naunyn-Schmiedeberg's Arch. Pharmacol.*, 346:243.

Deliganis, A.V. and Peroutka, S.J., 1991, 5-Hydroxytryptamine$_{1D}$ receptor agonism predicts antimigraine efficacy, *Headache*, 31:228.

Gupta, P., 1992, An endothelial 5-HT receptor that mediates relaxation in guinea-pig isolated jugular vein resembles the 5-HT$_{1D}$ subtype, *Br. J. Pharmacol.*, 106:703.

Hamel, E., Fan, E., Linville, D., Ting, V., Villemure, J.-G. and Chia, L.-S., 1993, Expression of mRNA for the serotonin 5-hydroxytryptamine$_{1D\beta}$ receptor subtype in human and bovine cerebral arteries, *Mol. Pharmacol.*, 44:242.

Hartig, P.R., Branchek, T.A. and Weinshank, R.L., 1992, A subfamily of 5-HT$_{1D}$ receptor genes, *Trends Pharmacol. Sci.*, 13:152.

Heuring, R.E. and Peroutka, S.J., 1987, Characterization of a novel ^3H-5-hydroxytryptamine binding site subtype in bovine brain membranes, *J. Neurosci.*, 7:894.

Humphrey, P.P.A., Hartig, P. and Hoyer, D., 1993, A proposed new nomenclature for 5-HT receptors, *Trends Pharmacol. Sci*, 14:233.

Limberger, N., Deicher, R., Starke, K., 1991, Species differences in presynaptic serotonin autoreceptors: mainly 5-HT$_{1B}$ but possibly in addition 5-HT$_{1D}$ in the rat, 5-HT$_{1D}$ in the rabbit and guinea-pig brain cortex, *Naunyn-Schmiedeberg's Arch. Pharmacol.*, 343:353.

Moskowitz, M.A., 1992, Neurogenic versus vascular mechanisms of sumatriptan and ergot alkaloids in migraine, *Trends. Pharmacol. Sci.*, 13:307.

Plassat, J.-L., Boschert, U., Amlaiki, N. and Hen, R., 1992, The mouse 5-HT$_5$ receptor reveals a remarkable heterogeneity within the 5-HT$_{1D}$ receptor family, *EMBO. J.*, 11:4779.

Schoeffter, P. and Hoyer, D., 5-Hydroxytryptamine (5-HT)-induced endothelium-dependent relaxation of pig coronary arteries is mediated by 5-HT receptors similar to the 5-HT$_{1D}$ receptor subtype, *J. Pharmacol. Exp. Ther.*, 252:387.

Van Sande, J., Allgeier, A., Massart, C., Czernolofski, A., Vassart, G., Dumont, J.E. and Maenhaut, C., 1993, The human and dog 5-HT$_{1D}$ receptors can both activate and inhibit adenylate cyclase in transfected cells, *Eur. J. Pharmacol.*, 247:177.

Zifa, E. and Fillion, G., 1992, 5-hydroxytryptamine receptors, *Pharmacol. Rev.*, 44:401.

Zgombick, J.M., Borden, L.A., Cochran, T.L., Kucharewicz, S.A., Weinshank, R.L. and Branchek, T.A., 1993, Dual coupling of cloned human 5-hydroxytryptamine$_{1D\alpha}$ and 5-hydroxytryptamine$_{1D\beta}$ receptors in stably transfected fibroblasts: inhibition of adenylate cyclase and elevation of intracellular calcium concentrations via pertussis toxin-sensitive G protein(s), *Mol. Pharmacol.*, 44:575.

REGULATION OF NEUROTRANSMITTER TRANSPORTERS

Cecilio Giménez, Francisco Zafra and Carmen Aragón

Centro de Biología Molecular "Severo Ochoa"
Facultad de Ciencias
Universidad Autónoma de Madrid-C.S.I.C.
28049-Madrid
Spain

INTRODUCTION

When an action potential arrives at the nerve ending, it triggers the release of a chemical transmitter from the terminal by opening calciun ion channels. The transmitter diffuses across the synaptic cleft to the postsynaptic membrane producing a change in ion permeability in the membrane and so a change in the postsynaptic membrane potential.

The overall process of chemical synaptic transmission can be divided in different stages: 1) release of transmitter into the synaptic cleft; 2) recognition and binding to a specific receptor; and 3) inactivation of the transmitter (Fig. 1).

With the exception of acetylcholine which is inactivated by enzymatic degradation, it has now become clear that termination of the signal occurs in most cases through sodium-coupled reuptake of the neurotransmitters into presynaptic neurons or the surrounding glial cells. Even in the case of acetylcholine mediated transmission which is hydrolyzed into acetate and choline by acetylcholine estearase, the choline moiety is removed by sodium-dependent reuptake.The reuptake is an active process which assures both constant and high levels of neurotransmitters in the neuron and low concentration in the cleft. The energy for the reuptake process comes from a sodium gradient generated by Na^+/K^+-ATPase acting in the plasma membrane of presynaptic and glial cells (for a review see Kanner and Shuldiner, 1987).

Cell Signal Transduction, Second Messengers, and Protein Phosphorylation in Health and Disease
Edited by A.M. Municio and M.T. Miras-Portugal, Plenum Press, New York, 1994

Amino acid neurotransmitters are involved in fast transmission processes where γ-aminobutyric acid (GABA), glutamate and glycine probably represent the principal fast chemical signallign agents between neurons and between neurons and glial cells in the central nervous system. During the past years, transport system for amino acid

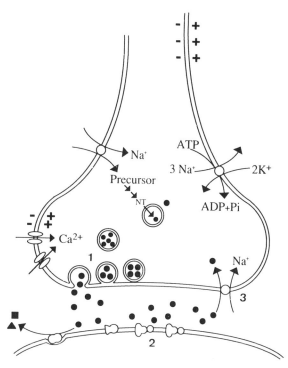

Figure 1. An overview of a chemical synapse. After the arrival of an action potential to the nerve terminal triggering Ca^{2+} entry, release of the transmitter occurs (1); recognition and binding to a specific receptor (2); and (3) inactivation of the transmitter.

neurotransmitters have been studied in detail by using different preparations as intact cells, synaptosomes or plasma membrane vesicles, and there is now a large body of experimental data on the ionic dependence, specificity, mechanistics and pharmacology of the transporters (see Amara and Kuhar, 1993). Most recently, GABA, glutamate and glycine transporters have been purified (Danbolt et al., 1990; López-Corcuera et al, 1991; Radian et al., 1986), and cDNA clones encoding transporters for GABA, catecholamines, serotonine and glycine have been isolated and a significant amino acid identity between these proteins has been found (Guastella et al., 1990; Hoffman et al., 1991; Kanai and Hediger, 1992; Pacholczyk et al, 1991; Shimana et al., 1991; Smith et al. 1992). Most of those transporters belong to a single family of membrane proteins that are homologous to each other in their amino acid sequences and share a similar structure that includes twelve transmembrane segments (Nelson, 1993). However, a second family of transporters devoted to glutamic acid uptake, sharing some structural features with bacterial transporters, has recently been identified (Pines et al., 1992).

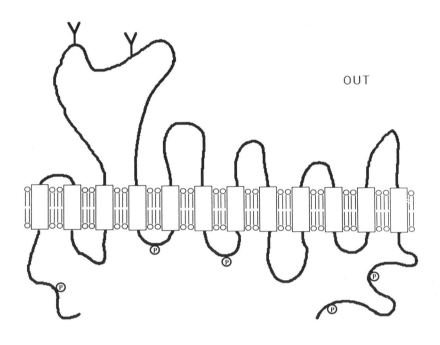

Figure 2. A structural model for a family of eukaryotic neurotransmitter transporters showing twelve putative membrane-spanning domains. Also shown are potential phosphorylation sites for different protein kinases and N-glycosilation sites.

In the past few years, our group has demonstrated a selective short-term regulation of glutamate and GABA transport by PKC-activation in glial cells through a mechanism independent of protein synthesis (Casado et al., 1991; Gomeza, et al., 1991). In these cases, whereas treatment of astrocytes with either phorbol esters or another agents acting through PKC, resulted in an up- and down-regulation of glutamate and GABA transporters respectively, they failed to modulate the same activities in neurons. Apart of the role of PKC on the GABA and glutamate transport, a similar PKC modulatory effect has been shown for the serotonin transporter function in endothelial cells (Myers et al, 1989). These data, together with the fact that all transporters so far cloned presented a variety of serine and threonine residues that could be substrates for phosphorylation, support the idea that these proteins can be very actively modulated by different kinases (Fig. 2). Recently we have demonstrated that a glial glutamate transporter can be phosphorylated both in vivo and in vitro and that this phosphorylation influence the transport of glutamate activity (Casado et al., 1993).

Results demonstrating that neurotransmitter transporters are regulated by second messengers are important because there is currently very little information available on the regulatory aspects of these proteins.

REGULATION OF GLUTAMATE AND GABA TRANSPORTERS THROUGH PKC ACTIVATION

The regulation of glutamate and GABA transporters by phorbol esters, calcium and phospholipase C has been studied in primary rat neuronal and glial cell cultures from rat brain cortex. Our neuronal and glial cell cultures were shown to be > 95% homogenous in each cell type by using immunofluorescent staining of specific neuronal (anti-Mß6-tubulin) and glial cell (anti-glial fibrillary acidic protein) markers (Chiu and Goldmam, 1984; Sullivan and Cleveland, 1986), and both types of cells were found to have active PKC (38.0 ± 0.2 and 13.3 ± 0.09 fmol of Pi/mg of protein for neurons and glial cells respectively).

We found that both neurons and glial cells from rat brain cortex actively accumulated glutamate and GABA. When glial cell cultures were incubated in the presence of the phorbol ester TPA, a dose and time activation of the glutamate transport was observed. On the contrary, the same treatment resulted in a significant inhibition of the transport of GABA. In both cases, neurons did not respond to the presence of phorbol esters. This modulation was a short-term effect, reaching a maximun after 30 min incubation. The action of TPA was not affected in any case in the presence of an inhibitor of protein synthesis as cycloheximide, thus suggesting that TPA actions were mediated by phosphorylation rather than by modification of the protein synthesis.

The possibility that TPA alters Na^+ gradients by a direct ionophoretic action was investigated by substrate influx studies in the presence or the absence of amiloride. Amiloride did not modified under any conditions GABA or glutamate influx in glial cells. So, the possibility that the observed TPA effect might be exerted through activation of sodium-proton exchange, a known effect of PKC activators in many cell classes, is ruled out.

In the case of glutamate, the main effect of TPA on the kinetic constants of glutamate transport is an increase of Vmax of the transport process whereas it has no significant effect on the Km for the substrate. According to enzyme kinetics, Vmax is an indicator of total enzyme activity; thus, in the case of uptake kinetics, an increase in the Vmax of the transport system could be explained either by an activation of preexisting transport systems or by an increase in the number of transport systems in the plasma membrane of the glial cells due to a stimulation of protein synthesis. Experiments performed in the presence of cycloheximide ruled out this possibility. Thus, TPA may enhance the turnover rate of the solute-loaded carrier sites or enhance the number of carrier sites available at the outer surface of the membrane.

The activatory or inhibitory effect observed on the glutamate and GABA transport respectively can be also achieved by OAG, a permeant DAG analoge which has been reported to stimulate PKC in several cells types (Nishizuka, 1984). Moreover, increased levels of endogenous DAG achieved by inhibition of DAG kinase with R59022 mimic the action observed with TPA. On the other hand, the effects exerted by TPA on glutamate and GABA transport in glial cells can be reverted by H7, an inhibitor of PKC (Matsui et al., 1986). It has been reported that phospholipase C generates IP3 and DAG from membrane phospholipids when added exogenously to a variety of cell types (Farese et al., 1988) . This would result in an increase in intracellular free Ca^{2+} and activation of PKC through the second messengers inositol trisphosphate and DAG respectively. Evidence is now accumulating that PKC activation and Ca^{2+} mobilization act synergistically to produce full cellular responses, particularly in short-term actions (Kaibuschi et al., 1983; Yamanishi et al., 1983: Nishizuka, 1984). Because TPA shows synergistic effects on modulating glutamate and GABA transport when added with the calcium ionophore A23187, which presumably works by elevating intracellular free Ca^{2+} concentration, it is likely that both arms of the phosphatydilinositol turnover are involved in the process.

PHOSPHORYLATION OF A GLIAL GLUTAMATE TRANSPORTER

Activation of glutamate transport by short-term exponsure of astroglial cells to phorbol esters suggest that the glutamate transporter is a substrate of PKC. Direct evidence for this was provided when the purified glutamate transporter was incubated with PKC and labelled ATP under optimal phosphorylation conditions. This resulted in the phosphorylation of the band corresponding to the glutamate transporter (66 kDa) in a preparative SDS-PAGE . The phosphoamino acid analysis of this band indicates that phosphorylation had predominantly occurred on serine residues. The functional consequences of the phosphorylation of the glutamate transporter was investigated in C6 cells, a cell line of glial origin which exhibits a highly active sodium-dependent L-glutamate transport activity (Schousboe, 1981). We phosphorylated and immunoprecipitated the glutamate transporter of glioma cells by using a polyclonal antibody raised against the purified transporter (Danbolt et al. 1992). The degree of labelling of the immunoprecipitated band increased with time upon exposure to TPA. In parallel experiments, the transport of glutamate in C6 cells exposed to TPA was increased concomitantly to the level of phosphorylation with similar kinetics. This strongly suggesting that the activation of glutamate transport is due to an increased phosphorylation of the transporter.

The phorbol ester TPA also stimulated glutamate transport in HeLa cells infected with a recombinant vaccinia virus expressing T7 RNA polymerase and transfected with a plasmid carrying a recently cloned rat brain glutamate transporter of glial origin (Pines et al., 1992). Site-directed mutagenesis clarified the mechanism of activation of TPA on glutamate transporter. Mutation of serine 113 to asparragine does not affect the levels of expressed transport but abolishes its stimulation by the phorbol ester, indicating that phosphorylation of a single serine residue by PKC causes the stimulation of glutamate transporter.

REGULATION BY cAMP OF THE GABA TRANSPORTER mRNA

There are likely to be other mechanisms that regulate the activities of neurotransmitter transporters. It has been reported the involvement of cAMP in long-term modulation of a high-affinity serotonin transporter in the human placental choriocarcinoma cell line (Cool et al., 1991). We have also found that forskolin, an agent know to increase cellular cAMP concentration, decreases the expression of the GAT-1 mRNA (a neuronal GABA transporter) in a dose- and time-dependent manner (Gomeza et al., 1993). The observed changes in GAT-1 mRNA caused by chronic exposure of the cells to forskolin can be also achieved by 8-Br-cAMP and dib-cAMP, suggesting that cAMP-dependent protein phosphorylation is involved in this effect. The inhibitory effect of cAMP is specific for the GABA transporter because under similar conditions, mRNA of glycine transporter is not affected. Interestingly, we have observed this regulatory effect by cAMP only in neuronal cultures but not in mixed glial cultures, thus supporting the existence of cell specific hetereogeneity in the regulation of GABA transport.

CONCLUSION

In conclusion, as it is now widely accepted the importance of the reuptake processes for neurotransmitters in the overal process of chemical neurotransmission, an understanding of the structure and function of neurotransmitter transporters may lead to a better understanding of synaptic function. Specific drugs acting on the neurotransmitter transporters can modulate neuronal function either by increasing or decreasing the duration of neurotransmitter action. In this context, neurotransmitter transporters for monoamines are the site of action for clinically important drugs as well as drugs abuse like cocaine or amphetamines, and transporters for inhibitory neurotransmitters as the GABA or the glycine transporters, are the potential target for drugs with anticonvulsant properties.

ACKNOWLEDGEMENTS

This work was supported by grants from the Spanish DGICYT (PB89-166 and PB92-0131), Boehringer Ingelheim España S.A. and an institutional grant from the Fundación Ramón Areces. We thank E.Núñez for his expert technical assistance.

REFERENCES

Amara S.G. and Kuhar, M.J. 1993. Neurotransmitter transporters: Recent progress. Annu. Rev. Neurosci. 16:73-93.

Casado M., Zafra F., Aragón C. and Giménez C. 1991. Activation of the high-affinity uptake of glutamate by phorbol esters in primary glial cell cultures. J. Neurochem. 57:1185-1190.

Danbolt N.C., Pines G. and Kanner B.I. 1990. Purification and reconstitution of the sodium- and potassium-coupled glutamate transport glycoprotein from rat brain. Biochemistry, 29:6734-6740.

Gomeza J., Casado M., Giménez C. and Aragón C. 1991. Inhibition of high-affinity γ-aminobutyric acid uptake in primary astrocyte cultures by phorbol esters and phospholipase C. Biochem. J. 275:435-439.

Guastella J., Nelson N., Nelson H., Czyzyk L., Keynan S., Miedel M., Davison N., Lester H.A. and Kanner B.I. 1990. Cloning and expression of a rat brain GABA transporter. Science. 249:1303-1306.

Hoffman B.J., Mezey E. and Browstein M.J. 1991. Cloning of a serotonin transporter affected by antidepressants. Science 254:579-580.

Kanay Y. and Hediger A. 1992. Primary structure and functional characterization of a high-affinity glutamate transporter. Nature. 360:467-471.

Casado M., Bendahan A., Zafra F., Danbolt N.C., Aragón C., Giménez C. and Kanner B.I. 1993. Phosphorylation and modulation of brain glutamate transporters by protein kinase C. J. Biol. Chem. (in press).

Cool D.R., Leibach F.H., Bhalla V.K., Mahesh V.B. and Ganapathy V. 1991. Expression and cyclic AMP-dependent regulation of a high-affinity serotonin transporter in the human placental choriocarcinoma cell line (JAR). J. Biol. Chem. 266:15750-15757.

Chiu F.-C. and Goldman J.E. 1984. Synthesis and turnover of cytoskeletal proteins in cultured astrocytes. J. Neurochem. 42:166-174.

Farese R.V., Cooper D.R., Konda T.S. Nair G., Standart M.L., Davis J.S. and Pollet R.J. 1988. Mechanisms whereby insulin increases diacylglycerol in BC34-1 myocytes. Biochem. J. 256:175-184.

Gomeza J., Giménez C. and Zafra F. 1993. Cellular distribution and regulation by cAMP of the GABA transporter (GAT-1) mRNA. Mol. Brain. Res. (in press).

Kaibuchi K., Takai Y., Sanamura H., Hoshijima M., Fujikna T. and Nishizuka Y. 1983. Synergistic functions of protein phosphorylation and calcium mobilization in platelet activation. J. Biol. Chem. 258:6701-6704.

Kanner B.I. and Shuldiner S. 1987. Mechanism of transport and storage of neurotransmitters. CRC Crit. Rev. Biochem. 22: 1-38.

López-Corcuera B., Vázquez J. and Aragón C. 1991. Purification of the sodium- and chloride-coupled glycine transport from central nervous system. J. Biol. Chem. 266:24809-24814.

Myers C.L., Lazo J.S. and Pitt B.R. 1989. Translocation of protein kinase C is associated with inhibition of 5-HT uptake by cultured endothelial cells. Am. J. Physiol. 257:L253-L258.

Nelson N. 1993. Presynaptic events involved in neurotransmission. J. Physiol. 87:171-178.

Nishizuka Y. 1984. The role of protein kinase C in cell surface signal transduction and tumor promotion. Nature. 388:693-698.

Pacholczyk T., Blakely R.D. and Amara S. 1991. Expression cloning of a cocaine and antidepressant-sensitive human noradrenaline transporter. Nature. 350:350-354.

Pines G., Danbolt N.C., Bjoras M., Zhang Y., Bendahan A., Eide L., Kolpsell H., Storn-Mathisen J., Seeberg E. and Kanner B.I. 1992. Cloning and expression of a rat brain L-glutamate transporter. Nature. 360:464-467.

Radian R., Bendahan A. and Kanner B.I. 1986. Purificacion and identification of the functional sodium- and chloride-coupled γ-amynobutyric acid transport glycoprotein from rat brain. J. Biol. Chem. 261:15437-15441.

Schousboe A. 1981. Transport and metabolism of glutamate and GABA in neurons and glial cells. Int. Rev. Neurobiol. 22:1-45.

Shimana S., Kitayama S., Lin C., Patel A., Nanthakumar E., Geagor P., Kuhar M. and Uhl G. 1991. Cloning and expresion of a cocaine-sensitive dopamine transporter complementary DNA. Science. 254:576-578.

Smith K.E., Borden L.A., Hartin P.R., Branchek T. and Weinshank R.L. 1992. Cloning and expression of a glycine transporter reveal colocalization with NMDA receptors. Neuron. 8:927-935.

Sullivan K.F. and Cleveland D.W. 1986. Identification of conserved isotype-defining variable region sequence for four vertebrate β-tubulin peptide classes. Proc. Natl. Acad. Sci. USA 83:4327-4331.

Yamanishi J., Takai Y., Kaibuchi K., Samo K., Castagna M. and Nishizuka Y. 1983. Synergistic action of phorbol esters and calcium in serotonin release from human platelets. Biochem. Biophys. Res. Comm. 112:778-786.

EXTRACELLULAR SIGNALS AND TRANSDUCTION MECHANISMS CONTROLLING THE ADENOSINE TRANSPORT IN NEURAL CELLS

Esmerilda G. Delicado, Raquel P. Sen, Teresa Casillas, M. Dolores Fideu and M.Teresa Miras-Portugal

Departamento de Bioquímica
Facultad de Veterinaria
Universidad Complutense
28040 Madrid, Spain

SUMMARY

The adenosine transport in bovine chromaffin cells is highly regulated by different extracellular signals, both at short and long-term. The short-term regulation by extracellular signals such as secretagogues or direct activation of protein kinases A and C inhibit the adenosine transport capacity. Binding studies with nitrobenzylthioinosine show that the inhibition of the transport runs in parallel with a decrease in the number of binding sites for this ligand. The adenosine transport inhibition is mediated by a phosphorylation-dephosphorylation process. Extracellular signals that bind to nuclear receptors are responsible for the long-term regulation. Thyroid hormones increase the adenosine transport by increasing the number of adenosine transporters at the plasma membrane level. This transport activation requires the protein-synthesizing mechanism. On the contrary, dexamethasone and retinoic acid inhibit the adenosine transport capacity and revert the action of thyroid hormones.

INTRODUCTION

Adenosine is an important endogenous modulator with relevant actions on nervous and vascular systems[1,2]. The extracellular actions of this nucleoside are mediated by membrane-bound receptors, designed P_1 purinergic receptors[3]. The adenosine receptors have been identified in neural and non-neural tissues and classified pharmacologically in different subtypes: A_1, A_2 and A_3 receptors, also taking into account the intracellular mechanisms involved in the physiological response. Some of the receptors have been recently cloned[4,5]. The extracellular actions of adenosine are

Cell Signal Transduction, Second Messengers, and Protein Phosphorylation in Health and Disease
Edited by A.M. Municio and M.T. Miras-Portugal, Plenum Press, New York, 1994

213

terminated by its reuptake into the cells. However, the origin and release of adenosine remains unclear. In neural tissues, adenosine is mainly originated from the hydrolysis of extracellular nucleotides[6]. ATP and adenine dinucleotides Ap_4A (diadenosine tetraphosphate), Ap_5A (diadenosine pentaphosphate) and Ap_6A (diadenosine hexaphosphate), are coreleased by exocytosis with classical neurotransmitters, such as acetylcholine or catecholamines, depending on the neural preparations.

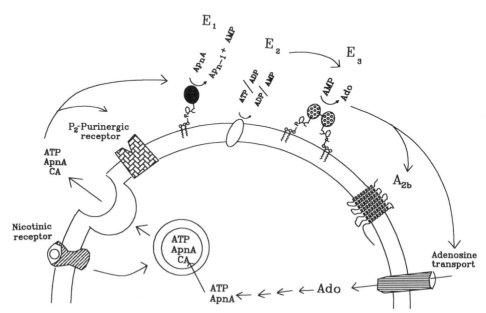

Figure 1.- The ATP-adenosine cycle in chromaffin cells.

Upon nicotinic-receptor stimulation, ATP and dinucleotides are exocytotically released together with catecholamines from the chromaffin granules. As they reach the extracellular space, they exert specific actions through the P_2-purinergic membrane receptors. The duration of these actions is then terminated by their degradation by the ectoenzyme-cascade (E_1 = ecto-Ap_4A-hydrolase; E_2 = ecto-ATPase and ecto-ADPase; E_3 = ecto-5'-nucleotidase), to yield adenosine as the final product. Adenosine can bind to its membrane receptors (A_{2b}-type in the case of the chromaffin cells), and finally, adenosine is reuptaked into the cell by the specific adenosine transporter. Once adenosine is inside the cell it is rapidly metabolized into nucleotides in order to replenish the intragranular nucleotide-pool.

These nucleotides can modulate the neurotransmission acting through the P_2 purinergic receptors. They are degraded to adenosine by ectonucleotidases to finalize their action[6-11]. As occurs with other neurotrasmitters or the degradation products, their transport is necessary to recover the synaptic functionality. Thus, adenosine can be considered as the last extracellular product of purinergic transmission (fig. 1). Adenosine is effectively transported into the cells by a high affinity transport system and then incorporated into the nucleotide pool by the action of adenosine kinase[12,13]. It is interesting to mention that in hypoxic and isquemic conditions, the extracellular levels of adenosine are significantly increased. In these conditions, the adenosine is released from the neurons by the reversion of the nucleoside transporter[14,15]. Since the adenosine transport represents the last step in the purinergic transmission and the main step regulating the extracellular adenosine levels, it is very interesting to know the possible regulation of this process. A good approach is to study the regulation of adenosine transport in chromaffin cells. These cells represent an homogeneous neural cell model, being similar to peripheral sympathetic neurons and can be obtained in sufficient number to carry out the transport regulation studies. The relevant physiological signals in this model are secretagogues that trigger the exocytosis, such as acetylcholine or nicotinic agonists; and also substances that have been released during the exocytotic event and can have a feed-back control on exocytosis, such as catecholamines, ATP and dinucleotides. The effect of hormones acting on gene expression, such as thyroid hormone and steroid hormones is also studied in long-term regulation of adenosine transport.

SHORT-TERM REGULATION OF ADENOSINE TRANSPORT IN CHROMAFFIN CELLS

The adenosine transport in chromaffin cells is well characterized[16]. No-sodium-dependent transport has been found. All the nucleoside transporters present in these cells take up adenosine by a facilitated diffusion process, exhibiting a high affinity (in the μmolar range) and high sensitivity to NBTI inhibition, as described for neural tissues, indicating the presence of only one transporter type[17,18]. This fact facilitates the regulation studies. In these neural cells this transport mechanism is a very important step for the salvage of nucleosides, that have been previously released as nucleotides by exocytosis[11,19].

The stimulation of chromaffin cells with secretagogues that bind to nicotinic receptors, such as carbachol and nicotine or depolarization of the cells with high K^+ concentrations, significantly inhibits the adenosine transport (Table I). This inhibition corresponds to a decrease of 50% in the transport capacity (V_{max}), without changing the transport affinity (K_m)[20]. Recently, the presence of dinucleoside polyphosphates, Ap_4A and Ap_5A, in the chromaffin granules and their release togheter with ATP and catecholamines during the exocytosis has been described[7]. These extracellular dinucleotides acting through purinergic receptors are also able to modulate the catecholamine secretion[10]. The purinergic receptors present in chromaffin cells show a P_{2y} pharmacological profile[21]. Taking into account these findings, the effect of Ap_4A on adenosine transport is analyzed. The treatment of chromaffin cells with Ap_4A and Ap_5A inhibits the adenosine transport. Similar results are obtained when cells are preincubated with bradykinin, that produces an increase in the cytosolic calcium concentration[22]. The inhibition of adenosine transport corresponds to a decrease in the V_{max} value (Table I).

TABLE I.- Modification of kinetic parameters of adenosine transport by several effectors.

Effector	Km (μM)	Vmax	%
Control	1.6 $\pm$ 0.2	40 $\pm$ 5.0	--
Carbachol (50 μM)	2.0 $\pm$ 0.3	20 $\pm$ 5.0*	50
K$^+$ (60 μM)	1.6 $\pm$ 0.3	25 $\pm$ 4.5*	37
Ap$_4$A (100 μM)	1.3 $\pm$ 0.4	22 $\pm$ 4.7*	45
Bradykinin (10μM)	1.6 $\pm$ 0.6	25 $\pm$ 5.5*	37
NECA (100 nM)	3.5 $\pm$ 3.0*	102 $\pm$ 17*	255

Transport experiments were carried out with chromaffin cells cultured in 24-well-Costar cluster dishes at a density of 250,000 cells/well. The kinetic parameters were determined measuring the transport during the first minute of incubation at 37°C with [^{3}H]adenosine, which corresponded to the linear period. Experiments in the presence of the above mentioned effectors were done by preincubating the cells at 37°C with these effectors before starting the transport incubation period. Cells were preincubated for 1 min with carbachol and K$^+$, for 2 min with bradykinin and Ap$_4$A, and for 10 min with NECA at the indicated concentrations.
The transport capacity (Vmax) is expressed as pmol/10^{6}cells/min.
% represents the percentage of inhibition or increase in the Vmax value with respect to control.
* $p \leq 0.001$
Values are the means $\pm$ SD of five experiments performed in quadruplicate.

As previously mentioned, the adenosine transport is the main mechanism that terminates the physiological actions of this nucleoside through its membrane receptors. However, several studies have shown that there is not a good correlation between the distribution of adenosine receptors and adenosine transporters in rat brain[23]. Since no work has been done on the possible interaction between the adenosine transporters and receptors, studies on the modulation of adenosine transport by the stimulation of adenosine receptors were carried out in chromaffin cells. The incubation of chromaffin cells in the presence of A_1 agonists had no effect on adenosine transport. On the contrary, the treatment of cells with the A_2 agonist, NECA (5'-N-ethylcarboxamidoadenosine), increased the adenosine transport. The NECA stimulation was antagonized by XAC (1,3-dipropyl-8-{4-[(2-aminoethyl)amino]-carbonylmethyl oxyphenyl} xanthine), which is a broad adenosine receptor antagonist, but not by DPCPX (8-cyclopentyl-1,3-dipropylxanthine), which is a specific A_1 antagonist[24]. These results show that the adenosine is able to modulate its own transport as has been reported for another neurotransmitter, such as dopamine[25]. These results do not exclude the possibility that A_1 or A_3 receptors can modulate the adenosine transport in other cellular models, in which the adenosine actions are mediated by these receptor types. Even the adenosine transport can be modulated by other extracellular signals of physiological relevance in different tissues.

Intracellular mechanisms that mediate the adenosine transport regulation

In bovine chromaffin cells, the main receptor clearly linked to the stimulation of the exocytosis of catecholamines is the nicotinic cholinergic receptor. It is also well established that the stimulation with nicotine or carbachol produces a rise in the intracellular calcium concentration, $[Ca^{+2}]_i$. The increase in the cytosolic calcium causes a translocation of protein kinase C (PKC) to the membranes[26,27]. The treatment of chromaffin cells with the calcium ionophore, A-23187, or direct activators of protein kinase C (phorbol esters) also inhibited the adenosine transport. Futhermore, when calcium ionophore was combined with phorbol esters showed synergistic effects on adenosine transport inhibition (Table II). These compounds inhibited the adenosine transport capacity with minor changes in the affinity of the transporter.

TABLE II.- Effect of several intracellular effectors on the kinetic parameters of adenosine transport.

Effector	Km (μM)	Vmax	%
Control	1.6 ± 0.2	40 ± 5.0	
A-23187 (0.5 μM)	1.9 ± 0.3	$30 \pm 7.3*$	23
PDBu (0.1 μM)	1.6 ± 0.2	$32 \pm 4.0*$	20
PDBu + A-23187	1.7 ± 0.3	$15 \pm 4.0*$	60
Forskolin (0.5μM)	2.2 ± 0.3	$18 \pm 3.0*$	55
ClPh$_c$AMP (0.1μM)	2.0 ± 0.1	$21 \pm 4.0*$	47

Cultured chromaffin cells were preincubated in the presence or absence of the indicated effectors for 10 min at 37°C before starting the transport experiments. The Vmax values are expressed as pmol/10^6 cells/min and the % represents the percentage of inhibition in the Vmax values with respect to control.
* $p \leq 0.001$
Values are the means $\pm$ SD of five experiments performed in quadruplicate.

The intracellular mechanism involved in the activation of adenosine transport due to NECA response was studied. The A_2 adenosine receptors are coupled to adenylate cyclase activation in the majority of the studied tissues[3]. But, in chromaffin cells, NECA did not stimulate the cAMP production. However, when the effect of a cAMP elevation on adenosine transport was analyzed by treating the cells with forskolin and the cAMP analogue, ClPhcAMP, a significant inhibition was observed in the transport capacity (Table II). These results support the hypothesis that the A_{2b} receptors present in chromaffin cells are not coupled to the stimulation of adenylate cyclase and protein kinase A activation. Nevertheless, these findings showed that the activation of PKC and protein kinase A (PKA) inhibit the adenosine transport in these neural cells, suggesting that a phosphorylation of the transporter can take place[20,28]. A possible explanation for the stimulatory effect of NECA on adenosine transport would be that the A_{2b} receptors mediated the protein phosphatase activation. Thus, this protein phosphatase could dephosphorylated the adenosine transporter explaining the transport stimulation. When similar experiments were carried out in the presence of a specific inhibitor of protein phosphatase (okadaic acid) the effect of NECA decreased, indicating that these phosphatases were involved in the activation process of the transport. Recent studies confirmed that NECA activated a cytosolic protein phosphatase in chromaffin cells[29].

Modulation of adenosine transport studied by NBTI binding experiments

The inhibitors of adenosine transport, such as dipyridamole, dilazep and nitrobenzylthioinosine (NBTI), have been employed therapeutically to prolong the extracellular actions of adenosine[13]. These inhibitors are commonly used to identify and quantify the nucleoside transporter in different tissues. Nitrobenzylthioinosine is the most specific inhibitor of the facilitated diffusion transport system, which has also been used for the following classification: i) NBTI-sensitive nucleoside transport that is inhibited by nM concentration of NBTI and ii) NBTI-resistant nucleoside transport that is inhibited by μM concentration of this ligand. The distribution of the two classes of facilitated nucleoside transport depends on the cell type which can even coexist in the same cell[12,17,18].

The nucleoside transporters present in chromaffin cells have been characterized by using [³H]NBTI for binding and photoaffinity labelling experiments. These studies have shown that chromaffin cells present a homogenous population of high affinity binding sites for NBTI (K_d = 0.6 nM) in a density of about 32,000 transporters per cell. As NBTI is a lipid soluble compound, it was necessary to determine whether the ligand is only bound to the plasma membrane or, on the contrary, it enters the cells and is also bound to the intracellular binding sites. The first approach was to carry out the [³H]NBTI binding experiments with permeabilized chromaffin cells with digitonin, allowing the total accessibility of the ligand to the cytosol. In these conditions, the subcellular structures were maintained and there were no changes in the number of high affinity binding sites for NBTI. However, when similar experiments were undertaken in hypoosmotic lysed cells, the density of the transporters significantly rose to 115,000 transporters/cell. Thus, the number of NBTI binding sites measured in intact cells corresponded to the binding sites present at the plasma membranes[30]. Taking into account that the adenosine transporters have been identified in different subcellular preparations from chromaffin tissue, these results indicate that the nucleoside transporter of intracellular organella membranes can not bind NBTI in physiological conditions, because of their internal acidic environment. Only when the intracellular organellas are lysed, the nucleoside transporters were exposed and recognized the ligand (fig. 2).

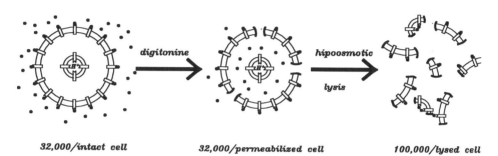

32,000/intact cell   32,000/permeabilized cell   100,000/lysed cell

Figure 2.- Number of NBTI binding sites in chromaffin cells.
The number of high-affinity NBTI binding sites in chromaffin cells were determined under several experimental conditions. In intact chromaffin cells, the NBTI only binds to the adenosine transporters located at the plasma membrane. Similarly, when the cells are permeabilized with digitonin, the NBTI has access to the cytosolic compartment but cannot bind to the transporters located in the subcellular organella, because of their internal acidic environment. The access to these organella binding sites is achieved when the cells are lysed in hypoosmotic medium as is revealed by the increase in the number of NBTI binding sites.

When chromaffin cells are stimulated with secretagogues or substances that induce the activation of PKA and PKC, a decrease in the NBTI binding sites was also observed, without any changes in the affinity for this ligand. Thus, the decrease in the transport capacity mentioned above can be explained by a similar decrease in the number of transporters at the plasma membrane level. These results indicate that in chromaffin cells the changes in the adenosine transport capacity runs in parallel with changes in the number of transporters present at the plasma membrane level. Besides, the efficiency of the transporter was not modified by the effectors, 12-13 molecules of adenosine being internalized per second and per transporter. These results also suggest the possibility that a phosphorylation of the transporter could take place and the modified transporter does not recognize the NBTI ligand and does not internalize adenosine. To verify this possibility the plasma membrane preparations of chromaffin cells were treated with purified PKA and PKC in the presence of $[\gamma^{32}P]ATP$ and the NBTI binding sites were quantified (Table III). The incorporation of $[^{32}P]$phosphate to the membranes corresponded to a 50% inhibition of the NBTI binding. These experiments provided additional evidences that the phosphorylation of the nucleoside transporters or a regulatory protein was responsible for the inhibition of the adenosine transport induced by secretagogues and other extracellular signals related to protein kinase activation.

TABLE III.- Effect of purified protein kinases A and C on [³H]NBTI binding to chromaffin cells plasma membranes.

	[³H]NBTI bound (fmol/mgprotein)
Control	140 ± 10
+ PKC	75 ± 6 *
+ PKA	73 ± 9 *

Chromaffin cell plasma membranes were incubated at 37°C for 10 min in the presence of the purified protein kinases A and C, and then the NBTI binding was performed by incubating at 37°C for 30 min with 0.5 nM [³H]NBTI. The assays with PKC were performed with 1 unit/assay of purified PKC from bovine brain[20]. The assays with PKA were carried out with comercially available PKA from bovine heart (33 μg/assay).
Results are the means ± SD of five experiments.
* $p \leq 0.01$ compared with result in the absence of protein kinases.

Similar effects of protein kinases on other facilitated transport processes have been described, such as the facilitated diffusion glucose transporter. Phosphorylation sites for protein kinases A and C have been identified at the cytoplasmic domains of glucose transporter (GLUT 4), able to modify the functionality of the transporter[31,32]. The glucose transport shares common features with the nucleoside transport, being the first step in the complex metabolic pathways of the two substances. Because of these similarities, the structure of the nucleoside transporter has been proposed to be similar to that of the glucose transporter with 12 membrane-spanning segments. On the other hand, other types of transporters, such as the neurotransmitter transporters that take up the released neurotransmitters into presynaptic terminals and vesicles, have also been proposed as proteins with 12 transmembrane region topologies. They mediate the reuptake of neurotransmitter by an active transport process coupled to ionic gradients, finishing the synaptic transmission. Some of these transporters are submitted to a short-term regulation by protein kinases actions. The activation of protein kinase C produces an increase in the sodium-dependent glutamate transport in neurons and glial cells[33]. Similar effects of protein kinase C activation on the uptake of GABA in brain synaptosomes has been reported[34]. On the contrary, in endothelial cells and platelets, the activation of protein kinase C is associated with the inhibition of serotonin transport[35,36]. All these findings show the relevance of protein kinase C modulating the functionality of transport proteins.

LONG-TERM REGULATION OF ADENOSINE TRANSPORT IN CHROMAFFIN CELLS

As previously mentioned, the adenosine transport is submitted to an acute regulation by signals that mediate rapid responses through phosphorylation-dephosphorylation. In addition, in peripheral sympathetic neurons it has been reported that catecholamine levels and their neural uptake is decreased in hypothyroid status[37,38]. Since adenosine is the ATP precursor, which is co-stored with catecholamines, a long-term regulation of the nucleoside transport could be expected. In this sense, a long-term regulation of adenosine transport has also been described[39]. It was demonstrated that nerve growth factor (NGF) added to the culture media produced an increase in the capacity of the adenosine transport in chromaffin cells in the following hours. Thus, the study of this type of regulation was undertaken by the use of substances that bind to nuclear receptors such as triiodo-L-thyronine (T_3), glucocorticoids and retinoic acid. The long-term incubation (24 hours) of chromaffin cells in the presence of these substances significantly modify the adenosine transport (Table IV). The treatment of chromaffin cells with T_3 increased the transport capacity, without changes in the transport affinity constant. The modification in adenosine transport is due to a parallel increase in the number of plasma membrane transporters quantified by [³H]NBTI binding experiments, raising about 40,000 binding sites/cell. The K_d for NBTI was not modified by the presence of T_3. Thyroid hormone treatment did not show short-term responses to adenosine transport; all the effects observed required the incubation with T_3 for several hours, suggesting that the protein synthesis was required. Cycloheximide, an inhibitor of protein synthesis, blocks the stimulatory effect of T_3 in these cells, and as expected the half-life values obtained were about 24 hours both in the presence or absence of T_3. The new transporters synthetized had the same affinity for adenosine, and also the same efficiency, but only the maximal transport capacity was modified[40].

TABLE IV.- **Effect of thyroid hormone (T₃), Dexamethasone (Dex) and Retinoic acid (RA) on adenosine transport and NBTI binding in chromaffin cells.**

	Adenosine transport		NBTI binding	
	K_m (μM)	V_{max}	K_d (nM)	B_{max}
Control	0.8 ± 0.10	36.21 ± 2.1	0.72 ± 0.07	$33,500 \pm 3,000$
T₃ 1 μM	0.7 ± 0.09	$44.17 \pm 3.5*$	0.79 ± 0.08	$40,200 \pm 3,700 *$
Dex 1 μM	1.0 ± 0.11	$30.05 \pm 4.8*$	0.70 ± 0.09	$33,100 \pm 3,200$
RA 1 μM	0.8 ± 0.07	$28.90 \pm 3.9*$	0.71 ± 0.07	$34,500 \pm 3,090$

Cultured chromaffin cells were treated in the presence or absence of the different compounds for 24 h of incubation at 37°C, prior to assesment the transport and binding experiments. Vmax are expressed as pmol/10⁶ cells/min and Bmax as the number of binding sites/cell. The values are the means $\pm$ SD of four experiments performed in quadruplicate.
*$p < 0.05$

To continue the long-term regulation studies on adenosine transport, the effect of synthetic glucocorticoids, such as dexamethasone and retinoic acid were analyzed. The long-term treatment (24 h) of chromaffin cells, in the presence of these compounds inhibited the adenosine transport (Table IV). No effect was observed at short-term incubation period (10 min). The maximal velocity was modified with a decrease about 20% in both cases, but no significant changes were observed for the affinity constant (K_m). Surprisingly, the NBTI binding to chromaffin cells was not modified by these compounds. Thus the efficiency of the transporters decreased to 9 molecules of adenosine per transporter and per second. These results can be explained by a modification of the transporter itself or by the effect of other intracellular components that affect exclusively the transport capacity, but not the binding site for the NBTI recognition. It is interesting to mention, that the activatory effect of T₃ on adenosine transport was antagonized by dexamethasone and retinoic acid[41]. In this regard, the retinoic acid also antagonized the effects of T₃ on the TRß₂ gen repression[42].

Recent studies on adenosine transport have shown that other regulatory possibilities exist. Adenosine at very low concentrations (nM), was able to induce allosteric changes on the transporter, showing kinetic and allosteric cooperativity[43]. Furthermore, it has revealed that in spite of the independence of adenosine transport of a sodium-gradient, it is coupled to the intracellular energetic state, that assures the metabolism to phosphorylated derivatives. An intracellular ATP binding site on adenosine transporter similar to that described for other membrane transporters has been described[44]. All these findings show that adenosine transport in these neural cells is a highly regulated process (fig. 3), where different mechanisms are necessary to establish a coupling between the extracellular signals and the intracellular metabolic state.

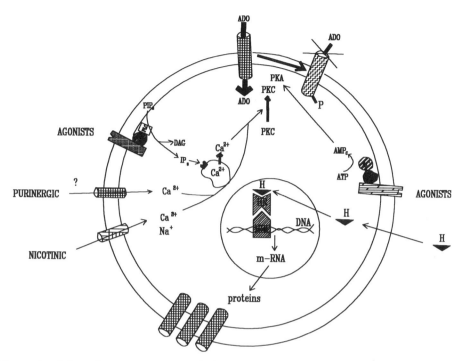

Figure 3.- Adenosine transport regulation by different types of membrane receptors present in chromaffin cells.
The stimulation of receptors in chromaffin cells that are coupled to the activation of PKC, either by activating the entry of extracellular calcium, such as the nicotinic receptors, or by elevating the $[Ca^{2+}]_i$ and generating diacylglycerol, such as the purinergic receptors, has an inhibitory effect on adenosine transport that is evident in the decrease of the transport capacity. The same effect is observed by increasing the cAMP levels and the subsequent activation of the PKA, mediated by not yet determined extracellular agonists. Therefore, the inhibition of adenosine transport by the activation of protein kinases A and C suggests the possibility that a phosphorylation process at the transporter level takes place. Another type of regulation of adenosine transport is mediated by several hormones, T_3, steroid hormones and retinoic acid, that binds to specific nuclear receptors. Long-exposure of chromaffin cells to the action of these hormones results in an increase in adenosine transport capacity in the case of T_3, and an inhibition in adenosine transport capacity for retinoic acid and steroid hormones, both effects being dependent on protein synthesis.

REFERENCES

1. S.M. Snyder. "Adenosine as a neuromodulator". Annu. Rev. Neurosci. 8:103 (1985).

2. R.A. Olsson and J.D. Pearson. "Cardiovascular purinoceptores". Physiol. Rev. 70:761 (1990).

3. M. Williams. "Purine receptors in mammalian tissues: pharmacology and functional significance". Annu. Rev. Pharmacol. Toxicol. 27:315 (1987).

4. J. Linden, A.L. Tucker and K.R. Lynch. "Molecular cloning of adenosine A_1 and A_2 receptors. Trends Pharmacol. Sci. 12: 326 (1991).

5. Q-Y Zhou, C-Y Li, M. Olah, R. Johnson, G. Stiles and O. Civelli. "Molecular cloning and characterization of a novel adenosine receptor, the A_3 adenosine receptor". Proc. Natl. Acad. Sci. USA 89:7432 (1992).

6. P.J. Richardson, S.J. Brown, E.M. Brailyes and J.P. Luzio. "Ectoenzymes control adenosine modulation of inmunoisolated cholinergic synapses". Nature 327:232 (1987).

7. J. Pintor, M. Torres and M.T. Miras-Portugal. "Carbachol induces release of diadenosine polyphosphates-AP_4A and Ap_5A-from perfused bovine adrenal medulla and isolated chromaffin cells". Life Sci. 48:2317 (1991).

8. J. Pintor, M.A. Díaz-Rey, M. Torres and M.T. Miras-Portugal. "Presence of diadenosine polyphosphates-Ap_4A and Ap_5A-in rat brain synaptic terminals. Ca^{+2}-dependent release evoked by 4-aminopyridine and veratridine". Neurosci. Lett. 136:141 (1992).

9. J. Pintor, H.J. Kowalewski, M. Torres, M.T. Miras-Portugal and H. Zimmermann." Synaptic vesicle storage of diadenosine polyphosphates in the Torpedo electric organ". Neurosci. Res. Commun. 10:9(1992).

10. E. Castro, M. Torres, M.T. Miras-Portugal and M.P. González. "Effect of diadenosine polyphosphates on catecholamine secretion from isolated chromaffin cells". Br. J. Pharmacol. 100:360 (1990).

11. M. Torres, J. Pintor and M.T. Miras-Portugal. "Presence of ectonucleotidases in cultured chromaffin cells: hydrolysis of extracellular adenine nucleotides". Arch. Biochim. Biophys. 279:37 (1990).

12. P.H. Wu and J.W. Phillis. "Uptake by central nervous tissues as a mechanism for the regulation of extracellular adenosine concentrations". Neurochem. Int. 6:613 (1984).

13. J. Deckert, P.F. Morgan and P.J. Marangos. "Adenosine uptake site heterogeneity in the mammalian CNS? Uptake inhibitors as probes and potencial neuropharmaceuticals". Life Sci. 42:1331 (1988).

14. J.W. Phillis and P.H. Wu. "The role of adenosine and its nucleotides in central synaptic transmission". Prog. Neurobiol. 16:87 (1981).

15. K.A. Rudolphi, P. Schubert, F.E. Parkinson and B.B. Fredholm. "Adenosine and brain ischemia". Cerebrovasc. Brain Metab. Rev. 4:346 (1992).

16. M.T. Miras-Portugal, M. Torres, P. Rotllán and D. Aunis. "Adenosine transport in bovine chromaffin cells in culture". J. Biol. Chem. 261:1712 (1986).

17. M. Torres, M.D. Fideu and M.T. Miras-Portugal. "All nucleoside transporters in bovine chromaffin cells are nitrobenzylthioinosine sensitive". Neurosci. Lett. 112:343 (1990).

18. R.P. Shank and W.J. Baldy. "Adenosine transport by rat and guinea pig synaptosomes: basis for differential sensitivity to transport inhibitors". J. Neurochem. 55:541 (1990).

19. M.T. Miras-Portugal, P. Rotllán and D. Aunis. "Incorporation of adenosine into nucleotides of chromaffin cells maintained in primary cultures". Neurochem. Int. 7:89 (1985).

20. E.G. Delicado, R.P. Sen and M.T. Miras-Portugal." Effects of phorbol esters and secretagogues on nitrobenzylthioinosine binding to nucleoside transporters and nucleoside uptake in cultured chromaffin cells". Biochem. J. 279:651 (1991).

21. J. Pintor, M. Torres, E. Castro and M.T. Miras-Portugal. "Characterization of diadenosine (Ap_4A) binding sites in cultured chromaffin cells: evidence for a P_{2y}-site". Br. J. Pharmacol. 103:1980 (1991).

22. R.P. Sen, E.G. Delicado, E. Castro and M.T. Miras-Portugal. "Effect of P_{2y} agonists on adenosine transport in cultured chromaffin cells". J. Neurochem. 60: 613 (1993).

23. J. Deckert, J.C. Bisserbe, E. Klein and P.J. Marangos. "Adenosine uptake sites in brain: regional distribution of putative subtypes in relationship to adenosine A_1-receptors". J. Neurosci. 8:2338 (1988).

24. E.G. Delicado, A. Rodrigues, R.P. Sen, A.M. Sebastiao, J.A. Ribeiro and M.T. Miras-Portugal."Effect of 5'-(N-ethylcarboxamido)adenosine on adenosine transport in cultured chromaffin cells". J.Neurochem. 54:1941 (1990).

25. S.M. Meiergerd, T.A.M. Patterson and J.O. Schenk. "D_2 receptors may modulate the function of the striatal transporter for dopamine: kinetic evidence from studies in vitro and in vivo". J. Neurochem. 61:764 (1993).

26. D.R. Terbush and R.W. Holz. "Effects of phorbol esters, diglyceride and cholinergic agonists on the subcellular distribution of protein kinase C in intact or digitonin-permeabilized adrenal chromaffin cells". J. Biol. Chem. 261:17099 (1986).

27. R.D. Burgoyne."Control of exocytosis in adrenal chromaffin cells". Biochim. Biophys. Acta. 1071: 174 (1991).

28. R.P. Sen, E.G. Delicado and M.T. Miras-Portugal."Effect of forskolin and cyclic AMP analog on adenosine transport in cultured chromaffin cells". Neurochem. Int. 17(4): 523 (1990).

29. J. Mateo, E. Castro, J. Zwiller, D. Aunis y M.T. Miras-Portugal. "5'(N-etilcarboxamido)adenosina (NECA) inhibe el influjo de Ca^{2+} y activa una proteina fosfatasa en células cromafines bovinas". XIV Reunión Nacional del grupo español de la célula cromafín. Madrid (1993).

30. M. Torres, E.G. Delicado, M.D. Fideu and M.T. Miras-Portugal. "Down-regulation and recycling of the nitrobenzylthioinosine-sensitive nucleoside transporter in cultured chromaffin cells". Biochim. Biophys. Acta 1105:291 (1992).

31. D.E. James, J. Hiken and J.C. Lawrence. "Isoproterenol stimulates phosphorylation of the insulin-regulatable glucose transporter in rat adipocytes". Proc. Natl. Acad. Sci. USA 86:8368 (1989).

32. L.A. Witters, C.A. Vater and G.E. Lienhard. "Phosphorylation of the glucose transporter in vitro and in vivo by protein kinase C". Nature 315:777 (1985).

33. M. Casadó, F. Zafra, C. Aragón and C. Giménez. "Activation of the high-affinity uptake of glutamate by phorbol esters in primary glial cell cultures". J. Neurochem 57:1185 (1991).

34. A. Cupello, B. Gasparetto, P. Mainardi, L. Vignolo and M.P Robello." Effect of protein kinase C activators on the uptake of GABA by rat brain synaptosomes". Int. J. Neurosci. 69:131 (1993).

35. C.L. Myers, J.S. Lazo and B.R. Pitt. "Translocation of protein kinase C is associated with inhibition of 5-HT uptake by cultured endothelial cells". Am. J. Physiol. 257:L253 (1989).

36. G.M. Anderson and W.C. Horne. "Activators of protein kinase C decrease serotonin transport in human platelets". Biochim. Biophys. Acta 1137:331 (1992).

37. T.A. Slotkin and R.J. Slepetis."Obligatory role of thyroid hormones in development of peripheral sympathetic and central nervous system catecholaminergic neurons: effects of propylthiouracil-induced hypothyroidism on neurotransmitter levels, turnover and release".J. Pharmacol. Exp. Ther. 230: 53 (1984).

38. C. Lau and T.A. Slotkin. "Maturation of sympathetic neurotransmission in the rat heart.VIII. Slowed development of noradrenergic synapses resulting from hypothyroidism". J. Pharmacol. Exp. Ther. 220: 629 (1982).

39. M. Torres, M.F. Bader, D. Aunis and M.T. Miras-Portugal."Nerve growth factor effect on adenosine transport in cultured chromaffin cells". J. Neurochem. 48:233. (1987).

40. M.D. Fideu and M.T. Miras-Portugal."Long-term regulation of nucleoside transport by thyroid hormone (T_3) in cultured chromaffin cells". Neurochem. Res. 17(11): 1099 (1992).

41. M.D. Fideu and M. T. Miras-Portugal. "Steroid-Induced inhibition of adenosine transport in cultured chromaffin cells". Cell. Mol. Neurobiol. 13(5): 493 (1993).

42. K.D. Davis and M.A. Lazar. "Selective antagonism of thyroid hormone action by retinoic acid". J. Biol. Chem. 267: 3185 (1992).

43. T. Casillas, E.G. Delicado, F. García-Carmona and M.T. Miras-Portugal. "Kinetic and allosteric cooperativity in L-adenosine transport in chromaffin cells. A mnemonical transporter. Biochemistry 32: 14203 (1993).

44. T. Casillas, E.G. Delicado and M.T. Miras-Portugal. "Adenosine 5'-triphosphate modulation of nitrobenzylthioinosine binding sites in plasma membranes of bovine chromaffin cells". Neurosci. Lett. 164:51 (1993).

SIGNAL TRANSDUCTION THROUGH LAMININ RECEPTORS. EFFECTS OF EXTRACELLULAR MATRIX ON BCS-TC2 ADENOCARCINOMA CELLS

Nieves Olmo and Mª Antonia Lizarbe

Department of Biochemistry and Molecular Biology
Faculty of Chemistry, Complutense University
Madrid, 28040, Spain

SUMMARY

The alterations of the extracellular matrix play a critical role in the establishment, growth and invasiveness of tumors. We have reported an altered collagen metabolism in a human colon adenocarcinoma; the relative collagen content in the tumor was significantly lower than in normal tissue. This is due to a permanent phenotypic alteration in the tumor-associated fibroblast-like cells. The cell line BCS-TC2 established from this primary human colon adenocarcinoma maintains in culture several characteristics of the original tumor. We have studied the effects of extracellular matrix proteins on the behavior of BCS-TC2 cells. The modulation of the AMPase activity of 5´-nucleotidase by laminin and fibronectin suggest a possible function of the ectoenzyme as a cell receptor for matrix proteins. BCS-TC2 cell adhesion to laminin indicates the presence of different laminin receptors, among them, several integrins and the 67-kDa high-affinity laminin receptor. Integrins, talin and tyrosine phosphorylation activity, located into the focal adhesion sites could be implicated in the signal transduction by receptor from extracellular matrix. The expression of the 67-kDa laminin receptor, is modulated by laminin, and may contribute to BCS-TC2 cells interaction with the basement membranes, and it seems to be related to the invasive properties of these cells.

ARE EXTRACELLULAR MATRICES INERT SUPPORTING MATERIALS?

The extracellular matrix is a structurally stable association of biopolymers surrounding or supporting the cells. In addition to its role as a physical barrier, it is responsible for the body shape and organ compartmentalization, and maintains the tissue integrity. Even though the major function of the extracellular matrix is to provide physical

Cell Signal Transduction, Second Messengers, and Protein Phosphorylation in Health and Disease
Edited by A.M. Municio and M.T. Miras-Portugal, Plenum Press, New York, 1994

227

support to the tissues, this matrix is not simply an inert supporting material. It is a cement or a natural glue for promoting cell adhesion, it controls the expression of the normal cell phenotype, and it exhibits a close functional relationship with the tissue cells. Thus, there are specific cells and macromolecular structures defining particular extracellular matrices.

Extracellular matrices are structurally diverse, and vary in a tissue-specific fashion; they are clearly distinct in both molecular composition and function[1-5]. Thus, skin, bone, cartilage, basement membranes, etc., contain different matrix proteins which are optimal for their individual functions. The extracellular matrices can be divided into two major groups: those forming the interstitial connective tissue and those from the basement membranes. Both types of matrices are composed of many different macromolecular structures that can be grouped, according to their major components, into collagens, noncollagenous glycoproteins and proteoglycans. A schematic representation of the major extracellular matrix components as well as some of their characteristics is shown in Figure 1.

Extracellular matrix components

Collagens were one of the first extracellular molecules studied since they account for a large amount of the tissues (one-third of the total protein); they were initially thought to be a stable and a relative inactive constituent of connective tissues. Collagen is defined as "the structural protein of the extracellular matrix which contains in their structure one or several domains in triple helical conformation"[6]. Type I collagen is a well known protein whose chemical and tridimensional structures have been clearly elucidated. However, this type of collagen is only a representative structure of a broad family of proteins. This large family of multimeric structural glycoproteins is composed of, at least, 14 distinct genetic types of collagen[6,7].

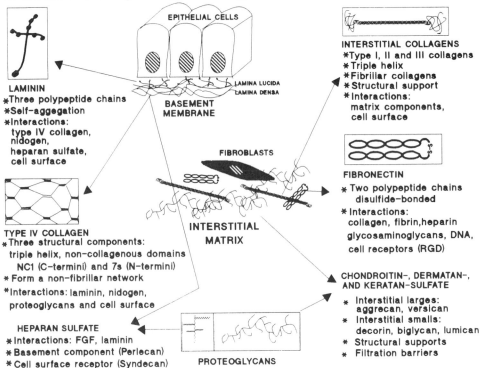

Figure 1. Extracellular matrices and their components.

Collagens are classified in different groups, according to the macromolecular aggregates they are able to form. The fibrillar collagens (types I, II, III, V and XI), are components of the homo- and heterotypic quarted-staggered fibrils. The FACITs (fibril-associated collagens with interrupted triple helices, types IX, XII and XIV) are non-fibrillar collagens, but contain one or two triple helical domains able to interact and adhere to collagen fibrils. And finally there are the nonfibrillar collagens which include the forming sheets collagens (the basement membrane type IV, and types VIII and X), type VI collagen which form beaded filaments, and type VII collagen, which is one component of the anchoring fibrils[6].

The second group of extracellular components are the noncollagenous glycoproteins[1,8-11]. They are multifunctional proteins, some of them not only found in the matrix but also in biological fluids. This family consists of several components but, due to their diversity, they are not yet subclassified as the collagen family. They are by alphabetical order as follows: fibronectin, laminin, nidogen/entactin, tenascin, thrombospondin, vitronectin, etc.. With respect to this presentation, attention is specially paid to fibronectin and laminin; the last one will be more fully discussed later.

Fibronectin is composed of two 250-kDa polypeptide chains joined near their C-terminal by disulfide bonds, and it has a modular structure[10,11]. Three kinds of repeats are present along the molecule: type I (45 amino acids; nine units comprise the N-terminal, and three the C-terminal), type II (60 amino acids; two repeats interrupt the N-terminal type I repeats) and type III (90 amino acids; a minimum of 15 repeats, in the middle of the fibronectin chain). Several binding sites have been described in this molecule: a domain for fibronectin matrix assembly, and binding sites for fibrin, heparin, gelatin, collagen, and cellular receptors. The major cell attachment site is located in the middle portion of fibronectin, in a type III repeat that contains the Arg-Gly-Asp (RGD) sequence, the cell adhesive recognition sequence. RGD is also found in other proteins (type I and type VI collagens, laminin, vitronectin, thrombospondin, von Willebrand factor, fibrinogen, and osteopontin). It is the basis of the biological activity of these proteins because this sequence is the ligand-recognition site of several cell surface receptors. Alternative splicing, which produces fibronectin variants, gives different properties to this molecule.

Proteoglycans are the third type of extracellular components[12,13]. They are macromolecules composed of sulfated polysaccharide chains linked covalently to a protein core. The glycosaminoglycans (i.e. long linear carbohydrate chains) are made up of repeating disaccharide subunits. They provide mechanical support to the tissues along with other extracellular matrix molecules. Proteoglycans are found: i) intracellularly in secretory granules (serglycine: chondroitin sulfate proteoglycan), ii) on the cell-surface membrane, where they could exhibit cellular-receptor properties (heparan sulfate type), and iii) in the extracellular matrices (aggrecan, decorin). Proteoglycans can bind growth factors such as FGF, EGF and TGF-β[12].

Non-structural functions of extracellular matrices

The architecture and physical properties of the tissues depends on the specific composition of the extracellular matrix (type of components, relative proportions of the different molecules, etc.) as well as on the interactions between these components. Thus, different cell types, and many biopolymers such as collagens, proteoglycans and glycoproteins, yield a highly specialized matrix structure. In addition, cell-matrix interactions are important in the control of cellular functions. Research on functional aspects have shown that extracellular matrix components play important roles in the regulation of cell behavior. Therefore cell-matrix interactions influence cell migration, morphology, growth, differentiation, morphogenesis, development, and metastasis of cancer cells. We now understand its role in the immobilization of growth factors and the mitogenic activity of some matrix proteins. The effects of matrix components on cells are thought to be mediated through specific cell surface receptors. These cell adhesion receptors present a broad distribution and a wide range of potential functions.

The following comments will try only to present the complexity of the extracellular matrix, and its active functional role through the cell-matrix interactions.

EXTRACELLULAR MATRIX AND METASTASIS

During the tumor growth, cancer cells invade the interstitial connective tissue, and migrate from the primary lesion. However, the basement membrane of the organs around the tumor is a physical continuos barrier that tumor cells must cross. Cancer cells break through the basal lamina, invade vessels, reach the blood stream and adhere to capillary walls at distant organs. Consequently, extracellular matrices play an important role in the establishment, growth, and invasiveness of tumors. Cancer cells typically secrete matrix degrading enzymes, show altered expression of matrix receptors as compared with their normal cell counterparts, and synthesize their own matrix constituents. The above observations suggest complex interactions of tumor cells with the host extracellular matrix (Fig. 2). These interactions trigger several cellular mechanisms inducing very different cellular responses, remodeling their surrounding to facilitate their invasive phenotype[14-16]. In addition, receptors for adhesive proteins on cancer cells are also other influential factors on cancer metastasis. After cell transformation, the number and type of receptors in the cell surface change; e.g. metastatic tumor cells have a great number of laminin receptors on the cell surface. Figure 2 illustrates some changes in cell surface after transformation of ephitelial cells.

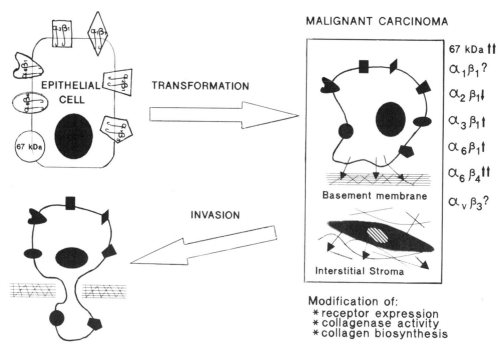

Figure 2. Schematic diagram showing hypothetical cellular receptor changes after malignat transformation of epithelial cells. The interaction of these receptors with basement membrane components triggers specific molecular mechanisms that modify the cell behavior; the cells are capable to produce matrix degrading enzymes contributing to tumor cell spreading.

In order to understand the molecular basis of malignant invasion and metastasis, biology of the extracellular matrix has been studied by using biopsies or well established cell lines as models. Being colorectal carcinoma one of the most common malignancies, with a high rate of occurence and the cause of a great number of deaths, our work was initially focussed on the study of the extracellular alterations in colon carcinoma.

The metabolism of collagen is usually involved in the development and growth of tumors. Thus, alterations of the metabolic pathways involved in either biosynthesis or degradation of collagen have been detected in a large number of neoplasms. We have analyzed the collagen content in a human colon adenocarcinoma. The overall decrease of collagen content in the tumor tissue was the first evidence of an alteration of the extracellular matrix in this system. For this reason, we have studied the collagen metabolism in different cell types derived from early subcultures obtained from tumor explants[17].

Cells affected by transformation in colorectal carcinoma exhibit an epithelial origin. Two different cell populations were obtained from the tumor tissue: stroma cells (fibroblast-like) and tranformed epithelial cells. Collagenase activity could be involved in the observed alteration. However, we have not detected any collagenase activity in conditioned media from the transformed epithelial cells. These cells show an almost negligible capability for collagen synthesis whereas normal epithelial cells are able to produce collagen; the absence of collagen biosynthesis must be related to the cell tranformation. Furthermore, in the tumor-associated fibroblast-like cells we have observed permanent altered phenotypic characteristics such as morphology, growth, adherence and collagen metabolism . Apparently the stroma cells maintained a normal ability for collagen biosynthesis. However, they exhibited a 2-fold increase in non-collagenous protein synthesis, when compared to control fibroblast from non-affected colon regions. An altered type I:type III ratio in collagens and procollagens as well as a decreased secretion of collagen with an abnormal deposition of procollagen in the cell layer were also observed[18]. The interstitial stroma alterations in fibroblast-like cells, which have been now studied for the first time, could be related to the invasiveness of the transformed epithelial cells.

From this primary and poorly differentiated human colon adenocarcinoma, we have established a cell line that we have designated as BCS-TC2 cells. The morphological, immunological, and ultrastructural features of these cells are related to the epithelial origin of the transformed cells observed in the tumoral tissue. BCS-TC2 cells maintain in culture several characteristics of the original tumor cells, therefore they represent a useful model for cell biology studies of primary tumors[19].

A relationship among metastatic potential and adhesive capacities, cell shape, and cytoskeletal organization of tumor cells has been suggested. These processes are regulated by cell-extracellular matrix interactions. Figure 3 shows the adhesion of BCS-TC2 cells to different substrates, extracellular matrices deposited by fibroblast cells and by themselves, as well as to individual matrix components. These cells highly increase their adhesion ability on both extracellular matrices[20]. In contrast, the adhesion increase with other epithelial cells is more relevant when the matrix has been deposited by endo- or epithelial cells than deposited by fibroblasts.

Cell adhesion on protein-coated subtrates is delayed, and the percentage of adherent cells is lower compared with the studied matrices. Both type I and type IV collagens cause significant cell adhesion in comparison to plastic. When cells are seeded on a laminin-coated substrate, the cells attach rapidly, and a high increase in the BCS-TC2 cell adhesion percentage is observed. Fibronectin also enhances the cell attachment but to a lower extent than laminin. Epithelial cells normally present a high affinity for basement membrane proteins such as laminin and type IV collagen but not for fibronectin and type I collagen. Nevertheless, BCS-TC2 cells exhibit affinity for interstitial molecules. This abnormal phenotype is possibly due to the cell dedifferentiation induced by malignant transformation, as it has been described for LoVo colon adenocarcinoma cells, which attach to type IV collagen, laminin, and fibronectin[21]. These results indicate that matrix components are able to induce changes in the behavior of BCS-TC2 cells[20]; the phenotype of these cells is expressed differently depending upon each matrix factor. The observed dualism, recognition of either interstitial or basement membrane constituents by these cells, is probably an important feature for tumor progression and metastasis. These results provide a functional basis for the identification of cell receptors, exhibited under certain situations, in reply to different matrix proteins.

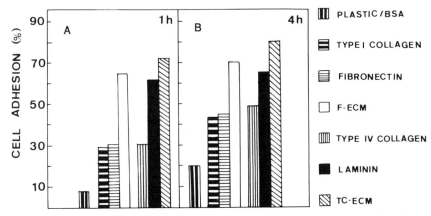

Figure 3. Effect of extracellular matrices and some of their components on BCS-TC2 cell attachment. Radiolabeled cells were added to each well and incubated for 1 (A) and 4 h (B); afterwards unattached cells were washed away. The remaining radiolabeled cells were solubilized and quantified. Cell adhesion is expressed as a percentage relative to the total number of cells seeded. F-ECM and TC-ECM are extracellular matrices deposited by fibroblast and BCS-TC2 cells, respectively.

Among the molecules that mediate adhesive interactions, a special interest has been focussed on the role of laminin. Interactions of epithelial and endothelial cells with laminin play an important role in the development of malignant tumors. High metastatic tumor cells in vitro present a rapid adhesion to laminin compared to their nonmetastatic counterparts[22]. Laminin also increases the metastatic phenotype of tumor cells since nonmetastatic cells become metastatic in the presence of laminin[23].

LAMININ

Laminins, multimeric glycoproteins and major constituents of basement membranes, are a family of adhesive proteins[3,4,8,9,24]. Members of this family (laminin A, s-laminin, merosin, etc.) have distinct polypeptide chains, chain composition and structure. Most of the studies on laminin used the protein isolated from the murine Engelbreth-Holm-Swarm tumor. Laminin was first purified and characterized in 1979 from this tumor[25] and studies related to the structure, function, and cellular receptors were one of the main fields in the last decade. The laminin molecule is a large and multidomain structure, which appears on electron microscopy as a four-armed structure. This multifunctional protein interacts with either basement membrane components or with cell surfaces. Thus, various biological activities have been described for this adhesive protein, and several binding sites have been mapped into the different structural domains (Fig. 4).

The laminin molecule

Laminin is a very large, cross-shaped protein with one long arm and three short arms, with a molecular mass of 900-kDa. The classical structure is composed of three disulfide-bonded polypeptide chains (A, B1, and B2). Protein fragment analyses, electron microscopic studies as well as sequences deduced from the cDNA indicate the presence of seven different structural domains. Each chain forms a different short arm, and all three chains together, in a coiled-coil structure, form the long arm of the asymmetric cross-structure. Globules are observed in the middle and at the end of each of the short arms, and a large globule (G-domain) is observed at the end of the long arm. B chains are homologous but not identical, and different interaction sites are described in each chain. A diagram of the laminin structure, some of its proteolitic fragments, and several of its biological activities are shown in Figure 4. Cysteine-rich repeats with a similarity to epidermal growth factor (EGF-like repeats, 8C) were found in each of the short arms.

Up to date, at least eight cell binding regions have been described (Fig. 5A). B1 chain includes half of these sites: the YIGSR sequence in domain III near the intersec-

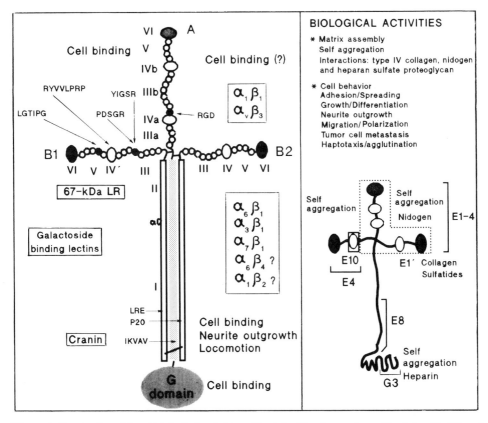

Figure 4. Structural model and biological activities of laminin. The three polypeptide chains (A, B1, and B2) form a cross-shaped structure. Each chain form a short arm of the cross, and all together form the long arm. Globular domains (◯ , ●) and EGF-like repeats (O) of the short arms are represented. The domains along the molecule are indicated by Roman numerals. Regions recognized by some laminin receptors, and sequences able to interact with cells are indicated. The diverse biological activities as well as the position of the different elastase-derived fragments are shown.

tion of the four arms of the laminin molecule, the PDSGR sequence which can act synergistically with the former, the LGTIPG elastin receptor binding sequence, and the site located in domain IV' of the short arm. Two sites are located on the A chain: a latent RGD-site in the short arm (domain IIIa), and the IKVAV sequence at the end of the long arm (domain I). The other two sequences have also been located on the E8 fragment[9,26].

Properties and functions of laminins

Laminin exhibits a variety of biological activities which are summarized in Figure 4. Laminin binds to nidogen, heparan sulfate proteoglycan, and type IV collagen, components of the basement membranes, but not to other matrix proteins. Due to these multiple interactions, it is likely that laminin plays an important role either in matrix assembly or maintaining the stability of the matrix. Other biological functions of laminin are related to its ability to anchor cells to the basement membrane. Laminin is involved in the regulation of cell attachment, spreading, migration, proliferation and differentiation of normal and neoplastic cells. Laminin primarily affects epithelial cells, and the response varies depending on the cell type.

Taking into account that laminin enhances BCS-TC2 cell attachment and that laminin contains multiple cell-adhesion-promoting domains, some of them cell-type specific, the effects of E1-4 and E8 laminin fragments were also examined. Data are summarized in Table 1. Both laminin fragments increase cell attachment to a similar extent as the intact molecule[20]. However, Clone A cells, derived from a poorly differentiated human colon carcinoma, which adhere avidly to a laminin substrate, only interact with the E8 laminin fragment[27]. In addition, a different behavior of BCS-TC2 cells is observed when cell spreading is studied. The E1-4 and E8 laminin fragments induce aggregation or spindle-shape morphology, respectively. Population doubling time is reduced in the presence of laminin from 39 to 32 h, indicating that the growth of these cells is also modified by laminin. The described behavior of BCS-TC2 cells on both laminin fragments strongly suggests the expression of at least two different types of laminin receptors in these cells, since the receptors described up to the present show affinity for only one of the laminin fragments.

Table 1. Effects of laminin and its fragments on the behavior of BCS-TC2 cells.

	PLASTIC	LAMININ	E1-4	E8
Adhesion[a]	15%	65%	52%	48%
Spreading[b]	-	+ +	+	+ + +
Aggregation[c]	-	+ +	+ + +	+
Morphology[d]	rounded	rounded-polygonal	rounded-polygonal	spindle
Growth[e]	39 h	32 h	nd	nd

[a] Cell adhesion, 1 hour after seeding, is expressed as the percentage relative to the total number of cell seeded. [b] Spreading, [c] aggregation, and [d] morphology of adhered cells 7 h after plating. [e] Population doubling time of cells growing on plastic and laminin substrates. (nd) not determined

LAMININ RECEPTORS

Considering the size, the presence of isoforms, the glycoprotein nature, and the structural and functional complexity of this protein, it should not seem unusual the large number of cell surface proteins and other macromolecules with affinity to laminin described so far[27-32]. Thus, a considerable number of receptors have been reported from a variety of cell types and tissues. They differ in apparent molecular masses, affinities for laminin, domains or sequences of laminin which are recognized, as well as in the cellular process in which they are involved. The increasing diversity of laminin-binding proteins was summarized and, in part, clarified with the establisment of protein families. One of the well-characterized group of laminin receptors are members of the integrin family. Others have been grouped as the non-integrin laminin receptors, a wide and diverse family of molecules. However, there are still some questions concerning the function as well as the possible role of all these molecules as true laminin receptors involved in the regulation of the cell behavior.

INTEGRINS

Cell-cell and cell-substrate interactions are mediated through several different families of receptors: integrins, cadherins, LEC-CAMs cell adhesion molecules, proteins from the immunoglobulin superfamily and homing receptors[33]. The integrins appear to be the primary mediators of cell-extracellular matrix adhesion. Integrins are heterodimers consisting of nonconvalently associated α and β subunits[33-36].

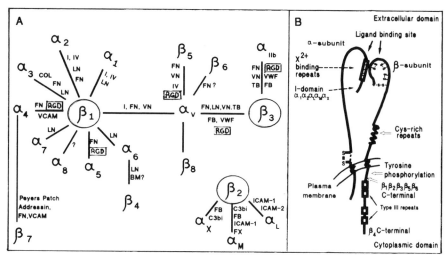

Figure 5. Schematic representation of the known integrin subunit associations. Structural characteristics of the integrins.
(A) Integrin family: one group of α subunits interact with a common β subunit generating distinct subfamilies. The ligands of the different integrins are also shown. Extracellular ligands: fibronectin (FN), fibrin(ogen) (FB), laminin (LN), type I (I) and type IV (IV) collagens, thrombospondin (TB), von Willebrand factor (VWF) and vitronectin (VN). Other ligands are intercellular adhesion molecules (ICAM), vascular cell adhesion molecules (VCAM), blood coagulation factor X (FX), complement component C3bi (C3bi) and basement membranes (BM). (B) Integrin structure. There are composed of one α and one β integral membrane subunits associated non-covalently. The extracellular domain contains the ligand binding site. The cysteine-rich repeats of the β subunit and cation binding sites (X^{2+}) in the α subunit are indicated. Different α subunits present a specific domain (I). The cytoplasmic domains of both subunits are relatively small. Only the β_4 subunit is large, presenting various fibronectin type III repeats.

These proteins are divided into subfamilies, each one with a common β subunit capable of associating with different α subunits. Up to date, 13 α subunits and 8 β subunits are known; they have been at least partially sequenced and have been shown to be distinct. The different combinations of the α and β subunits could form at least 19 integrins (Fig. 5A). A schematic representation, and the more important structural characteristics of a typical integrin, are shown in Figure 5B.

The integrin ligand-binding site appears to be formed by sequences from both subunits. The binding of integrins to their ligands is a cation-dependent process occurring with low affinities (10^{-6} M). One of the best studied recognition sites in the ligand is the RGD sequence. However, not all integrins bind to ligands via RGD-containing domains, for example some of the laminin integrins recognize the E8 fragment, which does not contain this sequence (Fig. 4).

The cytoplasmic domain of integrins establishes connections with the cytoskeleton. A tyrosine phosphorylation site appears to be located at the cytoplasmic sequences of the β_1, β_3 and other subunits, suggesting a potential site for integrin regulation. This makes integrins good candidates for transducing signals from the extracellular environment to the cell cytoplasm. However, the precise mechanism whereby signals from extracellular matrix proteins are transduced via integrins to the intracellular machinery controlling the cell behavior is not yet known. Considering the large number of integrin genes, the possible different combinations of α and β dimers, the variety of ligands, the alternative splicing of cytoplasmic domains as well as the existence of more than one affinity state for many integrins, the complexity of the integrin family could very well increase. As is shown in Figure 5, an integrin can recognize several matrix proteins. Fibronectin and laminin are recognized by at least ten and seven different integrin heterodimers, respectively (Fig. 5A). On the other hand, most of the other matrix ligands also have multiple receptors. Laminin is the specific ligand of the $\alpha_6\beta_1$, $\alpha_6\beta_4$, and $\alpha_7\beta_1$ heterodimers[27,28,31,32].

This complex view is enlarged by the fact that cells often display multiple integrins-capable of interacting with a particular matrix protein ligand. For example, in our model, we have observed that BCS-TC2 cells express different types of integrins. By flow cytometry and immunofluorescence analyses we have detected the expression of two types of β subunits, β_1 and β_4. They are combined with α subunits (α_1, α_2, α_6). This could indicate the presence of $\alpha_1\beta_1$, $\alpha_2\beta_1$, $\alpha_6\beta_1$ and $\alpha_6\beta_4$ integrins in these cells. All of them are able to recognize laminin, as well as other matrix components such as fibronectin and collagens. Figure 6 shows the integrin expression in BCS-TC2 cells by immunofluorescence and FACS analyses.

From what we have observed so far, several questions arise themselves. Why do the cells express different integrins upon the recognition of a single ligand? If the RGD sequence, present in a great number of ligands, is recognized by different integrins, which are the basis of the specificity? How is the signal transduced allowing different cellular responses? One possibility is that various α/β combinations might transduce different signals from the extracellular matrix to the cell interior. Although integrins lack the classical characteristics of a signal-generating receptor, such as kinase or phosphatase domains, or sequences suitable for interacting with G proteins, it is clear that integrins are able to transduce signals from the extracellular matrix to the cell interior, and that these signals may trigger changes in gene expression[37].

Do integrins transmit signals into the cells?

There is a great deal of experimental evidences showing that integrins mediate the transference of information into the cells. A mechanism of integrin-mediated signalling is based on the reorganization of the cytoskeleton, promoting the regulation of cell shape, motility and adhesive strength[38,39].

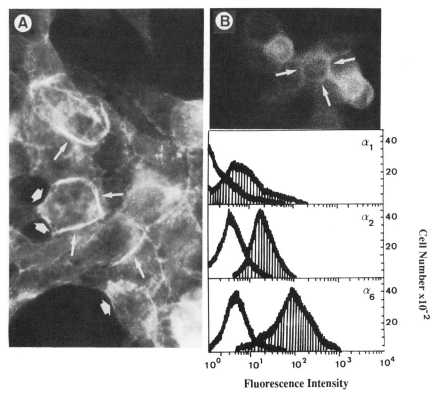

Figure 6. Study of cell surface expression of integrins by immunofluorescence and FACS analyses in BCS-TC2 cells.

For instance, as observed in Figure 7, BCS-TC2 cells show a very low degree of cytoskeletal organization; actin cable disorganization was also observed in other human colon ephitelial cells. This lack of cytoskeletal organization is probably due to malignant transformation, and it is normally associated with the invasive properties of tumoral cells, such as high motility, adaptability, and increased proliferation rate. On the other hand, some degree of actin microfilaments can be observed on BCS-TC2 cells growing on laminin. This suggests that cell-laminin interactions induce a reorganization of the cytoskeleton, and, as a result, cells have polygonal morphology. The glycoprotein also promotes the organization of the tubulin network on these cells. Besides, an interaction between microtubules and components of the cellular signal transduction pathways has been described[40]. Thus, microtubules could be involved in the signal transduction during the interaction of these cells with basement membranes.

The integrin cytoplasmic domain interacts with cytoskeletal components. Thus, the sites of integrin-mediated adhesion to the extracellular matrix could be a foci of nucleation for cytoskeletal assembly. An appropiate cell shape associated to the cytoskeletal organization can regulate the biosynthetic activity of the cell, and thus contribute to cell growth or differentiation.

The cytoplasmic domain of β_1 subunits carries the information needed for the movement of $\alpha\beta_1$ integrins to the focal contact sites. The function of the cytoplasmic domain of the α-subunit seems to be the regulation of the different interactions between the β_1 subunit cytoplasmic tail and cytoskeletal molecules[37]. Laminin is the ligand of five $\alpha\beta_1$ integrins (Fig. 5A). Considering that the cytoplasmic domains of the α-subunits in these heterodimers are different, each of these $\alpha\beta$ combinations could promote the organization of a distinct cytoskeletal framework which allows for different cellular responses.

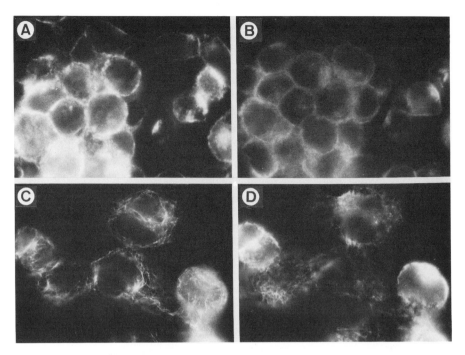

Figure 7. Tubulin and actin distribution in BCS-TC2 cells growing on glass coverslips and on laminin-coated sustrate. Fluorescence microscopy of cells growing for two days on glass coverslips (A & B) or on laminin (C & D). Tubulin (A & C) and actin (B & D) distribution was determined by indirect immunofluorescence (anti-tubulin antibodies) or by rhodamine-phalloidine staining, respectively. Original magnification x1000.

Clustering of β_1 integrins promoted by immobilized ligands or soluble antibodies has been described as sufficient to induce cytoskeletal associations in fibroblasts[41]. In contrast, it is known that the phosphorylation of the α_6 subunit is required to promote its association with the cytoskeleton and to induce macrophage attachment and spreading on laminin[42]. In epidermal cells, integrin $\alpha_3\beta_1$ interlinks extracellular matrix molecules with actin filaments in focal contacs, whereas integrin $\alpha_6\beta_4$ interconnects with intermediate filaments in hemidesmosome-like structures[43].

One of the best candidates for communication sites and signal transduction across the plasma membrane are the focal adhesions contacts, which are in close proximity to the extracellular environment[38,39,44]. The recognition of a ligand by an integrin develops a specific chain of protein-protein interactions in the cytoplasmic face. Involved in this chain are several cytoskeletal proteins and, as a result, the cytoskeleton is organized and cell behavior is regulated. Some of the mechanisms of the signal transduction that have been proposed are summarized in Figure 8. The major protein components situated on the cytoplasmic face of focal adhesions are α-actinin, talin and vinculin. Talin has been shown to be the first linking protein, binding to the integrin cytoplasmic tail and to vinculin. Vinculin is able to interact with α-actinin, an actin cross-linking protein. In addition to these major structural components, focal contact sites also contain several additional non-enzymatic proteins that are present at low levels, probably playing a regulatory role. These include paxillin, which is one of the major substrate for tyrosine kinases, and tensin, which has an *src* homology 2 (SH2) domain. Several proteins involved in signal transduction contain this SH2 domain, and they mediate interactions with other proteins by binding to specific phosphotyrosine residues. Thus, the presence of an SH2 domain in tensin suggests the performance of a regulatory function in focal adhesion organization. Additionally, focal contacts contain also enzyme regulatory molecules that become associated to the cytoskeleton. These include protein tyrosine kinase pp125[FAK] (FAK "focal adhesion kinase"[45]) and protein kinase C[44].

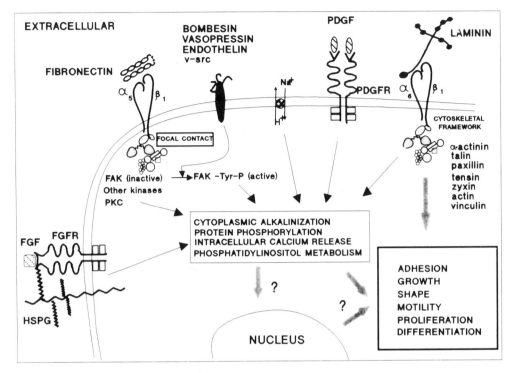

Figure 8. Signal transduction by integrins receptors.
The interaction of integrins with their ligands (e.g. laminin, fibronectin) promote different signals such as the organization of cytoskeletal framework and the tyrosine phosphorylation of focal adhesion kinase (FAK). The focal contact zone contains several cytoskeletal proteins as well as FAK, other tyrosine kinases and protein kinase C (PKC), which are implicated in protein phosphorylation. Integrin related signals are transduced in cooperation with other cell surface receptors for adhesion molecules and/or growth and differentiation factors (HSPG: heparan sulfate proteoglycan; FGF: fibroblast growth factor: FGFR: fibroblasts growth factor receptor; PDGF: platelet derived growth factor; PDGFR: platelet derived growth factor receptor).

Integrin tyrosine phosphorylation

Tyrosine phosphorylation of integrins by themselves has been proposed as a possible mechanism for the regulation of integrin affinity[36]. In addition, other kinds of tyrosine phosphorylations could be involved in signal transduction. The generation of different types of cystoskeletal frameworks seems to be able to transmit the extracellular information across integrins and give rise to a biochemical cellular response. It has been suggested that a potential transducer of integrin-generated signalling pathways is the tyrosine phosphorylation of cellular components[37,44,45]. High levels of tyrosine phosphate have been found concentrated at cell-substratum and cell-cell contact points. Also, the stimulation of tyrosine phosphorylation via integrins in platelets and adherent cells has been described[45].

In different cultured cells phosphorylation of pp125[FAK] has been observed in response to cell attachment on extracellular matrix proteins, such as fibronectin, laminin, collagen type IV or vitronectin. Cell adhesion to the extracellular matrix is also accompanied by tyrosine phosphorylation of paxillin, the 68- kDa focal-adhesion-associated protein[39]. In both platelets and NIH 3T3 cells, integrin-stimulated pp125[FAK] tyrosine phosphorylation correlates with an increase in the intrinsic kinase activity of pp125[FAK].

The pp125[FAK] is included in the tyrosine kinase family but it has special characteristics. The primary structure of pp125[FAK] indicates that it contains a tyrosine kinase catalytic domain. However, this molecule has at least three other unique features which distinguish it from other tyrosine kinases: i) the catalytic domain of this protein is flanked by large NH_2- and COOH-terminal domains of unknown function, ii) it seems to be a pure cytoplasmic kinase due to the absence of a transmembrane domain or any obvious sites for conjugation of lipid anchors, and iii) it lacks some domains that are usually found in cytoplasmic tyrosine kinases (SH2 and SH3)[45].

It has been suggested that integrin-mediated activation of pp125[FAK] is an early step in the cascade of signal transduction that flows the information from extracellular matrix to the cell inside. In light of several reports, it appears that activation of pp125[FAK] is related to anchorage dependent cell growth and that this kinase regulates the cytoskeleton organization[37]. Other possibility was also pointed out, it is the relation of integrin signalling and activation of pp125[FAK] to the control of gene expression and cell differentiation[46].

As observed in Figure 9, immunofluorescence studies reveal tyrosine phosphorylation in BCS-TC2 cells. The enzyme activity is found in areas of cell-to-cell interaction and also in some tiny focal contacts. This pattern is very similar to that observed for the β_1 integrin. A similar distribution is also observed after immunostaining of talin. Thus, β_1 integrins, talin and a tyrosine phosphorylation activity, which are located at the focal sites could be involved in the signal transduction in BCS-TC2 cells.

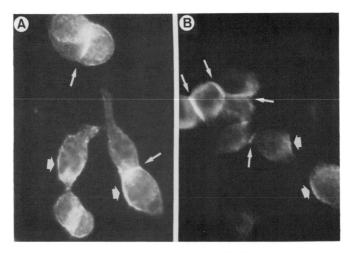

Figure 9. Distribution of β_1 integrins and phosphotyrosine in BCS-TC2 cells.
The distibution of β_1 integrins in the adherent cells is showed in (A). Cells were fixed at h after seeding. The cells were stained with mAB LIA1/4. The corresponding staining for phosphotyrosine (mAB PT-66) is shown in (B).

At first glance, signalling events in the cells are complex. Cytoskeletal changes and tyrosine phosphorylation, are only two examples from a great number of other integrin-related signal processes that have been described[37]. In a similar way, studies of cell binding to immobilized extracellular matrix molecules have also been shown to affect phosphatidylinositol metabolism, release intracellular calcium, alteration calcium-activated proteases, activation the Na^+/H^+ antiporter, and alteration the sub-cellular distribution of phosphoinositide-3-kinase. In addition, signalling events involving arachidonic acid metabolites have been associated with integrin-dependent cell adhesion[w]. Several studies are in progress in order to establish the actual pathways of integrin-mediated signalling. So far, our current knowledge suggests that integrins, like growth factor receptors, may bring about chemical signals.

Integrins are not isolated in the cell surface. One should keep in mind that they detect and transduce signals cooperatively with other types of adhesion and/or growth factor receptors[44]. In other words, the difference between the roles of growth factors and extracellular matrix receptors in signal transduction is not clearly established. Thus, the recognition of growth factors by cells can be modulated by extracellular matrix components and their receptors. Matrix components can regulate growth factor expression, bind growth factors and modify their activity or their interactions with specific receptors[47].

How is the information encoded in the extracellular matrix, and how is transduced into the cell by matrix receptors linked to the nuclear signalling machinery?. The answer to these questions are at present not clear.

NON-INTEGRIN LAMININ BINDING PROTEINS

In addition to the integrins which are involved in the adhesion of laminin, other potential laminin receptors or laminin binding proteins have been described so far. As can be observed in Table 2, a great number of molecules with affinity to laminin have been described[29,30,32].

Among these non-integrin laminin binding proteins, the 67-kDa protein family of cell-substrate adhesion receptors[33]. The high-affinity 67-kDa laminin receptor has been isolated from a variety of cell types and tissues. A correlation between cell malignancy and expression of this protein has been established although is not exclusively found in tumor cells. As a result, it is known the 67-kDa metastasis-associated laminin receptor. This protein binds to laminin through the sequence YIGSR located in the domain III of the B1 chain[48]. Recently a new binding site in laminin, located on the long arm and close to the intersection of the four arms has been described[49]. It is interesting that the 67-kDa laminin receptor seems to regulate the phenotypic characteristics of malignant cells such as rapid proliferation and metastatic potential. For instance, poorly differentiated colon carcinoma cells with high metastatic potential contains more laminin receptors than well-differentiated ones do.

The role of this protein as a laminin receptor has been questioned[30,32]. The cDNA from the 67-kDa laminin receptor codes a 37-kDa protein which has been proposed to be the precursor of the 67-kDa form. Figure 12 summarizes the main points under controversy, and shows the hypothetical mechanisms of the 67-kDa protein biosynthesis based on the proposed model[50-52]. The biological significance of this receptor as a true laminin receptor requires studies about the mechanism for signal transduction, and yet little is known about it. The 67-kDa laminin receptor has been shown to co-localize with actin microfilament bundles and recently the co-localization with α-actinin and vinculin has also been described[53].

The ^{125}I-laminin binding to BCS-TC2 cells reveals a high affinity, specific, and saturable binding. BCS-TC2 cells bind laminin with an estimated apparent K_D of around 4 nM and approximately 1-2 x 10^4 cell surface binding sites. The K_D value is close to that previously reported for the 67-kDa high-affinity laminin receptor from other sources. The integrin affinities for laminin are three orders of magnitude lower than the one of the 67-kDa laminin receptor. Thus, this result indicates that these cells express,

Table 2. Non-integrin laminin binding proteins[29,30,32,54]

Family of 67 kDa proteins:	Single-chain polypeptides. Bind elastin, laminin and collagen.
* High-affinity 67-kDa laminin-binding protein	Implicated in chemotaxis and metastasis. Precursor form of 37-kDa deduced from the cDNA sequence (see other details in text).
* 67-kDa elastin/laminin-binding protein	Similar to the high-affinity 67-kDa laminin receptor. It recognizes the VGVAPG and PGAIPG elastin sequences. Properties of peripheral membrane protein. Contains a lectin domain which controls the elastin binding. This receptor also binds to laminin although through a different region than the former, the LGTIPG sequence, and the carbohydrate side chains of laminin. The possible identity of this protein with the high affinity 67-kDa laminin receptor has been suggested. This elastin/laminin receptor was recently related to a 67 kDa catalytically inactive, alternative spliced form of β-galactosidase.
Galactoside-binding lectins:	Lectins from the S-lectin family. They bind with relatively high affinity poly-N-acetyllactosamine type oligosaccharides of laminin. The 67-kDa laminin-binding protein and the 67-kDa elastin/laminin receptor, which contain a lectin domain, can also be included in this group.
* 12-14 kDa	Lectins which are recognized by anti-67 kDa protein antibodies. Involved in the detachment of myoblasts from laminin during differentiation..
* 31-35 kDa	Macrophage antigen Mac-2, CBP-35, HLBP31. The 31-kDa β-galactoside-binding lectin can form dimers. The resulting 67-kDa homodimer is different from the 67-kDa laminin-binding protein.
Glicosylating enzymes	
* β-1,4 Galactosyltransferase	The cell surface enzyme binds to the side chain with N-acetylglucosamine. It functions as a laminin receptor during mesenchymal spreading and migration. It mediates processes of spreading and neurite outgrowth on laminin.
Other proteins:	Non antigenically related to 67-kDa laminin binding proteins.
* 36 and 38 kDa	Bind to laminin through the YIGSR and RGD sites, respectively.
* 55 kDa	Binding site unkown. Bind laminin with low affinity.
* Aspartactin (67 kDa) ?	Contains 33 consecutive aspartate residues at the COOH-terminal. Binding site: E3 fragment. Evidence against a laminin receptor role for this aspartactin /calsequestrin has recenly reported.
* Cranin (110 kDa)	Binds to IKVAV site. Present in neuronal cells and fibroblasts as a integral membrane protein.
* 5'-nucleotidase	The purified enzyme from chiken gizzard interacts with laminin and fibronectin (see other details in text).

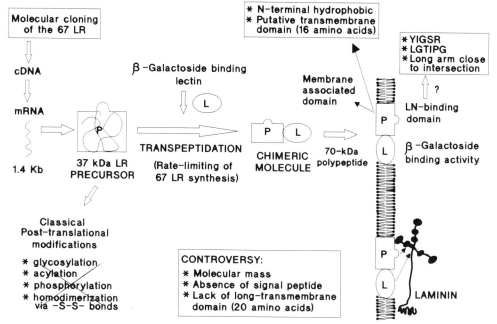

Figure 10.- Proposed model for the 67-kDa laminin receptor.
The 67-kDa laminin receptor (67 LR) is obtained by the association between the 37-kDa laminin receptor precursor (37-kDa LR) and a β-galactoside-binding lectin (L)[32,50]. Post-translational steps involved in the biosynthesis of this receptor, recognition sequences on laminin (LN), as well as the controversy[30,32] about its role as true laminin receptor are indicated.

in addition to integrins, other non-integrin laminin binding proteins[56].

Affinity chromatography on laminin-Sepharose of detergent-extracted surface-labeled proteins reveals the presence of a 67-kDa protein as well as other laminin binding proteins[55]. Besides the 67-kDa laminin receptor, some other components of the BCS-TC2 cell surface bind to laminin. Among them, we have found proteins with molecular masses of 74-kDa, 98-116 kDa, and two additional faint protein bands (52- and 43-kDa). In this context, proteins of 110-kDa with high-affinity for laminin have been reported in a variety of epithelial cells, and in neuronal and Sertoli cells. The 74 kDa-protein could be related to a 70-kDa precursor of the high affinity receptor.

Preincubation of BCS-TC2 cells with laminin increases the mRNA expression levels of the 67-kDa laminin receptor[55]. The over-regulation of this metastasis-associated laminin binding protein could play an important role in the invasive properties of malignant cells, and it could be related to an increased metastatic potential of these cells. It has been described that laminin increases the total cellular levels of the 67-kDa laminin receptor mRNA[56], and possibly regulates the expression of 67-kDa mRNA at different levels.

Turning to the 67-kDa laminin binding proteins, and considering the 67-kDa cell-substrate adhesion subfamily, several proteins which bind laminin, elastin and collagen are potential members of it[30,52,57]. Several characteristics of the integrins, such as structural related proteins, diversity of ligands, implication in a wide diversity of cell functions, and expression in different cell types, make a picture similar to the 67-kDa cell adhesion family. However, a great number of studies will be necessary to establish and clarify the funcion of this 67-kDa protein family.

Several cell lines recognize more than one site on laminin and express multiple functional receptors. The presence of all of these laminin-binding proteins in BCS-TC2 cells suggests that cells bind to multiple sites on laminin through a number of distinct surface proteins or receptors. This interaction with basement membranes is one of the primary requirement of metastasis; moreover, this multiple interaction could activate a number of different reactions involved in the subsequent steps of the metastatic process.

IS THE ECTO 5'-NUCLEOTIDASE A LAMININ/FIBRONECTIN RECEPTOR?

5'-Nucleotidase catalyzes the hydrolysis of nucleoside 5'-monophosphate to nucleosides and inorganic phosphate[58]. Molecular characteristics, kinetic parameters and properties of the ecto 5'-nucleotidase have been reported for the purified enzyme from different sources[58,59]. Nowadays, investigation on the ectoenzyme is mainly centered on structural aspects, such as its membrane topography, and on the involvement of the enzyme in the regulation of cellular functions[58]. The physiological role of the enzyme is not completely understood; it may be involved in the cellular uptake of nucleoside derivatives from nucleosides 5'-monophosphate. Studies concerning this enzyme have been stimulated by the growing evidence on the important role that adenosine plays in intercellular communication in several tissues. It is well known that the expression of 5'-nucleotidase is involved in processes such as maturation, migration, cell growth, and it appears modified in malignant cells in comparison to benign cells. We have analyzed the 5'-nucleotidase activity on different cell lines and its modification during cell proliferation[60].

In the last few years, two glycoproteins of the extracellular matrix, fibronectin and laminin, have been shown to interact directly and specifically with purified chicken gizzard 5'-nucleotidase and to regulate the AMPase activity of this ectoenzyme[61]. Taking into account the above mentioned modulation of 5'-nucleotidase activity by laminin and fibronectin, it is tempting to speculate that this ectoenzyme is involved in the regulation of cell behavior by extracellular matrix components.

Role of 5'-nucleotidase on cell-matrix interactions

To gain further support for the physiological relevance of the observed modulation, our group verified this effect on intact cells. In our studies we compare the effect of laminin and fibronectin on the 5'-nucleotidase activity using intact cells and plama membrane preparations[62]. The results are summarized in Figure 11.

We have observed that the effect of laminin and fibronectin is time-dependent and requires a preincubation of cells with the matrix proteins. The results obtained with BCS-TC2 cells are similar to those obtained with their plasma membrane preparations. An increase of the enzyme activity is observed in the presence of laminin while fibronectin exerts the opposite effect. This data suggest that the interaction of 5'-nucleotidase with the different matrix protein takes place in distinct sites of the ectoenzyme; also a possible role of the ectoenzyme as membrane receptor for both proteins can be proposed. The recognition and binding to different adhesive ligands, such as laminin and fibronectin, have been previously described for other cellular receptors as integrins. Similar experiments were performed with Rugli cells (rat glioblastoma cell line) and the effects were similar to those obtained with BCS-TC2 cells.

In order to discuss the relationship between cell-binding and AMPase-modulator domains in these two glycoproteins, the modulation of ecto-5'-nucleotidase by different laminin and fibronectin fragments has also been examined[62]. Laminin fragments E1-4 and E8, which are responsible for cell-laminin interactions, are able to induce activation of the AMPase activity present in BCS-TC2 cells, while E3 laminin fragment do not promote any effect upon the AMPase activity. Laminin fragments E1-4 and E8 probably enhance 5'-nucleotidase activity by inducing conformational changes in the ectoenzyme either by direct interaction or indirectly as a result of the laminin interaction with other cellular receptors.

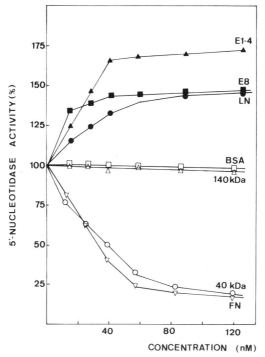

Figure 11. Effects of laminin, fibronectin and their fragments on the AMPase activity of BCS-TC2 cells. Cells were incubated in the presence of each effector.▲, E1-4;■, E8;●, laminin;□,BSA;△, 140 kDa;▽, fibronectin;○, 40-kDa. The activity is expressed as nmol of AMP hydrolyzed/cell in 15 min (x10⁵).

In contrast, preincubation of cells with the 140-kDa cell binding fibronectin fragment, which contains the RGD sequence, has no effect on the ecto-5'-nucleotidase activity. Nonetheless after preincubation of BCS-TC2 cells with other different fibronectin fragments, only the 40-kDa fragment shows AMPase activity inhibition. This fragment binds gelatin and collagen and it is involved in the fibrous polymerization of fibronectin, and is not recognized by integrin receptors. Preparations of the 40-kDa fragment incubated with gelatin retain the inhibitory activity. Thus, the binding of fibronectin to cells through this domain of the molecule seems to be independent from the gelatin-binding sequences. Is the 40-kDa fragment a cooperative site for the fibronectin-cell interaction? What is the meaning of this interaction? It has been suggested that the fibronectin cell-adhesive domain and an amino-terminal matrix assembly domain participate in the fibronectin assembly into fibroblast pericellular matrix. An hypothesis has been proposed in which two receptors could participate in the fibronectin-matrix deposition process: the cell attachment receptor (integrin, binding to the RGD site) and other matrix assembly receptor. This is supported by two facts: i) matrix assembly is inhibited by monoclonal antibodies that prevent the binding of the $\alpha_5\beta_1$ integrin to fibronectin, and ii) a fragment from the NH$_2$-terminal domain of fibronectin competes with fibronectin binding to the cell layer[63]. Taking this into account and considering the interaction between the 40-kDa fragment and 5'-nucleotidase, it is possible to speculate that this ectoenzyme can be the non-integrin receptor needed for the fibronectin-matrix assembly.

Our results point to a function of the ecto-5'-nucleotidase as a possible cell receptor for extracellular matrix proteins. Other reports also suggest that the ectoenzyme can function as a binding protein for laminin[64] and as a receptor for fibronectin[65]. A similar bifunctional role as ectoenzyme and cell adhesion molecule has been described for the rat liver ecto-ATPase. Moreover, a possible role of 5'-nucleotidase in cell-to-cell interaction can be suggested as it has been reported the presence of the HNK-1 carbohydrate epitope in the ectoenzyme[66]. This epitope is present in a variety of surface-located proteins involved in cell adhesion such as N-CAM, L1 or cytotactin[58].

The ectoenzyme from several sources is released from the plasma membranes by treatment with phosphatidylinositol-specific phospholipase C (PIPL-C), suggesting that phosphatidylinositol is involved in the anchoring of 5'-nucleotidase to the membrane[67]. Considering that transmembrane signalling may be an intrinsic and functionally important property of GPI-anchored membrane proteins, a new possible role of the 5'-nucleotidase is opened. Most of the current information on GPI-mediated signalling has been obtained by studies on lymphocytic activation. GPI-linked membrane proteins may play two major functions: interaction with other cells and initiation of biological responses[68].

Recent results indicate that the ectoenzyme from BCS-TC2 cells is partially liberated from plasma membranes and intact cells after treatment with PI-PLC. This demostrates that ecto-5'-nucleotidase is anchored through a phosphatidylinositol-glycan linker. However, a certain percentage of the total 5'-nucleotidase activity is resistant to PI-PLC. This suggest an heterogeneity in the anchoring of 5'-nucleotidase to the plama membrane of BCS-TC2 cells. In contrast, 5'-nucleotidase activity is totally released from Rugli cells. It is yet unknown how 5'-nucleotidase is able to transduce activation signals. Nevertheless, more studies are required to elucidate the bifunctional role of the ecto-5'-nucleotidase, the mechanisms involved in the modulation by laminin and fibronectin, and the enzyme ability for transduce activation signals.

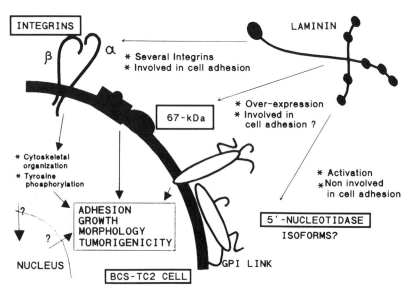

Figure 12. Schematic representation of the laminin binding proteins found on the surface of BCS-TC2 cells.

In summary, these colon adenocarcimona BCS-TC2 cells express different types of surface proteins involved in the recognition of laminin such as integrins, the 67-kDa laminin receptor, and 5'-nucleotidase (Fig.12). Integrins can mediated the adhesion of cells to laminin modifying the cytoskeletal framework. Tyrosine phosphorylation can also involved in the signal transduction. The 67-kDa laminin receptor over-expression induced by laminin could be an important regulatory mechanism of the metastatic potential of these cells. 5'-Nucleotidase activity is modulated by matrix glicoproteins and it can be an additional protein implicated in cell-cell and cell-matrix interactions. We conclude that BCS-TC2 cells through different surface receptors can recognize laminin, one of the major basement membrane component. This is one of the primary requirements during the metastatic course. The tumor cell environment is able to modulate the cell behavior by different signal transductions. Diverse cellular processes are activated and can be involved in the subsequent steps of the metastasis.

ACKNOWLEDGEMENTS

The original research described in this review was supported by grants from the spanish DGICYT (PB89/0129 and PB92/0552), and by an institutional grant from the Complutense University (PR180/91-3529).

REFERENCES

1. A.E. Postlethwaite and A.H. Kang, Fibroblasts and matrix proteins, in: "Inflammation: Basic Principles and Clinical Correlates," J.I. Gallin, I.M. Goldstein and R. Snyderman, ed., Raven Press, Ltd., New York (1992).
2. Ch. M. Lapierre and T. Krieg, "Connective tissue diseases of the skin", Marcel Dekker, Inc., New York (1993)
3. R. Timpl, Structure and biological activity of basement proteins, *Eur. J. Biochem.* 180:487 (1989).
4. M. Paulsson, Basement membrane proteins: structure, assembly, and cellular interactions, *Crit. Rew. Biochem. Mol. Biol.* 27:93 (1992).
5. E.D. Hay. "Cell Biology of Extracellular Matrix", Plenum Press, New York (1991).
6. M. van der Rest and R. Garrone, Collagen family of proteins, *FASEB J.* 5:2814 (1991).
7. R. Fleischmajer, B.B. Olsen and K. Kühn. "Structure, Molecular Biology, and Pathology of Collagen," New York Academic of Sciences, New York (1990).
8. P.M. Royce and B. Steinmann, "Connective tissue and its heritable disorders", Willey-Liss, New York (1993).
9. K. Beck, I. Hunter and J. Engel, Structure and function of laminin: anatomy of a multido main glycoprotein, *FASEB J.* 4:148 (1990).
10. R.O. Hynes and K.M. Yamada, Fibronectins: multifunctional modular glycoproteins, *J. Cell Biol.* 95:369 (1982).
11. R.O. Hynes. "Fibronectins," Springer-Verlag, New York (1989).
12. L. Kjellen and U. Lindahl, Proteoglycans: structures and interactions, *Annu. Rev. Bio chem.* 60:443 (1991).
13. T.E. Hardingham and A.J. Fosang, Proteoglycans: many forms and many functions, *FASEB J.* 6:861 (1992).
14. L.A Liotta, C.N Rao and U.M Wewer, Biochemical interactions of tumor cells with base ment membrane. Ann Rev Biochem 55:1037 (1986).
15. E. Ruoslahti, Control of cell motility and tumor invasion by extracellular matrix interac tions, *Br. J. Cancer* 66:239 (1992).
16. M. Nakajima and A.M. Chop, Tumor invasion and extracellular matrix degradative enzymes: regulation of activity by organ factors. Semin Cancer Biol 2:115 (1991).
17. J. Turnay, N. Olmo, J.G. Gavilanes and M.A. Lizarbe, Collagen metabolism in human colon adenocarcinoma. *Connect Tissue Res* 33:251 (1989).
18. J. Turnay, N. Olmo, J.G. Gavilanes and M.A. Lizarbe, Fibroblasts-like primary cells from human adenocarcinoma explants: collagen biosynthesis. *In vitro Cell Dev Biol* 27:447 (1991).
19. J. Turnay, N. Olmo, J.G. Gavilanes, J. Benitez and M.A. Lizarbe, Establishment and characterization of a new human colon adenocarcinoma cell line: BCS-TC2. *Cytotechnology* 3:75 (1990).

20. J. Turnay, N. Olmo, T. López, and M.A. Lizarbe, Influence of extracellular matrix com ponents on the behaviour of BCS-TC2 cells, a human adenocarcinoma cell line, *In vitro Cell Dev Biol* (Submitted)

21. S. Aznavoorian, L.A. Liotta and H.Z. Kupchika, Characteristics of invasive and non-inva sive human colorectal adenocarcinoma cells, *J. Natl. Cancer. Inst.* 82:1485 (1990).

22. G.W. Daneker, A.J. Piazza,G.D. Steele and A.M. Mercurio, Relationship between extra cellular matrix interactions and degree of differentiationin human colon carcinoma cell lines, *Cancer Res.* 49:681 (1989).

23. V.P. Terranova, E.S. Hujanen, D.M. Loeb, G.R. Martin, L. Thornburg and V. Glusko, Use a reconstituted basement membrane to measure cell invasiveness and select for highly invasive tumor cells, *Proc. Natl. Acad. Sci. USA* 83:465 (1986).

24. E. Engvall, Laminin variants: why, where and when?, *Kidney Int* 43:2 (1993)

25. R. Timpl, H. Rhode, P.G. Robey, S.I. Rennard, J.M. Foidart and G.R. Martin, Laminin - a glycoprotein from basement membranes, *J. Biol. Chem.* 254:9933 (1979).

26. T.M. Sweeney, M.C. Kibbey, M. Zain, R. Fridman and H.K. Kleinman, Basement mem brane and the SIKVAV laminin-derived peptide promote tumor growth and metastases, *Cancer Met. Rev.* 10:245 (1991).

27. E.C. Lee, M.N. Lotz, G.D. Steele and A.M. Mercurio, The integrin $\alpha6\beta4$ is a laminin re ceptor, *J. Cell Biol.* 117:671 (1992).

28. S.K. Akiyama, K. Nagata and K.M. Yamada, Cell surface receptors for extracellular matrix components, *Biochim. Biophys. Acta* 1031:91 (1990).

29. R.P. Mecham, Laminin receptors, *Annu. Rev. Cell Biol.* 7:71 (1991).

30. A.M. Mercurio and L.M. Shaw, Laminin binding proteins, *BioEssays* 13:469 (1991).

31. A. Sonnenberg, Laminin receptors in the integrin family, *Pathol. Biol.* 40:773 (1992).

32. V. Castronovo, Laminin receptors and laminin-binding proteins during tumor invasion and metastasis, *Invas. Metas.* 13:1 (1993).

33. S.M. Albelda and C.A. Buck, Integrins and other cell adhesion molecules, *FASEB J.* 4:2868 (1990).

34. S.M. Albelda, Role of integrins and other cell adhesion molecules in tumor progression and metastasis, *Lab Invest* 68:4 (1993).

35. E. Ruoslahti, Integrins, *J. Clin. Invest.* 87:1 (1991).

36. R.O. Hynes, Integrins: versatility, modulation, and signalling in cell adhesion, *Cell* 69:11 (1992).

37. R.L. Juliano and S. Kaskill, Signal transduction from the extracellular matrix, *J. Cell Biol.* 120:533 (1993).

38. D. Ingber, Integrins as mechanochemical transducers, *Curr. Opin. Cell Biol.* 3:841 (1991).

39. C.E. Turner and K. Burridge, Transmembrane molecular assemblies in cell-extracellular matrix interactions, *Curr. Opin. Cell Biol.* 3:849 (1991).

40. M.M. Rasenick, C.M. O´Callahan, C.A. Moore, and R.S. Kaplan, GTP-binding proteins which regulate neuronal adenylate cyclase interact with microtubule proteins, in: "Mi crotubules and microtubules inhibitors," M. De Brabander and J. De Mey, eds., Else vier Science Publishers B.V., Amsterdam (1985).

41. A. Horvath and S. Kellie, Regulation of integrin motility and cytoskeletal association in normal and RSV-transformed chick embryo fibroblasts, *J. Cell Sci.* 97:307 (1990).

42. L.M. Shaw, J.M. Messier and A.M. Mercurio, The activation dependent adhesion of macrophages to laminin involves cytoskeletal anchoring and phosphorylation of the $\alpha_6\beta_1$ integrin, *J. Cell Biol.* 110:2167 (1990).

43. W.G. Carter, P. Kauer, S.G. Gil, P.J. Gahr and E.A. Wayner, Distinct functions for inte grins $\alpha_3\beta_1$ in focal adhesions and $\alpha_6\beta_4$/bullous pemphigoid antigen in a new stable anchoring contact (SAC) of keratinocytes: relation to hemidesmosomes, *J. Cell Biol.* 111:3141 (1990).

44. C.H. Damsky and Z. Werb, Signal transduction by integrin receptors for extracellular matrix: cooperative processing of extracellular information, *Cuur. Opin. Cell Biol.* 4:772 (1992).

45. I. Zachary and E. Rozengurt, Focal adhesion kinase (p125[FAK]): a point of convergence in the action of neuropeptides, integrins, and oncogenes, *Cell*, 71:891 (1992).

46. A.D. Yurochko, D.Y. liu, D. Eierman and S. Haskill, Integrins as a primary signal trans duction molecule regulating monocyte-early gene induction, *Proc. Natl. Acad. Sci. USA* 89:9034-9038.

47. R. Flaumenhaft and D.B. Rifkin, Extracellular matrix regulation of growth factor and

protease activity, *Curr. Opin. Cell Biol.* 3:817 (1991).

48. J. Graf, R.C. Ogle, F.A. Robey, M. Sasaki, G.R. Martin, Y. Yamada and H.K. Kleimman, A pentapeptide from the laminin B1 chain mediates cell adhesion and binds the 67000 laminin receptor, *Biochemistry* 26:6896 (1987).

49. V. Cioce, I.M.K. Margulies, M.E. Sobel and V. Castronovo, Interaction between the 67 kilodalton metastasis-associated laminin receptor and laminin, *Kidney Int.* 43:30 (1993).

50. V. Castronovo, A.P. Claysmith, K.T. Barker, V. Cioce, H.C. Krutzsch and M.E. Sobel, Biosynthesis of the 67 kDa high affinity laminin receptor, *Biochem Biophys Res Commun* 177:177 (1991).

51. V. Castronovo, G. Taraboletti and M.E. Sobel, Functional domains of the 67-kDa lami nin receptor precursor, *J. Biol. Chem.* 266:20440 (1991).

52. M.E. Sobel ME, Differential expression of the 67 kDa laminin receptor in cancer. *Semin. Cancer Biol* 4:311 (1993).

53. S.P. Massia, S.S. Rao and J.A. Hubbel, Covalently immobilized laminin peptide Tyr-Ile-Gly-Ser-Arg (YIGSR) supports cell spreading and co-localization of the 67-kilodalton laminin receptor with α-actinin and vinculin, *J. Biol. Chem.* 268:8053 (1993).

54. A. Hinek, M. Rabinovitch, F. Keeley, Y. Okamura-Oho and J. Callahan, The 67-kD elastin/laminin binding protein is related to an enzymatically inactive, alternatively spliced form of β-galactosidase, *J. Clin. Invest.* 91:1198 (1993).

55. N. Olmo, J. Turnay, J.M. Navarro, T. Lopez, K. von der Mark and M.A. Lizarbe, Inte raction of BCS-TC2 cells (human adenocarcinoma cell line) with laminin. Identifica-tion of a 67-kDa laminin receptor, *J. Cell Biochem.* (Submitted)

56. V. Castronovo and M.E. Sobel, Laminin and fibronectin increase the steady state level of the 67 kD high affinity metastasis-associated laminin receptor mRNA in human cancer cells, *Biochem. Biophys. Res. Commun* 168:1110 (1990).

57. J.G. Tidball, Identification and distribution of a novel, collagen-binding protein in the developing subepicaradium and endomysium, *J. Biol. Chem.* 267:21211 (1992).

58. H. Zimmermann, 5´-nucleotidase: molecular structure and functional aspects. *Biochem. J.* 285:345 (1992).

59. J. Turnay, N. Olmo, J.M. Navarro, J.G. Gavilanes and M.A. Lizarbe, Isolation and cha racterization of the ecto-5´-nucleotidase from a rat glioblastoma cell line. *Mol. Cell Biochem.* 117:23 (1992).

60. J. Turnay, N. Olmo, G. Risse, K. von der Mark and M.A. Lizarbe, 5'-Nucleotidase acti vity in cultured cell lines. Effect of different assay conditions and implications of cell proliferation. *In vitro Cell Dev Biol* 25:1055 (1989).

61. J. Dieckhoff, J. Mollenhauer, U. Kühl, V. Niggemeyer, K. von der Mark and H.G. Mannherz, The extracellular matrix proteins laminin and fibronectin modify the AM-Pase activity of 5'-nucleotidase from chicken gizzard smooth muscle, *FEBS Lett.* 195:82 (1986).

62. N. Olmo, J. Turnay, G. Risse, R. Deutzmann, K. von der Mark and M.A. Lizarbe, M.A. Modulation of 5'-nucleotidase activity in plasma membranes and intact cells by extra-cellular matrix proteins laminin and fibronectin, *Biochem. J.* 282:181 (1992).

63. S.K. Akiyama, S.S. Yamada, W.T. Chen and K.M. Yamada, Analysis of fibronectin re ceptors function with monoclonal antibodies: roles in cell adhesion, migration, matrix assembly, and cytoskeletal organization, *J. Cell Biol.* 109:863 (1989).

64. U. Stochaj and H.G. Mannherz, Chicken gizzard 5'-nucleotidase functions as a binding protein for the laminin/nidogen complex, *Eur. J. Cell. Biol.* 59:364-372, 1992.

65. U. Stochaj, H. Richter and H.G. Mannherz, Chicken gizzard 5'-nucleotidase is a receptor for the extracellular matrix component fibronectin, *Eur. J. Cell Biol.* 51: 335 (1990).

66. M. Vogel, H.J. Kowalewski, H. Zimmermann, A. Janetzko, R.U. Margolis and H.E. Wollny, Soluble low-Km 5'-nucleotidase from electric-ray (*Torpedo marmorata*) elec-tric organ and bovine cerebral cortex is derived from the glycosyl-phosphatidyl-inosi-tol-anchored ectoenzyme by phospholipase C clevage, *Biochem. J.* 284:621 (1992).

67. U. Stochaj, F. Floke, W. Mathes and H.G. Mannherz, 5'-Nucleotidase of chicken gizzard and human pancreatic adenocarcinoma cells are anchored to the plasma membrane via a phosphatidylinositol-glycan, *Biochem. J.* 262:33 (1989).

68. P.J. Robinson, Signal transduction by GPI-anchored membrane proteins, in: "GPI mem brane anchors," M. Lucia Cardoso de Almeida, ed., Academic Press Ltd, London (1992).

CONTRIBUTORS

Allan, G
Aragón, C
Asaoka, Y
Avila, J
Bazan, N. G
Boscá, L
Braquet, P. G
Cano, E
Carceller, F
Casillas, T
Castro, E
Correas, I
Cuevas, P
Defontaine, N
Delicado, E. G
Diaz Nido, J
Diaz-Guerra, M.J.M
Diaz-Meco, M. T
Domínguez, J
Farooqui, A. A
Fideu, M. D
Fischer, E. H
Fresno, M
Garcia de Ancos, J
Garcia Rocha, M
Garcia-Higuera, I
Genaro, A. M
Giménez, C
Gimenez-Gallego, G
González, M. P
Herrero, I
Herrero, M. T
Horrocks, L. A
Hortelano, S
Hosford, D
Koltai, M
Ledesma M. D
Lizarbe, M. A

Manenti, S
Martín Municio, A
Mayor jr, F
Micheli, L
Miras-Portugal, M. T
Montejo, E
Moscat, J
Moskowitz, M. A
Muñoz-Fernandez, M. A
Murga, C
Nakamura, S-I
Nishizuka, Y
Ogita, K
Olmo, N
Oset-Gasque, M. J
Paillard, M
Parramon, M
Penela, P
Pintor, J
Poggioli, J
Reimers, D
Rodriguez-Pascual, F
Ruiz-Gomez, A
Sánchez, C
Sánchez-Prieto, J
Sen, R. P
Sorokine, O
Taniguchi, H
Torres, M
Ulloa, L
Van Dorsselaer, A
Vázquez, E
Villanueva, N
Waeber, C
Xiaobing, F
Zafra, F

INDEX

MARC kinase sustrate (MARCKS) *(cont'd)*
  demyristoylation, 77, 81
  expression, 78
  membrane interaction, 79
  nerve terminals, 65, 68
  purification, 77
  synaptosomes, 81
Microtubules
  associated proteins (MAPS), 149, 153
  cytomegalovirus, 149
  MAPS phosphorylation, 153
  neural processes, 153

Nerve growth factor (NGF)
  adenosine transport regulation, 221
Neurotransmitter transporters
  adenosine transport, 213
  aminoacid transporters, 206
  Na^+ dependent, 205
  PKC regulation, 207
  structural model, 207
  vesicular transport, 174
Nitric oxide (NO)
  Ca^{2+} dependent release, 123
  cyclic GMP, 54, 57
  effects, 51, 54
  phosphorylation of DARPP-32, 54
Nitric oxide synthase (NOS)
  cytokine-inducible, 53
  induction by cytokines, 54
  induction by PKC, 51, 55
  in endothelial tissues, 53
  in macrophages, 57
  in neural cells, 53
  isoenzymes, 52
  lipopolysaccharides, 54, 57
  PAF action, 104
  promoter region, 59
  reaction, 52
  second messengers in expression, 55
Nuclear factor
  gene activation, 45
  nuclear translocation, 46

Okadaic acid, 17

Phorbol esters
  adenosine transport, 217
  glutamate release, 65
  neurotransmitter transporters, 208
  PKC activation, 56, 67, 83
  proximal tubule, 83
Phosphatidylcholine
  hydrolysis, 33, 38, 43
Phosphatidylcholine-phospholipase C
  activation of PKC, 44
  antibodies, 43
  mitogenic signalling, 43
  ras p21, 44

Phosphatidylcholine-phospholipase C *(Cont'd)*
  tumor growth factor, 44
Phospholipases, 8
Phospholipase A_2 (PLA$_2$)
  arachidonic acid, 34, 114
  cellular injury, 114
  cytosolic, 114
  IL2-R, 139
  isoforms, 39
  link to PKC, 36
  PAF activation, 101
  reaction, 12, 34
  secretory enzyme, 36, 114
Phospholipase C (PLC)
  action, 12, 33
  diacylglycerol in synaptosomes, 69
  isoforms, 39
  linked to glutamate receptors, 69
  NOS expression, 55
Phospholipase D (PLD)
  activation, 37, 139
  isoforms, 39
  NOS expression, 55
  reaction, 12
Phospholipids
  degradation, 35, 36, 95, 143
  degradation in mitogenic signals, 43
  PAF effects, 103
Platelet-activating factor (PAF)
  arachidonic acid release, 101
  binding sites, 95
  brain receptors, 95, 104
  calcium signaling, 102, 106
  down-regulation of release, 106
  ginkgolide, 96
  hetrazepine, 96
  muscarinic acetylcholine receptor, 98
  neuronal gene expression, 97
  phospholipase A_2, 101
  synaptic membranes, 95
  synaptic transmission, 105
Protein Kinase A (PKA)
  adenosine transport, 217
  BAR regulation, 131
  cAMP dependent, 25
  GABA-b agonists, 192
  glycogen metabolism , 23, 25
  IL-2 activation, 142
Protein Kinase C (PKC)
  activation, 33, 34, 40, 45
  adenosine transport, 217
  angiotensin II, 87
  catalytic fragment PKM, 79
  diadenosine polyphosphates, 178
  family structure, 40
  glutamate release, 65
  IL-2R, 139
  induction of NO synthase, 51, 59
  intracellular signalling, 33